MATHEMATICS RESEARCH DEVELOPMENTS

EVOLUTION EQUATIONS

THEORIES, SOLUTIONS AND FUNCTIONS

MATHEMATICS RESEARCH DEVELOPMENTS

Additional books in this series can be found on Nova's website under the Series tab.

Additional E-books in this series can be found on Nova's website under the E-book tab.

MATHEMATICS RESEARCH DEVELOPMENTS

EVOLUTION EQUATIONS

THEORIES, SOLUTIONS AND FUNCTIONS

GASTON M. N'GUEREKATA
EDITOR

Nova Science Publishers, Inc.
New York

For permission to use material from this book please contact us:
Telephone 631-231-7269; Fax 631-231-8175
Web Site: http://www.novapublishers.com

Additional color graphics may be available in the e-book version of this book.

LIBRARY OF CONGRESS CATALOGING-IN-PUBLICATION DATA

Evolution equations : theories, solutions, and functions / [edited by] Gaston M. N'Guerekata.
p. cm.
Includes index.
ISBN 978-1-61324-090-8 (hardcover)
1. Evolution equations. I. N'Guerekata, Gaston M., 1953-
QA377.3.E955 2011
515'.39--dc22

2011007084

Published by Nova Science Publishers, Inc. † New York

CONTENTS

PREFACE

This new book presents and discusses new developments in the study of evolution equations. Topics discussed include a qualitative study of a perturbed critical semi-linear wave equation in variable metric; renormalized solution for a nonlinear anisotropic degenerate parabolic equation; periodic solutions of impulsive evolution equations; nonlinear spectral theory and controllability of semilinear evolution equations; uniform exponential stability of linear skew-evolution semifl ws and integral solutions for nondensely define evolution equations.

Chapter 1 deals with the global existence and blow-up of solutions to a polytropic fil tration equation with inner absorption and nonlinear boundary flux The authors establish critical exponent for the polytropic filtratio equation by constructing upper and lower solutions.

Chapter 2 is devoted to the description of bounded energy sequences of solutions to the equation

$$\textbf{(E)} \qquad \partial_t^2 u - \operatorname{div}(A(x)\nabla_x u) + |u|^4 u + |u|^{\alpha-1}u = 0 \qquad \text{in} \quad \mathbb{R}_t \times \mathbb{R}_x^3,$$

where $1 < \alpha < 5$, and A is a regular function valued in the space 3×3 positive definit matrix which is the identity outside a fi ed compact. The authors show that the subcritical perturbation does not induce any new variations on the dynamic of the equation $(\mathbf{E})$ according to the energy-Strichartz norm. The authors also prove that the description by means of linear concentrating waves given by S. Ibrahim about critical case still holds for the equation $(\mathbf{E})$.

In Chapter 3 the authors consider the anisotropic degenerate parabolic equation of the form:

$$\partial_t u + \nabla.F(u) - \sum_{i=1}^{N} \partial_{x_i}\left(a_{ii}(u)\partial_{x_i} u\right) = f(u) \text{ in } Q =]0, T[\times \mathbb{R}^N, \;\; T > 0 \qquad (1)$$

where $a(u) = (a_{ij}(u))_{1 \le i,j \le N}$ is a nonnegative diagonal matrix such that each coefficien is locally integrable in $\mathbb{R}$. The authors prove the well-posedness (existence and uniqueness) of renormalized solution of the associated Cauchy problem for L^1 initial data and source term. The novelty in the authors' paper is the fact that they work in an unbounded domain with u_0, f in L^1, the convection-diffusion flu functions are not necessarily Lipschitz on $\mathbb{R}$ and f depends on u that extends the results of Bendahmane and Karlsen and Chen and Perthame. The authors establish the comparison principle, the existence and uniqueness of renormalized solution under an assumption (H_2) on the modulus of continuity of F_i and $A_{ii}(r) = \int_0^r a_{ii}(\xi)d\xi$ for $i = 1, ..., N$. This assumption seems to be optimal for conservation laws included here and turns out to be the key-point of the uniqueness.

In Chapter 4 the authors study the following semilinear evolution equation

$$\begin{aligned} &u'(t) + A(t)u(t) = f(t, u(t)), \quad t > 0, \quad t \neq t_i, \\ &u(0) = u_0, \\ &\Delta u(t_i) = I_i(u(t_i)), \quad i = 1, 2, \cdots, \quad 0 < t_1 < t_2 < \cdots < \infty, \end{aligned}$$

in a Banach space $(X, \|\cdot\|)$ with unbounded operators $A(t)$. Here, $\Delta u(t_i) = u(t_i^+) - u(t_i^-)$ constitutes an impulsive condition, which can be used to model more physical phenomena than the traditional initial value problems $u(0) = u_0$. The authors assume that $f(t, u)$ and $A(t)$ are T–periodic in t and then prove with some compactness conditions that if solutions of the equation are ultimate bounded, then the equation has a T–periodic solution. The new results obtained here extend some earlier results in this area for equations without impulsive conditions.

Chapter 5 introduces the criteria for boundedness of solutions of set control differential equations(SCDEs).

In Chapter 6, the authors study the convergence of strong solutions of a Magneto-Hydro-Dynamic system. On the torus $\mathbb{T}^3$, the proof is based on Schochet's methods, whereas in the case of the whole space $\mathbb{R}^3$, the authors use Strichartz's type estimates and a product law's $2D \times 3D$.

In Chapter 7 the authors introduce and apply a new nonlinear spectrum to characterize the exact controllability for a semilinear evolution equation. To this end, they firs recall a necessary and sufficien condition for the exact controllability of the corresponding linear system and prove that, under suitable hypotheses on the nonlinear right-hand side, the exact controllability of the linear system carries over to the perturbated semilinear system. Afterwards the authors apply this result to a nonlinear damped wave equation.

In Chapter 8 the authors have considered a modifie version of the basic model of Herz et al., emulating the HIV dynamics in human blood. The modifie model includes a time delay in the death term of the infected CD4$^+$T-lymphocytic (immune) cells. as literature shows, there exists a positive discrete time lag between "the infection" and the death" of infected CD4$^+$T-cells and hence the inclusion of the delay. Delay differential model equations are solved theoretically as well as numerically. The authors results show that delay induced instability are nonexistent and asymptotically stable solution, characteristic of a non delayed system, exists. The (asymptotic) concentrations of healthy CD4$^+$T cells and virus populations may significantl be altered with some crucial model parameters and controlling of these parameters may be useful for better management of the HIV inflicte disease AIDS.

Recently, the fiel of differential equations has been studying in a very abstract method. Instead of considering the behaviour of one solution of a differential equation, one studies its sheaf-solution. Instead of studying a differential equation, one studies differential inclusion. Especially, one studies fuzzy differential equation (FDE) (a differential equation whose variables and derivative are fuzzy sets).In some of the references cited, the authors have studied the fuzzy control differential equation (FCDE). In Lakshmikantham,Gnana Bhaskar and VasundharaDevi,the authors have investigated the set differential equation (SDE). The authors have studied the set control differential equation (SCDE).

In this paper, the authors present the problem of global controllable (GC) for set control differential equation (SCDE). Chapter 9 is a continuation of some works in this direction.

In Chapter 10 the authors study the concept "linear skew-evolution semifl ws" and give several characterizations for the property of the uniform exponential stability of linear skew-evolution semifl ws on Banach and Hilbert spaces. By extending some of previous techniques, the authors obtain some generalization of notable results due to R. Datko, S. Rolewicz, Zabczyk and J.M.A.M van Neerven.

In Chapter 11 the authors study the initial boundary value problem of wave equations and reaction-diffusion equations with two nonlinear source terms of different sign. By introducing a family of potential wells W_δ and corresponding outside sets V_δ of W_δ the authors firs obtain the invariant sets and vacuum isolating of solutions. Then they get the threshold result of global existence and nonexistence of solutions. Finally the authors prove the global existence of solutions for the problem with critical initial conditions.

Chapter 12 is concerned with the existence for nondensely define evolution equations with nonlocal conditions. Using the techniques of fi ed point theory, existence results are obtained, for integral solutions, without the compactness assumption on nonlocal items.

In Chapter 13, under general conditions on the function f, the authors prove the existence of positive continuous solutions for the nonlinear elliptic equation $\Delta u(x) + f(x, u(x)) = 0$ with zero Dirichlet boundary condition on unbounded non-smooth domain D in $\mathbb{R}^n$, $n \geq 2$. In particular, the authors extend known results to more general classes of functions and domains and they provide the optimal class of nonnegative potentials V, which guarantees the existence of such solutions for the semi-linear equation $\Delta u(x) + V(x)u^p(x) = 0$, $p > 1$ studied by several authors.

In Chapter 14 the authors study the boundary value problem $u - div(a(x, \nabla u)) = f$ in Ω, $u = 0$ on $\partial\Omega$, where Ω is a smooth bounded domain in $\mathbb{R}^N$ and $div(a(x, \nabla u))$ is a $p(x)$-Laplace type operator. The authors obtain existence and uniqueness of an entropy solution for L^1 data f and for a variable exponent not necessarily continuous.

In: Evolution Equations: Theories, Solutions and Functions ISBN 978-1-61324-090-8
Editor: Gaston M. N'Guerekata

Chapter 1

Critical Exponent for a Polytropic Filtration Equation with Absorption and Nonlinear Boundary Flux

Zhongping Li*
Department of Mathematics, China West Normal University,
Nanchong 637002, P. R. China

Abstract

This paper deals with the global existence and blow-up of solutions to a polytropic filtratio equation with inner absorption and nonlinear boundary flux We establish critical exponent for the polytropic filtratio equation by constructing upper and lower solutions.

Keywords: critical exponent, polytropic filtratio equation, inner absorption, nonlinear boundary flux global existence, blow-up

2000 AMS Subject Classification: 35K50; 35K55; 35K65; 35B33.

1. Introduction

In this paper, we consider the following polytropic filtratio equation with absorption and nonlinear boundary flu

$$\begin{cases} (u^k)_t = \triangle_p u - au^\alpha, & x \in \Omega, \quad t > 0, \\ \nabla_p u \cdot \nu = u^\beta, & x \in \partial\Omega, \quad t > 0, \\ u(x,0) = u_0(x), & x \in \overline{\Omega}, \end{cases} \tag{1.1}$$

where

$$\triangle_p u = \operatorname{div}(|\nabla u|^{p-1}\nabla u) = \sum_{i=1}^{N}(|\nabla u|^{p-1}u_{x_i})_{x_i},\ \nabla_p u = (|\nabla u|^{p-1}u_{x_1}, ..., |\nabla u|^{p-1}u_{x_N}),$$

*E-mail address: zhongping-li@sohu.com

$\Omega \subset R^N$ is a bounded domain with smooth boundary $\partial\Omega$, ν is the outward normal vector on the boundary $\partial\Omega$, $k, p, a > 0$, $\alpha, \beta \geq 0$ and $u_0(x)$ is a positive C^1 function and satisfie the compatibility conditions.

Equations (1.1) is well-known non-Newtonian polytropic filtratio equation, which is used to describe the nonstationary fl w in a porous medium of fluid with a power dependence of the tangential stress on the velocity of displacement under polytropic condition. It has been intensively studied(see [12], [15]-[17] and the references therein). For the Neumann problem (1.1), the local in time existence have been established, see the survey [17].

Our main concern of this paper is the blow-up phenomenon, a subject that has deserved a great deal of attention in recent years, see for example book [13] and the surveys [5, 7]. The problem of determining critical exponent is an interesting one in the general theory of blowing-up solutions to different non-linear evolution equations of mathematical physics.

Recently, Li et al. [9] studied the nonlinear diffusion equation

$$\begin{cases} (u^k)_t = \Delta u - au^\alpha, & x \in \Omega, \quad t > 0, \\ \frac{\partial u}{\partial \nu} = u^\beta, & x \in \partial\Omega, \quad t > 0, \\ u(x,0) = u_0(x), & x \in \overline{\Omega}, \end{cases} \tag{1.2}$$

with $k \geq 1, a > 0, \alpha, \beta \geq 0$, and determined that the solutions of (1.2) blow up in finit time for large initial value if $\alpha < k, \beta > (k+1)/2$ or $\alpha > k, \beta > (\alpha+1)/2$. For the case of $0 < k < 1$, Andreu et al. [4] proved that the solutions of (1.2) blow up in finit time for large initial value if $\beta > 1, \beta > (\alpha+1)/2$ or $k < \beta \leq 1, \beta > \alpha$.

In [15], S. Wang considered the following equation

$$\begin{cases} (u^k)_t = \Delta_p u, & x \in \Omega, \quad t > 0, \\ \nabla_p u \cdot \nu = u^\beta, & x \in \partial\Omega, \quad t > 0, \\ u(x,0) = u_0(x), & x \in \overline{\Omega}, \end{cases} \tag{1.3}$$

and they proved that all positive solutions of (1.3) exist globally if and only if $\beta \leq k$ when $k \leq p$ and if and only if $\beta \leq \frac{(k+1)p}{p+1}$ when $k > p$.

As it is well known that degenerate and singularity equations need not possess classical solutions, we give a precise definitio of a weak solution for the problem (1.1).

Definition 1.1. *Let $T > 0$ and $Q_T = \Omega \times (0,T]$. A function $u(x,t)$ is called an upper (lower respectively) solution to (1.1) in Q_T if $u(x,t) \in L^\infty(0,T; W^{1,\infty}(\Omega)) \cap W^{1,2}(0,T; L^2(\Omega)) \cap C(\overline{Q_T})$ and satisfies*
(i) $u(x,0) \geq (\leq) u_0(x)$;
(ii) for any positive function $\psi \in L^1(0,T; W^{1,2}(\Omega)) \cap L^2(Q_T)$,

$$\int\int_{Q_T} [(u^k)_t\psi + \nabla_p u \cdot \nabla\psi + au^\alpha\psi]dxdt \geq (\leq) \int_0^T \int_{\partial\Omega} u^\beta \psi dsdt.$$

In particular, $u(x,t)$ is called a weak solution of (1.1) if it is both a weak upper and a weak lower solution. For every $T < \infty$, if u is a solution of (1.1) in Q_T, we say u is global.

By Theorem 2.1 and the proof of Theorem 3.1 for the uniqueness (pp. 118-123 in [3]), we have the following proposition:

Proposition 1.1. (Comparison Principle) *Let $\underline{u}$ and $\overline{u}$ be nonnegative sub and supersolutions of (1.1) in Q_T respectively. Then $\underline{u} \leq \overline{u}$ a.e. in Q_T if $\underline{u}(x,0) \leq \overline{u}(x,0)$ with $\underline{u} \geq \delta$ (or $\overline{u} \geq \delta$) for some $\delta > 0$.*

To describe the critical exponent for the problem (1.1), we defin the parameter ρ as follows

$$\rho = \begin{cases} \beta - \frac{p}{p+1}[k+1+(\alpha-k)_+], & k \geq p, \\ \beta - [k+(\alpha-k)_+ - \frac{1}{p+1}(\alpha-p)_+], & k < p. \end{cases} \tag{1.4}$$

To state our results, we need some notation for Ω (see [10]- [12]). Let λ and ϕ be respectively the firs eigenvalue and the corresponding eigenfunction of

$$-\Delta_p \phi = \lambda \phi^p \text{ in } \Omega, \quad \phi = 0 \quad \text{on} \quad \partial\Omega \tag{1.5}$$

with the normalization $\phi(x) > 0$ in Ω and $||\phi||_\infty = 1$. There exist positive constants $c_i (i = 1, 2, 3)$ such that

$$c_1 \leq -\frac{\partial \phi}{\partial \nu} \leq c_2 \quad \text{on} \quad \partial\Omega, \qquad |\nabla \phi| < c_3 \quad \text{on} \quad \overline{\Omega}.$$

Defin

$$\Omega_1 = \{x \in \Omega : dist(x, \partial\Omega) < \epsilon\}, \qquad \Omega_2 = \Omega \setminus \overline{\Omega}_1.$$

It is easy to see that there exist small $\epsilon > 0$ and some $c_4 \in (0,1)$ such that

$$|\nabla \phi| \geq \frac{c_1}{p+1} \quad \text{on } \Omega_1, \qquad \phi \geq c_4 \quad \text{on } \overline{\Omega}_2.$$

Let $d = \max\{||x||, \ x \in \overline{\Omega}\}$, $h(x) = \sum_{i=1}^N x_i + Nd + 1$ and c_5 be the positive constant satisfying $c_5^p e^{\frac{c_5 p^2 (2Nd+1)}{(p+1)\sqrt{N}}} = 1$.

Our main result is stated as follows.

Theorem 1.1 *If $\rho > 0$, then there are no nonnegative global solutions of (1.1) for large initial data.*

Remark 1.1 The known results for nonlinear diffusion equation (1.2) (in [4, 9]) cited in this section can be covered by Theorem 1.1. Indeed, when one take $p = 1$, Theorem 1.1 agree with those in [7] for the case $k \geq 1$ and for the case $k < 1$, Theorem 1.1 improve the results of [4].

Theorem 1.2 *If $\rho < 0$, the solutions of (1.1) are global.*

Remark 1.2 Theorems 1.1 and 1.2 show that the global existence exponent of (1.1) is described by $\rho = 0$.

Theorem 1.3 *Assume $\rho = 0$. (i) If $\alpha > k$ and*

$$a < \begin{cases} \min\{\frac{(\alpha+1)p}{p+1}[\frac{c_1}{(p+1)(c_3^{p-1}c_2)^{\frac{1}{p}}}]^{p+1}, \ \lambda c_4^{p+1}\frac{\alpha-p}{p+1}(\frac{1}{c_3^{p-1}c_2})^{\frac{p+1}{p}}\}, & \alpha \geq p, \\ \frac{c_5^p N^{\frac{1}{2}}}{2Nd+1}, & \alpha < p, \end{cases}$$

then the the solutions of (1.1) are non-global under large initial data;
(ii) If $\alpha > k$ and

$$a > \frac{2\alpha}{c_1^{p+1}}[\lambda c_1(p+1)^{p-1} + 3pc_3^{p+1}(p+1)^{\frac{p^2-1}{p}}],$$

then the solutions of (1.1) are globally bounded.
(iii) If $\alpha \leq k$, then the solutions of (1.1) are global.

Remark 1.3 For the critical case $\rho = 0$, the solutions of (1.1) can be both global and non-global, depending on the geometry and the size of domain Ω.

We will prove Theorems 1.1-1.3 in the following three sections respectively. The main technique employed is findin suitable sub or supersolutions blowing up in finit time or existing for all time, for which a precise analysis is necessary to determine the complicated interactions among the multi-nonlinearities included in the equation.

2. The Proof of Theorem 1.1

Lemma 2.1 *If $\alpha > k$, $k \geq p$, $\rho > 0$, then there are no nonnegative global solutions of (1.1) for large initial data.*

Proof Set

$$\underline{u}(x,t) = A[M\phi A^{\frac{\alpha-p}{p+1}} + (1-ct)^L]^{-\frac{p+1}{\alpha-p}}, \quad (x,t) \in \overline{\Omega} \times [0, \frac{1}{c}),$$

where $L \geq \max\{1, \frac{\alpha-p}{(p+1)(\alpha-k)}\}$, and $A, M > 1$, $c > 0$ to be determined. A simple computation shows

$$\begin{aligned}
(\underline{u}^k)_t &\leq \frac{(p+1)LkcA^k}{\alpha-p}[M\phi A^{\frac{\alpha-p}{p+1}} + (1-ct)^L]^{-\frac{(p+1)\alpha}{\alpha-p}}[MA^{\frac{\alpha-p}{p+1}} + 1]^{\frac{(p+1)(\alpha-k)}{\alpha-p} - \frac{1}{L}}, \\
\nabla\underline{u} &= -\frac{(p+1)M}{\alpha-p}A^{\frac{\alpha+1}{p+1}}[M\phi A^{\frac{\alpha-p}{p+1}} + (1-ct)^L]^{-\frac{\alpha+1}{\alpha-p}}\nabla\phi, \\
\Delta_p\underline{u} &= \lambda[\frac{(p+1)M}{\alpha-p}]^p A^{\frac{(\alpha+1)p}{p+1}}[M\phi A^{\frac{\alpha-p}{p+1}} + (1-ct)^L]^{-\frac{(\alpha+1)p}{\alpha-p}}\phi^p \\
&\quad + \frac{\frac{p(\alpha+1)}{\alpha-p}(\frac{p+1}{\alpha-p})^p M^{p+1}A^\alpha|\nabla\phi|^{p+1}}{[M\phi A^{\frac{\alpha-p}{p+1}} + (1-ct)^L]^{\frac{\alpha(p+1)}{\alpha-p}}}, \\
a\underline{u}^\alpha &= \frac{aA^\alpha}{[M\phi A^{\frac{\alpha-p}{p+1}} + (1-ct)^L]^{\frac{\alpha(p+1)}{\alpha-p}}}.
\end{aligned}$$

We have

$$\begin{aligned}
\tfrac{1}{2}\Delta_p\underline{u} - a\underline{u}^\alpha &\geq \frac{A^\alpha}{[M\phi A^{\frac{\alpha-p}{p+1}} + (1-ct)^L]^{\frac{\alpha(p+1)}{\alpha-p}}}[\frac{p(\alpha+1)}{2(p+1)}(\frac{Mc_1}{\alpha-p})^{p+1} - a] \quad \text{for } x \in \Omega_1, \\
\tfrac{1}{2}\Delta_p\underline{u} - a\underline{u}^\alpha &\geq \frac{A^\alpha}{[M\phi A^{\frac{\alpha-p}{p+1}} + (1-ct)^L]^{\frac{\alpha(p+1)}{\alpha-p}}}[\frac{\lambda}{2}(\frac{p+1}{\alpha-p})^p(Mc_4)^{p+1} - a] \quad \text{for } x \in \Omega_2.
\end{aligned}$$

Choose

$$\begin{aligned}
M^{p+1} &\geq \max\{[\frac{2a(p+1)}{p(\alpha+1)}](\frac{\alpha-p}{c_1})^{p+1}, \ \frac{2a}{\lambda}(\frac{\alpha-p}{p+1})^p(\frac{1}{c_4})^{p+1}\}, \\
c &\leq \frac{a(\alpha-p)A^{\alpha-k}}{(p+1)Lk[MA^{\frac{\alpha-p}{p+1}} + 1]^{\frac{(p+1)(\alpha-k)}{\alpha-p} - \frac{1}{L}}}.
\end{aligned}$$

Then

$$(\underline{u}^k)_t \le \Delta_p \underline{u} - a\underline{u}^\alpha \quad \text{in } \Omega \times (0, \frac{1}{c}). \tag{2.1}$$

On the other hand, by choosing

$$A^{\beta - \frac{p(\alpha+1)}{p+1}} \ge c_2 c_3^{p-1} [\tfrac{(p+1)M}{\alpha - p}]^p,$$

on the boundary, we have

$$\nabla_p \underline{u} \cdot \nu \le c_2 c_3^{p-1} [\tfrac{(p+1)M}{\alpha-p}]^p A^{\frac{p(\alpha+1)}{p+1}} (1-ct)^{-\frac{(\alpha+1)pL}{\alpha-p}} \le \underline{u}^\beta = A^\beta (1-ct)^{-\frac{(p+1)\beta L}{\alpha-p}}. \tag{2.2}$$

(2.1) and (2.2) implies that $\underline{u}$ is a non-global subsolution of (1.1) if $u_0(x) \ge \underline{u}(x, 0)$. □

Lemma 2.2 *If $\alpha > k,\ 0 < k < p,\ \rho > 0$, then there are no nonnegative global solutions of (1.1) for large initial data.*

Proof Construct

$$\underline{u}(x,t) = \begin{cases} A[M\phi A^{\frac{\alpha-p}{p+1}} + (1-ct)^L]^{-\frac{p+1}{\alpha-p}}, & \alpha > p, \\ \frac{A^p}{(1-ct)^{Lp}} e^{Mh^{1+\frac{1}{p}} A^{\alpha-p}(1-ct)^{-L(\alpha-p)}}, & k < \alpha \le p, \end{cases}$$

where $L \ge \max\{1, \frac{\alpha-p}{(p+1)(\alpha-k)}\}$, and $A, M > 1,\ c > 0$ to be determined.

The subcase $\alpha > p$ has been proven in Lemma 2.1. In the subcase $k < \alpha \le p$, a simple computation shows

$$\begin{aligned} (\underline{u}^k)_t &\le \frac{LpkcA^{kp}}{(1-ct)^{Lpk+1}} e^{kMh^{1+\frac{1}{p}} A^{\alpha-p}(1-ct)^{-L(\alpha-p)}}, \\ \nabla \underline{u} &= \frac{(p+1)MA^\alpha}{p(1-ct)^{L\alpha}} e^{Mh^{1+\frac{1}{p}} A^{\alpha-p}(1-ct)^{-L(\alpha-p)}} h^{\frac{1}{p}} (1, ..., 1), \\ \Delta_p \underline{u} &\ge [\frac{(p+1)MA^\alpha}{p(1-ct)^{L\alpha}}]^p e^{pMh^{1+\frac{1}{p}} A^{\alpha-p}(1-ct)^{-L(\alpha-p)}} N^{\frac{p+1}{2}}, \\ a\underline{u}^\alpha &= \frac{aA^{p\alpha}}{(1-ct)^{Lp\alpha}} e^{\alpha Mh^{1+\frac{1}{p}} A^{\alpha-p}(1-ct)^{-L(\alpha-p)}}. \end{aligned}$$

By taking

$$M^p \ge a(\tfrac{p}{p+1})^p N^{-\frac{p+1}{2}}, \qquad 0 < c < \frac{A^{(\alpha-k)p}}{Lpk} [(\tfrac{p+1}{p} M)^p N^{\frac{p+1}{2}} - a],$$

then we have

$$(\underline{u}^k)_t \le \Delta_p \underline{u} - a\underline{u}^\alpha \quad \text{in } \Omega \times (0, \frac{1}{c}). \tag{2.3}$$

On the other hand, by choosing

$$A^{(\beta-\alpha)p} \ge [(1+\tfrac{1}{p})MN^{\frac{1}{2}}]^p (2Nd+1) e^{pM(2Nd+1)^{1+\frac{1}{p}}},$$

on the boundary, we have

$$\nabla_p \underline{u} \cdot \nu \le [\frac{(p+1)MA^\alpha N^{\frac{1}{2}}}{p(1-ct)^{L\alpha}}]^p (2Nd+1) e^{pM(2Nd+1)^{1+\frac{1}{p}}} \le \frac{A^{p\beta}}{(1-ct)^{Lp\beta}} \le \underline{u}^\beta. \tag{2.4}$$

We know from (2.3) and (2.4) that $\underline{u}$ is a non-global subsolution of (1.1) if $u_0(x) \ge \underline{u}(x, 0)$.
□

Lemma 2.3 *If* $\alpha \le k,\ \rho > 0$, *then there are no nonnegative global solutions of (1.1) for large initial data.*

Proof. We fis consider the following problem

$$\begin{array}{lll} (w^k)_t = \Delta_p w - a w^{\alpha_0}, & x \in \Omega, & t > 0, \\ \nabla_p w \cdot \nu = w^\beta, & x \in \partial\Omega, & t > 0, \\ w(x,0) = w_0(x), & x \in \overline{\Omega}, & \end{array} \tag{2.5}$$

where $\alpha_0 > k \ge \alpha$. The solution w of (2.5) is non-global for large initial data by Lemmas 2.1-2.2. Let T be blow-up time of w. It suffice to prove that $w(x,t) \ge 1$ for all $(x,t) \in \overline{\Omega} \times [0,T)$ provided w_0 is large enough, which guarantees that w can be a subsolution of (1.1). To see this, fi T_1 such that

$$[\tfrac{k}{a(\alpha_0-k)T_1}]^{\frac{1}{\alpha_0-k}} > 1. \tag{2.6}$$

Let w_0 be sufficientl large that $T \le T_1$. Consider the O.D.E. problem

$$(\xi^k(t))_t = -a\xi^{\alpha_0}(t), \qquad \xi(0) = M$$

with solution

$$\xi(t) = (\tfrac{k}{a(\alpha_0-k)t + kM^{k-\alpha_0}})^{\frac{1}{\alpha_0-k}},$$

where $M > 1$ is to be determined. By the comparison principle, we have that

$$w(x,t) \ge \xi(t) \ge (\tfrac{k}{a(\alpha_0-k)T_1 + kM^{k-\alpha_0}})^{\frac{1}{\alpha_0-k}} \ge 1, \quad (x,t) \in \overline{\Omega} \times [0,T),$$

provided M is large enough due to (2.6). The proof is complete with $w_0 \ge M$. □

Combing Lemmas 2.1-2.3 we get Theorem 1.1 directly.

3. The Proof of Theorem 1.2

Lemma 3.1 *If* $\alpha > k,\ k \ge p,\ \rho < 0$, *then the solutions of (1.1) are globally bounded.*

Proof Set

$$\overline{u}(x,t) = A[2 - e^{-\epsilon\phi A^{\frac{\alpha-p}{p+1}}}]^{-1},$$

where $0 < \epsilon \le 1,\ A \ge 1$ to be determined. A simple computation shows

$$\begin{aligned} \nabla\overline{u} &= -\epsilon A^{\frac{\alpha+1}{p+1}} e^{-\epsilon\phi A^{\frac{\alpha-p}{p+1}}} [2 - e^{-\epsilon\phi A^{\frac{\alpha-p}{p+1}}}]^{-2}\nabla\phi, \\ \Delta_p\overline{u} &= A^\alpha e^{-p\epsilon\phi A^{\frac{\alpha-p}{p+1}}} \Big\{ [2 - e^{-\epsilon\phi A^{\frac{\alpha-p}{p+1}}}]^{-2p}[\lambda\epsilon^p\phi^p A^{\frac{p-\alpha}{p+1}} + p\epsilon^{p+1}|\nabla\phi|^{p+1}] \\ &\quad + 2pe^{-\epsilon\phi A^{\frac{\alpha-p}{p+1}}} [2 - e^{-\epsilon\phi A^{\frac{\alpha-p}{p+1}}}]^{-(2p+1)} \epsilon^{p+1}|\nabla\phi|^{p+1} \Big\} \\ &\le A^\alpha(\lambda\epsilon^p + 3p\epsilon^{p+1}c_3^{p+1}), \\ a\overline{u}^\alpha &\ge aA^\alpha 2^{-\alpha}. \end{aligned}$$

Let

$$\epsilon^p \le \{1, \frac{a}{2^\alpha(\lambda+3pc_3^{p+1})}\}.$$

Consequently,

$$(\overline{u}^k)_t \ge \Delta_p\overline{u} - a\overline{u}^\alpha, \qquad (x,t)\in\Omega\times(0,\frac{1}{c}). \tag{3.1}$$

On the other hand, let

$$A^{\frac{(\alpha+1)p}{p+1}-\beta} \ge (\epsilon c_1)^{-p}(p+1)^{p-1} \quad \text{and} \quad \frac{A}{2} \ge ||u_0||_\infty.$$

On the boundary, we have

$$\nabla_p\overline{u}\cdot\nu \ge (\epsilon c_1)^p(p+1)^{1-p}A^{\frac{(\alpha+1)p}{p+1}} \ge A^\beta = \overline{u}^\beta. \tag{3.2}$$

We follow from (3.1) and (3.2) that $\overline{u}$ is a time-independent supersolution of (1.1). □

Lemma 3.2 *If* $\alpha > k,\ k < p,\ \rho < 0$, *then the solutions of (1.1) are globally bounded.*

Proof Construct

$$\overline{u}(x,t) = \begin{cases} A[2-e^{-\epsilon\phi A^{\frac{\alpha-p}{p+1}}}]^{-1}, & \alpha > p, \\ A^p[2-e^{-\epsilon\phi A^{\alpha-p}}]^{-1}, & k < \alpha \le p, \end{cases}$$

where $0<\epsilon\le 1,\ A\ge 1$ to be determined.
The subcase $\alpha > p$ has been proven in Lemma 3.1.
Now, we prove the subcase $k<\alpha\le p$. A simple computation shows

$$\begin{aligned}
\nabla\overline{u} &= -\epsilon A^\alpha e^{-\epsilon\phi A^{\alpha-p}}[2-e^{-\epsilon\phi A^{\alpha-p}}]^{-2}\nabla\phi, \\
\Delta_p\overline{u} &= A^{\alpha p}e^{-p\epsilon\phi A^{\alpha-p}}\Big\{[2-e^{-\epsilon\phi A^{\alpha-p}}]^{-2p}[\lambda\epsilon^p\phi^p + pA^{\alpha-p}\epsilon^{p+1}|\nabla\phi|^{p+1}] \\
&\quad +2pA^{\alpha-p}e^{-\epsilon\phi A^{\alpha-p}}[2-e^{-\epsilon\phi A^{\alpha-p}}]^{-(2p+1)}\epsilon^{p+1}|\nabla\phi|^{p+1}\Big\} \\
&\le A^{\alpha p}(\lambda\epsilon^p + 3p\epsilon^{p+1}c_3^{p+1}), \\
a\overline{u}^\alpha &\ge aA^{\alpha p}2^{-\alpha}.
\end{aligned}$$

Let

$$\epsilon^p \le \{1, \frac{a}{2^\alpha(\lambda+3pc_3^{p+1})}\}.$$

Consequently,

$$(\overline{u}^k)_t \ge \Delta_p\overline{u} - a\overline{u}^\alpha \quad \text{in } \Omega\times(0,\frac{1}{c}). \tag{3.3}$$

Let

$$A^{(\alpha-\beta)p} \ge (\epsilon c_1)^{-p}(p+1)^{p-1} \text{ and } \frac{A^p}{2} \ge ||u_0||_\infty.$$

In addition, we have on the boundary that

$$\nabla_p\overline{u}\cdot\nu \ge (\epsilon c_1)^p(p+1)^{1-p}A^{\alpha p} \ge A^{\beta p} = \overline{u}^\beta. \tag{3.4}$$

By (3.3) and (3.4), we know that $\overline{u}$ is a time-independent supersolution of (1.1). □

Lemma 3.3 *If* $\alpha \le k,\ \rho \le 0$, *then the solutions of (1.1) are globally bounded.*

Proof. We consider the equation without inner absorption

$$\begin{array}{lll}(w^k)_t = \Delta_p w, & x \in \Omega, & t > 0, \\ \nabla_p w \cdot \nu = w^\beta, & x \in \partial\Omega, & t > 0, \\ w(x,0) = w_0(x), & x \in \overline{\Omega}, & \end{array} \tag{3.5}$$

where $w_0(x)$ is positive C^1 function satisfying the compatibility conditions on $\partial\Omega$. Clearly, the solution w of (3.5) is a supersolution of (1.1). Since $\alpha \le k$, we have $\beta \le \frac{(k+1)p}{p+1}$ for $k > p$; $\beta \le k$ for $k \le p$. According to [15], the solution of (3.5) is global. □

Combing Lemmas 3.1-3.3, we obtain Theorem 1.2.

4. The Proof of Theorem 1.3

Lemma 4.1 *If $\alpha > k,\ k \ge p,\ \rho = 0$ and*

$$a < \min\left\{ \frac{(\alpha+1)p}{p+1}\left[\frac{c_1}{(p+1)(c_3^{p-1}c_2)^{\frac{1}{p}}}\right]^{p+1},\ \lambda c_4^{p+1}\frac{\alpha-p}{p+1}\left(\frac{1}{c_3^{p-1}c_2}\right)^{\frac{p+1}{p}} \right\},$$

then there are no nonnegative global solutions of (1.1) for large initial data.

Proof Construct

$$\underline{u}(x,t) = \left[\frac{\alpha-p}{(p+1)(c_3^{p-1}c_2)^{\frac{1}{p}}}\phi + (1-ct)^L\right]^{-\frac{p+1}{\alpha-p}},$$

where $L \ge \max\{1, \frac{\alpha-p}{(p+1)(\alpha-k)}\}$, and $c > 0$ to be determined.
As for $(x,t) \in \Omega \times (0, \frac{1}{c})$, a simple computation shows

$$\begin{aligned}
(\underline{u}^k)_t &\le \frac{(p+1)kLc}{\alpha-p}\left[\frac{\alpha-p}{(p+1)(c_3^{p-1}c_2)^{\frac{1}{p}}}\phi + (1-ct)^L\right]^{-\frac{(p+1)\alpha}{\alpha-p}}\left[\frac{\alpha-p}{(p+1)(c_3^{p-1}c_2)^{\frac{1}{p}}} + 1\right]^{\frac{(p+1)(\alpha-k)}{\alpha-p}-\frac{1}{L}}, \\
\nabla\underline{u} &= -\frac{1}{(c_3^{p-1}c_2)^{\frac{1}{p}}}\left[\frac{\alpha-p}{(p+1)(c_3^{p-1}c_2)^{\frac{1}{p}}}\phi + (1-ct)^L\right]^{-\frac{\alpha+1}{\alpha-p}}\nabla\phi, \\
\Delta_p\underline{u} &= \frac{\lambda}{c_3^{p-1}c_2}\left[\frac{\alpha-p}{(p+1)(c_3^{p-1}c_2)^{\frac{1}{p}}}\phi + (1-ct)^L\right]^{-\frac{(\alpha+1)p}{\alpha-p}}\phi^p \\
&\quad + \frac{(\alpha+1)p}{p+1}\left(\frac{1}{c_3^{p-1}c_2}\right)^{\frac{p+1}{p}}\left[\frac{\alpha-p}{(p+1)(c_3^{p-1}c_2)^{\frac{1}{p}}}\phi + (1-ct)^L\right]^{-\frac{(p+1)\alpha}{\alpha-p}}|\nabla\phi|^{p+1}.
\end{aligned}$$

When $x \in \Omega_1$, we have $\nabla\phi \ge \frac{c_1}{p+1}$, and

$$\Delta_p\underline{u} - a\underline{u}^\alpha \ge \left[\frac{\alpha-p}{(p+1)(c_3^{p-1}c_2)^{\frac{1}{p}}}\phi + (1-ct)^L\right]^{-\frac{(p+1)\alpha}{\alpha-p}}\left[\frac{(\alpha+1)p}{p+1}\left(\frac{c_1}{(c_3^{p-1}c_2)^{\frac{1}{p}}(p+1)}\right)^{p+1} - a\right].$$

When $x \in \Omega_2$, we have $\phi \ge c_4$, and

$$\Delta_p\underline{u} - a\underline{u}^\alpha \ge \left[\frac{\alpha-p}{(p+1)(c_3^{p-1}c_2)^{\frac{1}{p}}}\phi + (1-ct)^L\right]^{-\frac{(p+1)\alpha}{\alpha-p}}\left[\lambda c_4^{p+1}\frac{\alpha-p}{p+1}\left(\frac{1}{c_3^{p-1}c_2}\right)^{\frac{p+1}{p}} - a\right].$$

By using the assumption on a and letting

$$c \le \min\left\{ \frac{(\alpha-p)\left[\frac{(\alpha+1)p}{p+1}\left(\frac{c_1}{(c_3^{p-1}c_2)^{\frac{1}{p}}(p+1)}\right)^{p+1}-a\right]}{(p+1)kL\left[\frac{\alpha-p}{(p+1)(c_3^{p-1}c_2)^{\frac{1}{p}}}+1\right]^{\frac{(p+1)(\alpha-k)}{\alpha-p}-\frac{1}{L}}},\ \frac{(\alpha-p)\left[\lambda c_4^{p+1}\frac{\alpha-p}{p+1}(\frac{1}{c_3^{p-1}c_2})^{\frac{p+1}{p}}-a\right]}{(p+1)kL\left[\frac{\alpha-p}{(p+1)(c_3^{p-1}c_2)^{\frac{1}{p}}}+1\right]^{\frac{(p+1)(\alpha-k)}{\alpha-p}-\frac{1}{L}}} \right\},$$

we have in summary that

$$(\underline{u}^k)_t \le \Delta_p \underline{u} - a\underline{u}^\alpha \quad \text{in } \Omega\times(0,\ \frac{1}{c}).$$

Moreover, we have on the boundary that

$$\nabla_p \underline{u}\cdot\nu \le (1-ct)^{-\frac{(\alpha+1)pL}{\alpha-p}} = (1-ct)^{-\frac{(p+1)\beta L}{\alpha-p}} = \underline{u}^\beta.$$

If in addition the initial data is large enough $u_0(x) \ge \underline{u}(x,0)$, then $\underline{u}$ is a blowing up subsolution of (1.1). □

Lemma 4.2 *If* $\alpha > k,\ k < p,\ \rho = 0$ *and*

$$a < \begin{cases} \min\{\frac{(\alpha+1)p}{p+1}[\frac{c_1}{(p+1)(c_3^{p-1}c_2)^{\frac{1}{p}}}]^{p+1},\ \lambda c_4^{p+1}\frac{\alpha-p}{p+1}(\frac{1}{c_3^{p-1}c_2})^{\frac{p+1}{p}}\}, & \alpha > p, \\ \frac{c_5^p N^{\frac{1}{2}}}{2Nd+1}, & \alpha \le p, \end{cases}$$

then there are no nonnegative global solutions of (1.1) for large initial data.

Proof The subcase $\alpha > p$ can be treat with the same proof as was used for Lemma 4.1.
In subcase $k < \alpha \le p$, construct

$$\underline{u}(x,t) = \frac{1}{(1-ct)^{Lp}} e^{\frac{pc_5}{p+1}[N^{\frac{p}{2}}(2Nd+1)]^{-\frac{1}{p}}h^{1+\frac{1}{p}}(1-ct)^{-L(\alpha-p)}}, \quad (x,t)\in\overline{\Omega}\times[0,\frac{1}{c}),$$

where $L \ge \frac{1}{p(\alpha-k)}$, and $c > 0$ to be determined.
For $(x,t)\in\Omega\times(0,\frac{1}{c})$, a simple computation shows

$$\begin{aligned}
(\underline{u}^k)_t &\le \frac{kLpc}{(1-ct)^{kLp+1}} e^{\frac{kpc_5}{p+1}[N^{\frac{p}{2}}(2Nd+1)]^{-\frac{1}{p}}h^{1+\frac{1}{p}}(1-ct)^{-L(\alpha-p)}}, \\
\nabla\underline{u} &= [N^{\frac{p}{2}}(2Nd+1)]^{-\frac{1}{p}}\frac{c_5 h^{\frac{1}{p}}}{(1-ct)^{L\alpha}} e^{\frac{pc_5}{p+1}[N^{\frac{p}{2}}(2Nd+1)]^{-\frac{1}{p}}h^{1+\frac{1}{p}}(1-ct)^{-L(\alpha-p)}}(1,\ldots,1), \\
\Delta_p\underline{u} &\ge \frac{c_5^p N^{\frac{1}{2}}}{(2Nd+1)(1-ct)^{Lp\alpha}} e^{\frac{p^2c_5}{p+1}[N^{\frac{p}{2}}(2Nd+1)]^{-\frac{1}{p}}h^{1+\frac{1}{p}}(1-ct)^{-L(\alpha-p)}}, \\
a\underline{u}^\alpha &= \frac{\alpha}{(1-ct)^{Lp\alpha}} e^{\frac{\alpha pc_5}{p+1}[N^{\frac{p}{2}}(2Nd+1)]^{-\frac{1}{p}}h^{1+\frac{1}{p}}(1-ct)^{-L(\alpha-p)}}.
\end{aligned}$$

According to assumption $a < \frac{c_5^p N^{\frac{1}{2}}}{2Nd+1}$ with

$$c \le \frac{1}{Lkp}(\frac{c_5^p N^{\frac{1}{2}}}{2Nd+1} - a),$$

we have that

$$(\underline{u}^k)_t \le \Delta_p \underline{u} - a\underline{u}^\alpha \quad \text{in } \Omega \times (0, \frac{1}{c}).$$

Further, we have on the boundary that

$$\nabla_p \underline{u} \cdot \nu \le (1-ct)^{-L\alpha p} = (1-ct)^{-L\beta p} \le \underline{u}^\beta.$$

Thus, $\underline{u}$ is a subsolution of (1.1) if $u_0(x) \ge \underline{u}(x,0)$. □

Lemmas 4.1 and 4.2 complete the proof for part (i) of Theorem 1.3.

Lemma 4.3 *If* $\alpha > k,\ k \ge p,\ \rho = 0$ *and*

$$a > \frac{2^\alpha}{c_1^{p+1}}[\lambda c_1(p+1)^{p-1} + 3pc_3^{p+1}(p+1)^{\frac{p^2-1}{p}}],$$

then the solutions of (1.1) are globally bounded.

Proof Construct time-independent functions

$$\overline{u}(x,t) = A[2 - e^{-c_1^{-1}(p+1)^{\frac{p-1}{p}}\phi A^{\frac{\alpha-p}{p+1}}}]^{-1}$$

with $A \ge 1$ to be determined. Then

$$\begin{aligned}
\nabla\overline{u} &= -c_1^{-1}(p+1)^{\frac{p-1}{p}} A^{\frac{\alpha+1}{p+1}}[2 - e^{-c_1^{-1}(p+1)^{\frac{p-1}{p}}\phi A^{\frac{\alpha-p}{p+1}}}]^{-2} e^{-c_1^{-1}(p+1)^{\frac{p-1}{p}}\phi A^{\frac{\alpha-p}{p+1}}}\nabla\phi,\\
\Delta_p\overline{u} &= A^\alpha e^{-pc_1^{-1}(p+1)^{\frac{p-1}{p}}\phi A^{\frac{\alpha-p}{p+1}}}\\
&\times\Big\{[2 - e^{-c_1^{-1}(p+1)^{\frac{p-1}{p}}\phi A^{\frac{\alpha-p}{p+1}}}]^{-2p}[\lambda c_1^{-p}(p+1)^{p-1}\phi^p A^{\frac{p-\alpha}{p+1}} + pc_1^{-(p+1)}(p+1)^{\frac{p^2-1}{p}}|\nabla\phi|^{p+1}]\\
&+2pe^{-c_1^{-1}(p+1)^{\frac{p-1}{p}}\phi A^{\frac{\alpha-p}{p+1}}}[2 - e^{-c_1^{-1}(p+1)^{\frac{p-1}{p}}\phi A^{\frac{\alpha-p}{p+1}}}]^{-(2p+1)} c_1^{-(p+1)}(p+1)^{\frac{p^2-1}{p}}|\nabla\phi|^{p+1}\Big\}\\
&\le A^\alpha(\lambda c_1^{-p}(p+1)^{p-1} + 3pc_3^{p+1}c_1^{-(p+1)}(p+1)^{\frac{p^2-1}{p}}),\\
a\overline{u}^\alpha &\ge aA^\alpha 2^{-\alpha}.
\end{aligned}$$

By the assumption on a, it is easy to see that

$$(\overline{u}^k)_t \ge \Delta_p\overline{u} - a\overline{u}^\alpha \quad \text{in } \Omega \times (0, \frac{1}{c}).$$

We have further on the boundary that

$$\nabla_p\overline{u} \cdot \nu \ge A^{\frac{(\alpha+1)p}{p+1}} = A^\beta = \overline{u}^\beta.$$

We choose A large enough such that $\overline{u}(x,0) \ge u_0(x)$. So $\overline{u}(x,t)$ is a time-independent supersolution of (1.1). □

Lemma 4.4 *If* $\alpha > k,\ k < p,\ \rho = 0$ *and*

$$a > \frac{2^\alpha}{c_1^{p+1}}[\lambda c_1(p+1)^{p-1} + 3pc_3^{p+1}(p+1)^{\frac{p^2-1}{p}}],$$

then the solutions of (1.1) are globally bounded.

Proof Construct

$$\overline{u}(x,t) = \begin{cases} A[2 - e^{-c_1^{-1}(p+1)^{\frac{p-1}{p}}\phi A^{\frac{\alpha-p}{p+1}}}]^{-1}, & \alpha > p, \\ A^p[2 - e^{-c_1^{-1}(p+1)^{\frac{p-1}{p}}\phi A^{\alpha-p}}]^{-1}, & k < \alpha \leq p, \end{cases}$$

where $A \geq 1$ to be determined.
The subcase $\alpha > p$ has been proven in Lemma 4.3.
In subcase $k < \alpha \leq p$, a direct calculation in $\Omega \times (0, +\infty)$ illustrates that

$$\begin{aligned} \nabla\overline{u} &= -c_1^{-1}(p+1)^{\frac{p-1}{p}} A^\alpha e^{-c_1^{-1}(p+1)^{\frac{p-1}{p}}\phi A^{\alpha-p}}[2 - e^{-c_1^{-1}(p+1)^{\frac{p-1}{p}}\phi A^{\alpha-p}}]^{-2}\nabla\phi, \\ \Delta_p\overline{u} &= A^{\alpha p} e^{-pc_1^{-1}(p+1)^{\frac{p-1}{p}}\phi A^{\alpha-p}} \\ &\times\Big\{[2 - e^{-c_1^{-1}(p+1)^{\frac{p-1}{p}}\phi A^{\alpha-p}}]^{-2p}[\lambda c_1^{-p}(p+1)^{p-1}\phi^p + pA^{\alpha-p}c_1^{-(p+1)}(p+1)^{\frac{p^2-1}{p}}|\nabla\phi|^{p+1}] \\ &+2pA^{\alpha-p}e^{-c_1^{-1}(p+1)^{\frac{p-1}{p}}\phi A^{\alpha-p}}[2 - e^{-c_1^{-1}(p+1)^{\frac{p-1}{p}}\phi A^{\alpha-p}}]^{-(2p+1)}c_1^{-(p+1)}(p+1)^{\frac{p^2-1}{p}}|\nabla\phi|^{p+1}\Big\} \\ &\leq A^{\alpha p}(\lambda c_1^{-p}(p+1)^{p-1} + 3pc_3^{p+1}c_1^{-(p+1)}(p+1)^{\frac{p^2-1}{p}}), \\ a\overline{u}^\alpha &\geq aA^{\alpha p}2^{-\alpha}, \end{aligned}$$

and hence, due to the assumptions on a, we have

$$(\overline{u}^k)_t \geq \Delta_p\overline{u} - a\overline{u}^\alpha \quad \text{in } \Omega \times (0, \frac{1}{c}).$$

Moreover, for $(x,t) \in \partial\Omega \times (0, +\infty)$, we have

$$\nabla_p\overline{u} \cdot \nu \geq A^{p\alpha} = \overline{u}^\beta.$$

If A are large enough that $\overline{u}(x,0) \geq u_0(x)$, then the determined $\overline{u}(x,t)$ is a time-independent supersolution of (1.1). □

The proof for part (ii) of Theorem 1.3 is covered by Lemmas 4.3 and 4.4.

By Lemma 3.3, we see that part (iii) of Theorem 1.3 holds.

Acknowledgments

This work was supported by the Natural Science Foundation Project of China West Normal University (07B047).

References

[1] H. W. Alt and S. Luckhaus, Quasilinear elliptic-parabolic differential equations, *Math. Z.* **183** (1983), 311-341.

[2] H. Amann, Parabolic evolutions equations and nonlinear boundary conditions, *J. Differential Equations* **72** (1988), 201-269.

[3] J. R. Anderson, Local existence and uniqueness of solution of degenerate parabolic equations, *Comm. Partial Differential Equations* **16** (1991), 105-143.

[4] F. Andreu, J. M. Mazón, J. Toledo and J. D. Rossi, Porous medium equation with absorption and a nonlinear boundary condition, *Nonlinear Analysis* **49** (2002), 541-563.

[5] K. Deng and H. A. Levine, The role of critical exponents in blow-up theorems: the sequel, *J. Math. Anal. Appl.* **243** (2000), 85-126.

[6] J. Escher, Global existence and nonexistence for semilinear parabolic systems with nonlinear boundary conditions, *Math. Ann.* **284** (1989), 285-305.

[7] H. A. Levine, The role of critical exponents in blow up theorems, *SIAM Rev.* **32** (1990), 262-288.

[8] H. A. Levine, Stability and instability for solutions of Burder's equation with a nonlinear boundary conditions, *SIAM J. Math. Anal.* **19** (1988), 312-336.

[9] H. L. Li, M. X. Wang and M. Yang, Fujita type profil and blow-up rate for a fast-diffusion equation with nonlinear absorption and boundary flux (preprint).

[10] P. Lindqvist, On the equation $div(|\nabla u|^{p-2}\nabla u) + \lambda|u|^{p-2}u = 0$, *Pro. Amer. Math. Soc.* **109** (1990), 157-164.

[11] P. Lindqvist, On the equation $div(|\nabla u|^{p-2}\nabla u) + \lambda|u|^{p-2}u = 0$, *Pro. Amer. Math. Soc.* **116** (1992), 583-584.

[12] Y. X. Li and C. H. Xie, Blow-up for p-Laplacian parabolic equations, *J. Differential Equations* **20** (2003), 1-12.

[13] A. A. Samarskii, V. A. Galaktionov, S. P. Kurdyumov and A. P. Mikhailov, "Blow-up in Quasilinear Parabolic Equations", *Walter de Gruyter, Berlin*, 1995.

[14] X. Song and S. Zheng, Blow-up and blow-up rate for a reaction-diffusion model with multiple nonlinearities, *Nonlinear Analysis* **54** (2003), 279-289.

[15] W. J. Sun and S. Wang, Nonlinear degenerate parabolic equation with nonlinear boundary condition, *Acta Mathematica Sinica, English Series* **21** (2005), 847-854.

[16] S. Wang and J. Deng, Doubly nonlinear parabolic equation with nonlinear boundary conditions, *J. Math. Anal. Appl.* **255** (2001), 109-121.

[17] Z. Q. Wu, J. N. Zhao, and J. X. Yin el al, "Nonlinear Diffusion Equations", *World Scientific Publishing Co. Inc. River edge, NJ, 2001.*

[18] S. N. Zheng, Nonexistence of positive solutions to a semilinear elliptic system and blow up estimates for a reaction-diffusion system, *J. Math. Anal. Appl.* **232** (1999), 293-311.

In: Evolution Equations: Theories, Solutions and Functions ISBN 978-1-61324-090-8
Editor: Gaston M. N'Guerekata

Chapter 2

QUALITATIVE STUDY OF A PERTURBED CRITICAL SEMI-LINEAR WAVE EQUATION IN VARIABLE METRIC

Olfa Mahouachi*
Faculty of Sciences of Tunis, Department of Mathematics

Abstract

This work is devoted to the description of bounded energy sequences of solutions to the equation

$$(\mathbf{E}) \qquad \partial_t^2 u - \operatorname{div}(A(x)\nabla_x u) + |u|^4 u + |u|^{\alpha-1} u = 0 \qquad \text{in} \quad \mathbb{R}_t \times \mathbb{R}_x^3,$$

where $1 < \alpha < 5$, and A is a regular function valued in the space 3×3 positive defi nite matrix which is the identity outside a fi ed compact. We show that the subcritical perturbation does not induce any new variations on the dynamic of the equation $(\mathbf{E})$ according to the energy-Strichartz norm. We also prove that the description by means of linear concentrating waves given in [13] about critical case still holds for the equation $(\mathbf{E})$.

Keywords: nonlinear wave equations, Stichartz inequalities, geometric-optics, scattering

2000 AMS Subject Classification: 35Q55, 35L25.

1. Introduction

In this paper, we focus on the equation

$$\partial_t^2 u - \operatorname{div}(A(x)\nabla_x u) + |u|^4 u + |u|^{\alpha-1} u = 0 \qquad \text{in} \quad \mathbb{R}_t \times \mathbb{R}_x^3, \tag{1}$$

*E-mail address: olfa.mahouachi@fst.rnu.tn. The author is grateful to the Laboratory of PDE and Applications at the Faculty of Sciences of Tunis.

where $1 < \alpha < 5$ and A is a regular function, valued in the space 3×3 positive definit matrix satisfying

$$(\mathcal{H}) \qquad \begin{cases} c_0|\xi|^2 \leq A(x)\xi.\xi \leq c_0^{-1}|\xi|^2; \quad \xi, x \in \mathbb{R}^3, \\ \\ A(x) \equiv Id, \quad |x| \geq R_0, \end{cases}$$

for some positive constants c_0 and R_0. The operator $\operatorname{div}(A(x)\nabla_x.)$ is the Laplace-Beltrami operator associated to the Riemannian metric $A^{-1}(x)$. In the constant case, namely when $A(x) = Id$, the operator $\Box_A := \partial_t^2 - \operatorname{div}(A(x)\nabla_x)$ is the usual wave operator on $\mathbb{R}_t \times \mathbb{R}_x^3$ define by $\Box := \partial_t^2 - \Delta_x$. The firs condition in $(\mathcal{H})$ guarantee the hyperbolicity of the equation (1), while, the second condition means that the metric associated to A is fla at the infinit .

In the spirit of the works of Bahouri-Gérard [2], Ibrahim [13] and Majdoub [18], our main concern consists firs on proving that the subcritical perturbation in (1) does not induce any new variations in the dynamic of the equation (1) in an appropriate sense. Second, we shall give a description of sequences of solutions to (1) via linear concentrating waves which we will give the definitio in the next section. More precisely, we show that the profil decomposition of sequences of solutions given by Ibrahim in [13] stills valid for sequences of solutions of (1).

First, we recall the global existence and uniqueness result that we know about the Cauchy problem related to the equation (1). An easy adaptation of the arguments used in [16] shows that, for every (φ, ψ) in $(\dot{H}^1(\mathbb{R}^3) \cap L^{\alpha+1}(\mathbb{R}^3)) \times L^2(\mathbb{R}^3)$, there exists a unique solution

$$u \in C(\mathbb{R}, \dot{H}^1(\mathbb{R}^3) \cap L^{\alpha+1}(\mathbb{R}^3)) \cap L^5_{loc}(\mathbb{R}, L^{10}(\mathbb{R}^3)), \quad \partial_t u \in C(\mathbb{R}, L^2(\mathbb{R}^3)),$$

to the equation (1) satisfying

$$(u, \partial_t u)_{/t=0} = (\varphi, \psi).$$

Moreover, this solution satisfie the conservation of the energy

$$\begin{aligned} E(u,t) &:= \int_{\mathbb{R}^3} [\frac{1}{2}(|\partial_t u|^2 + |A^{\frac{1}{2}}\nabla_x u|^2) + \frac{1}{\alpha+1}|u|^{\alpha+1} + \frac{1}{6}|u|^6](t,x)dx \\ &:= E_0(u,t) + \int_{\mathbb{R}^3} [\frac{1}{\alpha+1}|u|^{\alpha+1} + \frac{1}{6}|u|^6](t,x)dx \\ &= E(u,0). \end{aligned}$$

The equation (1) is a subcritical perturbation of the critical wave equation

$$\Box_A v + |v|^4 v = 0, \tag{2}$$

which belongs to a large class of equations of type

$$\Box_A v + |v|^{p-1} v = 0, \tag{3}$$

where $1 < p < +\infty$. The critical growth related to the equation (2) is the exponent $p_c = 5$. The equation (2) is critical in the sense that no global solutions are constructed yet with growth greater than 5 and with arbitrary initial data in the energy space $\mathcal{E} := \dot{H}^1(\mathbb{R}^3) \times L^2(\mathbb{R}^3)$. The equation (3) has been studied by Gérard [6] and Bahouri and Gérard [2] by comparison with the linear wave equation

$$\square_A w = 0. \tag{4}$$

In what follows, we recall some few facts dealing with the linearization and the description by means of linear concentrating waves. In order to proceed in a more easy way, we give several definitions
Let (φ_n, ψ_n) be a sequence in $\dot{H}^1(\mathbb{R}^3) \times L^2(\mathbb{R}^3)$ such that

$$(\varphi_n, \psi_n) \rightharpoonup (0,0) \quad \text{in} \quad \dot{H}^1(\mathbb{R}^3) \times L^2(\mathbb{R}^3) \tag{5}$$

$$\exists\, R_0 > 0, \quad \forall\, n, \quad \text{supp}(\varphi_n) \cup \text{supp}(\psi_n) \subset \{\, |x| \leq R_0 \,\}. \tag{6}$$

We denote by (v_n) (respectively (w_n)) the sequences of finit energy solutions to (3) (respectively (4)) such that

$$(v_n, \partial_t v_n)_{/t=0} = (w_n, \partial_t w_n)_{/t=0} = (\varphi_n, \psi_n).$$

Definition 1.1 *Let T be a positive real number. We say that the sequence (v_n) is linearizable on $[0,T]$ if*

$$\sup_{t\in[0,T]} E_0(v_n - w_n, t) \longrightarrow 0 \quad \textit{as} \quad n \to +\infty.$$

In the constant case, Gérard showed in [6] that in the subcritical case $(p < 5)$, the sequences of solutions of (3) are linearizable on every time bounded interval. But, in the critical case; $p = 5$, those solutions can present a nonlinear behavior due essentially to the lack of compactness in the Sobolev embedding

$$\dot{H}^1(\mathbb{R}^3) \hookrightarrow L^6(\mathbb{R}^3),$$

which can create some concentrations effects.

In this case (p=5), the sequence (v_n) is linearizable on $[0,T]$ if and only if the sequence (w_n) satisfie

$$\|w_n\|_{L^\infty([0,T],L^6(\mathbb{R}^3))} \longrightarrow 0 \quad \text{as} \quad n \to +\infty. \tag{7}$$

Remark 1.2 *This result illustrates the critical character of the exponent $p = 5$. In fact, we notice in the case of critical growth a qualitative modification in the dynamic of the equation.*

The same result was proved by Ibrahim-Majdoub in [15] in the variable critical case for dimensions $d \geq 3$. In this case, the equation takes the following form

$$\square_A u + |u|^{p-1} u = 0 \qquad \text{in} \quad \mathbb{R}_t \times \mathbb{R}^d_x, \tag{8}$$

where $1 < p \leq p_c$ and $p_c = \frac{d+2}{d-2}$ is the critical exponent. They also showed that the necessary and sufficien condition (7) can be reduced to a sufficien condition satisfie by the initial Cauchy data. The idea is to exhibit a transport equation showing that the propagation of the defect of compactness is done along the radius of the Hamiltonian $H_{\sqrt{A(x)\xi.\xi}}$.

As regards the total description by profil decomposition, we recall that in the constant coefficien case, Bahouri and Gérard [2] give a total description of sequences of solutions (v_n) to the equation (2). This result can be seen as a stability of the class of Shatah-Struwe solutions under weak convergence in the energy space. The so-called structure Theorem in [2] basically states that any such sequence (v_n) with bounded initial data can be decomposed as a superposition of nonlinear concentrating waves $(v_n^{(j)})$ (see [2] for the detailed definitio of concentrating wave and the properties therein) with

$$\Box v_n^{(j)} + v_n^{(j)5} = 0, \quad (v_n^{(j)}, \partial_t v_n^{(j)})(t_n^{(j)}, .) = \frac{1}{\sqrt{h_n}}\Big(\varphi, \frac{\psi}{h_n}\Big)\Big(\frac{\cdot - x_n}{h_n}\Big).$$

Therefore, these nonlinear concentrating waves can be reduced, using scattering theory, to their corresponding linear concentrating waves, namely

$$\Box w_n^{(j)} = 0, \qquad (w_n^{(j)}, \partial_t w_n^{(j)})(t_n^{(j)}) = (v_n^{(j)}, \partial_t v_n^{(j)})(t_n^{(j)}).$$

Recently, Majdoub proves in [18] that the description given by Bahouri and Gérard [2] of the sequence (v_n) of solutions of (2) remains valid for the sequences (u_n) of solutions of (1) in the constant case by proving that the remainder terms $r_n = u_n - v_n$ satisfy $|||r_n|||_I \to 0$ as $n \to +\infty$, where $|||.|||_I$ corresponds to the energy-Strichartz norm on an interval of time I containing 0 and given by

$$|||g|||_I := \sup_{t\in I} (\|\nabla_{t,x} g\|_{L^2(\mathbb{R}^3)}) + \|g\|_{L^5(I,L^{10}(\mathbb{R}^3))}. \tag{9}$$

In the variable case, Ibrahim [13] has investigated the equation (8) in the absence of focus. In order to give a total description to sequences of solutions to (8), he showed that by fixin the coefficient on concentrating points, nonlinear geometrical optics of (8), is reduced to linear geometrical optics. The proof was carried out by the use of the scattering related to the critical equation in the constant case.
In this paper, we are invited to show firs that the perturbation term in (1) has no impact on the dynamic of the equation (1) on every bounded interval of time I containing zero, and according to the energy-Strichartz norm define by (9). Concretely, we shall compare in the energy-Strichartz space the sequences of solutions of (1) with those of (2) stemming from the same initial Cauchy data (φ_n, ψ_n) satisfying the conditions (5) and (6). Second, we study the particular case of linear concentrating wave proving that the profil decomposition given in [13] still holds for sequences of solutions of (1).

The rest of the paper is organized as follows: in the next section we give some defi nitions and notations in order to state our main result. The third section is devoted to the proof of our firs result (see Theorem 2.1 below). The last section deals with the description of sequences of solutions of (1) via linear concentrating waves. We mention that these results use the so-called Strichartz and energy estimates and absorptions Lemmas. We also mention that all the results remains true in dimensions greater than 3.

2. Definitions and Main Results

Our firs result deals with the qualitative study of bounded energy sequences of solutions to the equation (1).

Theorem 2.1 *Let I be a bounded interval of time containing zero. Let (φ_n, ψ_n) be a bounded sequence in the energy space $\mathcal{E}$ satisfying the conditions (5) and (6). Let $\underline{u} = (u_n)$ be the sequence of solutions of (1) and $\underline{v} = (v_n)$ the sequence of solutions of (2) satisfying*

$$(u_n, \partial_t u_n)_{/t=0} = (v_n, \partial_t v_n)_{/t=0} = (\varphi_n, \psi_n), \tag{10}$$

then, we have

$$\overline{\lim_n} \ |||u_n - v_n|||_I = 0. \tag{11}$$

Remark 2.2 *We notice that the subcritical perturbation does not induce any new oscillations or energy concentrations.*

In what follows and before giving our second result, we introduce the definitio of a linear concentrating wave and recall the description given in [13]. We mention that in [13], Ibrahim's results are given in dimensions $d \geq 3$. We begin by recalling some geometric definition and notations. We refer to [5] for the geometric tools.
We denote by $a_{i,j}(x)$ the coefficient of the matrix $A(x)$ and $a^{i,j}(x)$ those of $A^{-1}(x)$. We endow $\mathbb{R}^3$ by the metric g define by

$$g = \sum_{i,j} a^{i,j} dx^i \otimes dx^j.$$

For all $x \in \mathbb{R}^3$, and $u, v \in T_x\mathbb{R}^3$, we defin the inner product

$$g_x(u, v) = < A^{-1}(x)u, v >,$$

and $|\cdot|_x$ the corresponding norm. We denote by $r_0 > 0$ the lower bound for the injectivity radius on $(\mathbb{R}^3, A^{-1}(x))$, $B(x, r)$ is the euclidean ball with center x and radius r, and $B'(x, r)$ the geodesic ball with center x and radius r,

$$B'(x, r) := \{y \in \mathbb{R}^3 \quad \text{such that} \quad |\exp_x^{-1}(y)|_x < r\}.$$

Let $\theta \in C_0^\infty(B(0, r_0))$ be a cut-off function satisfying $\theta(x) \equiv 1$ in $B(0, \frac{r_0}{2})$. For all $x_1, x_2 \in \mathbb{R}^3$, we set $\theta_{x_2}(x_1) = \theta(\exp_{x_2}^{-1}(x_1))$. Note that θ_{x_2} is well define because θ is supported in the ball $B(0, r_0)$.

Definition 2.3 [Linear concentrating wave in the variable case]
A linear concentrating data is a triple $\mathcal{W} = [(\varphi, \psi), \underline{\varepsilon}, (\underline{x}, \underline{t})]$ such that $(\varphi, \psi) \in \mathcal{E}$, $\underline{\varepsilon} = (\varepsilon_n)$ is a sequence of positive numbers going to 0 as n goes to the infinit and $(\underline{x}, \underline{t}) = (x_n, t_n)$ is a sequence of $\mathbb{R}^3 \times \mathbb{R}$ such that $\lim_n (x_n, t_n) = (x_\infty, t_\infty)$. The linear concentrating wave associated to the linear concentrating data $\mathcal{W}$ is the unique sequence of solutions (w_n) of (4) satisfying

$$\begin{cases} w_n(t_n, .) &= \frac{1}{\sqrt{\varepsilon_n}} \theta(\exp_{x_n}^{-1} .)\varphi\left(\frac{\exp_{x_n}^{-1} .}{\varepsilon_n}\right) + o(1) \quad \text{in} \quad \mathcal{E}(B'(x_n, r_0)) \\ \partial_t w_n(t_n, .) &= \frac{1}{\sqrt{\varepsilon_n}} \theta(\exp_{x_n}^{-1} .)\frac{1}{\varepsilon_n}\psi\left(\frac{\exp_{x_n}^{-1} .}{\varepsilon_n}\right) + o(1) \quad \text{in} \quad \mathcal{E}(B'(x_n, r_0)) \\ (w_n, \partial_t w_n)(t_n, .) &= o(1) \quad \text{in} \quad \mathcal{E}(\overline{B'^c}(x_n, r_0)). \end{cases}$$

Remark 2.4 *1. We write* $\underline{w} = [(\varphi, \psi), \underline{\varepsilon}, (\underline{x}, \underline{t})]$ *the linear concentrating wave associated to the concentrating data* $[(\varphi, \psi), \underline{\varepsilon}, (\underline{x}, \underline{t})]$*, and from now, we will not distinguish between the linear concentrating data and its linear concentrating wave.*

2. *In the constant case,* $\exp_{x_n}^{-1}(x) = x - x_n$ *and* $r_0 = +\infty$.

3. *Under the hypothesis satisfied by the metric* g*, we still use the same definition of a linear concentrating wave as in [2]. Indeed, there is a local equivalence between the geodesic distances and the Euclidean ones according to the following inequality*

$$c_\infty |x - y| \leq |\exp_x^{-1}(y)| \leq {c_\infty}^{-1} |x - y| \quad \forall \quad x, y \in \mathbb{R}^3,$$

where $0 < c_\infty \leq 1$*. Thus, one may be tempted to show that in the geodesic ball* $B'(x_n, r_0)$*, we have*

$$\underline{w}(t_n, .) = \frac{1}{\sqrt{\varepsilon_n}} \theta(\exp_{x_n}^{-1} .) \varphi\Big(\frac{\exp_{x_n}^{-1} .}{\varepsilon_n}\Big) = \frac{1}{\sqrt{\varepsilon_n}} \varphi\Big(\frac{. - x_n}{\varepsilon_n}\Big) + R_n(.),$$

with,

$$\lim_n \quad \|R_n\|_{L^2(\mathbb{R}^3)} = 0.$$

Definition 2.5 *Let* $\underline{w} = (w_n)$ *be a linear concentrating wave, the nonlinear profile associated to* $\underline{w}$ *corresponds to the solution* $\underline{v} = (v_n)$ *to the cauchy problem*

$$\begin{cases} \Box_A v_n & + & |v_n|^4 v_n = 0 \\ (v_n, \partial_t v_n)_{/t=0} & = & (w_n, \partial_t w_n)_{/t=0}. \end{cases}$$

We denote by $S^*(\mathbb{R}^3)$ the unit co-sphere bundle and, $|\Gamma|$ is the surface measure of a subset Γ of $S^*(\mathbb{R}^3)$ induced by the metric g on $T_x^*\mathbb{R}^3$.
Notation We consider

$$\mathcal{F}_{x_\infty}(x, t) := \{\xi \in S_x^*(\mathbb{R}^3) \quad \text{such that} \quad \exp_x(t\xi) = x_\infty\},$$

the set of directions of geodesics stemming from a point x and reaching x_∞ in a time t. Recalling that by fixin x, x_∞ and the time t, the existence of such geodesics is due to the Cauchy Theorem.

Definition 2.6 *The point* $x_\infty \in \mathbb{R}^3$ *is said to be focus if there exist a point* $x \in \mathbb{R}^3$ *and a time* t *such that* $\mathcal{F}_{x_\infty}(x, t)$ *has a positive surface measure.*

Before coming to the second main Theorem, we recall that in [13] a description of sequences (v_n) of solutions to (2) by linear concentrating waves (w_n) was given. Concretely, it was shown the existence of some compact terms r_n such that $r_n = v_n - w_n$ goes to zero as n goes to infinit according to the energy-Strichartz norm on some interval of time I containing 0 and not t_∞. The above result was established under the following geometric condition

$$\mathbf{H}(I, x_\infty): \qquad |\mathcal{F}_{x_\infty}(y, t_\infty - t)| = 0 \qquad \forall t \in I \quad \text{and} \quad \forall y \in \mathbb{R}^3.$$

Remarks 2.7 1) *The choice that I doesn't contain t_∞ is made to avoid concentrations effects. Indeed, at $t = t_\infty$, we have*

$$\mathcal{F}_{x_\infty}(y,0) := \{\xi \in S^*_x(\mathbb{R}^3) \quad \textit{such that} \quad \exp_y(0) = x_\infty\}$$

which doesn't have a null surface measure.
2) *The geometric condition $\mathbf{H}(I, x_\infty)$ is made to avoid the lack of compactness in some Sobolev embeddings needed in the sequel and takes into account the fact that in the variable case the linear solution $\underline{w}$ is able to concentrate several times between 0 and t_∞. However, when the metric is flat, this condition is global, because the geodesics are the right-hand sides stemming from $t = t_\infty$ which never intersect.*

For more details, the main result in [13] reads,

Theorem 2.8 *Let $\underline{w} = [(\varphi,\psi), \underline{\varepsilon}, (\underline{x},\underline{t})]$ be a linear concentrating wave, $\underline{v} = (v_n)$ its nonlinear profile. The initial data are supposed to be bounded in the energy space. There exists two linear concentrating waves*

$$\underline{w}^{\pm} = [(\varphi^\infty,\psi^\infty)_\pm, \underline{\varepsilon}, (\underline{x},\underline{t})]$$

such that for every interval of time I containing zero and not t_∞, satisfying the condition $\mathbf{H}(I, x_\infty)$, we have

$$v_n = [(\varphi^\infty,\psi^\infty)_\pm, \underline{\varepsilon}, (\underline{x},\underline{t})] + \rho^{\pm}_{n,\lambda} \qquad \textit{on} \quad I^{\pm}_{n,\lambda}, \tag{12}$$

such that

$$\overline{\lim_n} \; |||\rho^{\pm}_{n,\lambda}|||_{I^{\pm}_{n,\lambda}} \longrightarrow 0 \quad \textit{as} \quad \lambda \longrightarrow +\infty, \tag{13}$$

where K is a compact subset of I and

$$I^{\pm}_{n,\lambda} := \{t \in K \quad \textit{such that} \quad \pm(t_\infty - t \leq -\lambda\varepsilon_n)\}.$$

The key Lemma used in the proof of the above Theorem is the following.

Lemma 2.9 *Assuming that the geometric condition is satisfied on the point of concentration x_∞, then for all compact subset K of I, the linear concentrating wave satisfies*

$$\overline{\lim_n} \sup_{t \in K \setminus [t_\infty - \lambda\varepsilon_n, t_\infty + \lambda\varepsilon_n]} \|w_n(t,.)\|_{L^6(\mathbb{R}^3)} \longrightarrow 0 \qquad \textit{as} \qquad \lambda \longrightarrow +\infty. \tag{14}$$

We refer to [13] for the proof of this Lemma. Now, we are ready to state our next result.

Theorem 2.10 *Let $\underline{w} = [(\varphi,\psi), \underline{\varepsilon}, (\underline{x},\underline{t})]$ be a linear concentrating wave, $\underline{u} = (u_n)$ its nonlinear perturbed profile; that is $\underline{u}$ is the sequence of solutions of (1) satisfying*

$$(u_n, \partial_t u_n)_{/t=0} = (w_n, \partial_t w_n)_{/t=0},$$

where the initial data are supposed to be bounded in the energy space. For every interval of time I containing zero and not t_∞, satisfying the geometric condition $\mathbf{H}(I, x_\infty)$, and under the same notations of $I^{\pm}_{n,\lambda}$ in Theorem 2.8, we have

$$u_n = [(\varphi^\infty,\psi^\infty)_\pm, \underline{\varepsilon}, (\underline{x},\underline{t})] + \gamma^{\pm}_{n,\lambda} \qquad \textit{on} \quad I^{\pm}_{n,\lambda}, \tag{15}$$

such that

$$\overline{\lim_n} \quad |||\gamma_{n,\lambda}^{\pm}|||_{I_{n,\lambda}^{\pm}} \longrightarrow 0 \quad as \quad \lambda \longrightarrow +\infty, \tag{16}$$

where

$$[(\varphi^{\infty}, \psi^{\infty})_{\pm}, \underline{\varepsilon}, (\underline{x}, \underline{t})]$$

are the two linear concentrating waves in the Theorem 2.8 given by

$$\underline{w}^{-} = [(\varphi, \psi), \underline{\varepsilon}, (\underline{x}, t_{\infty})] \quad and \quad \underline{w}^{+} = [S^{\infty}(\varphi, \psi), \underline{\varepsilon}, (\underline{x}, t_{\infty})].$$

We mention that in Theorem 2.8, the existence of the two linear concentrating waves $\underline{w}^{\pm}$ is due to the scattering operator S^{∞}. We give the details in the fourth section.
This description shows that the subcritical term in the equation (1) does not modify the dynamic of the equation in the energy-Strichartz space for bounded intervals of time far away from the time of concentration t_{∞}. Where the times are suficentl close to the concentration time t_{∞}, we show that $\underline{u}$ takes the same behavior as in the constant coefficien case.

2.1. Tools of the Proofs

We recall in this paragraph some a priori estimates and some absorption Lemmas that are needed in several proofs.

Lemma 2.11 [Energy estimate]
Let h be a solution in $I = [a, b]$ of $\Box_A h = f \in L^1(I, L^2(\mathbb{R}^3))$. Assume that

$$(h, \partial_t h) \in C(I, \dot{H}^1(\mathbb{R}^3)) \times C(I, L^2(\mathbb{R}^3)),$$

then,

$$\sup_{t\in I} \|\nabla_{t,x} h\|_{L^2(\mathbb{R}^3)} \leq C\{\|\nabla_{t,x} h_{/t=a}\|_{L^2(\mathbb{R}^3)} + \|f\|_{L^1(I,L^2(\mathbb{R}^3))}\}. \tag{17}$$

Lemma 2.12 [Strichartz estimate]
Under the same notations and hypothesis of Lemma 2.11, the function h satisfie

$$\|h\|_{L^q(I,L^r(\mathbb{R}^3))} \leq C_q\{\|\nabla_{t,x} h_{/t=a}\|_{L^2(\mathbb{R}^3)} + \|f\|_{L^1(I,L^2(\mathbb{R}^3))}\}, \tag{18}$$

with

$$\frac{1}{q} + \frac{3}{r} = \frac{1}{2} \quad \text{and} \quad q > 2.$$

We mention that every finit energy solution to the free wave equation $\Box_A w = 0$ satisfie the additional property $w \in L^q(I, L^r(\mathbb{R}^3))$, with

$$\|w\|_{L^q(I,L^r(\mathbb{R}^3))} \leq \tilde{C}_q E_0^{1/2}(w, 0).$$

We note also that in the constant coefficien case the inequalities (18) are global.

Strichartz inequalities were obtained by Ginibre-Velo [10] in the constant case and by Kapitanski [17] when the coefficient are smooth. We also mention that these inequalities will often be combined with the following classical absorptions Lemmas. We leave the proofs to the reader.

Lemma 2.13 [Boot-strap Lemma]
Let $X : [0,T] \longrightarrow \mathbb{R}^+$ be a continuous function such that

$$X(t) \le a + bX^\theta(t) \qquad \text{where,} \quad a,b>0 \quad \theta>1$$

$$a < (1-\frac{1}{\theta}).\frac{1}{(\theta b)^{\frac{1}{\theta}}} \qquad \text{and} \qquad X(0) \le \frac{1}{(\theta b)^{\frac{1}{\theta-1}}},$$

then,

$$\forall t \in [0,T] \qquad X(t) \le \frac{\theta}{\theta-1}a.$$

Lemma 2.14 *Let $X = X(t)$ be a nonnegative continuous function on $[0,T]$ satisfying*

$$X(0)=0 \quad \text{and} \qquad X(t) \le C(a + X^\beta(t) + X^5(t)),$$

where C is a positive constant, $1<\beta<5$ and $0<a<a_0(C,\beta)$, then, there exists $\tilde{C}>0$ such that

$$X(t) \le \tilde{C}a.$$

3. Proof of Theorem 2.1

We start by proving the Theorem 2.1. For the sake of simplicity, set $I = [0,T]$ for any positive time T. The proof will be made in several steps. First we apply Strichartz estimates on the remainder term $u_n - v_n$. Then we prove the existence of a decomposition of time up to small intervals I_k such that the quantity $\|v_n\|_{L^5(I_k,L^{10}(\mathbb{R}^3))}$ is small enough. This fact enables us to absorb the linear term in the Strichartz inequality. Then, using Lemma 2.14, then the result given by Lemma 3.1 below, our result holds. We need firs the following result.

Lemma 3.1 *Let (v_n) be a sequence of solutions of (2), then*

$$\overline{\lim_n} \quad \|v_n\|_{L^\alpha([0,T],L^{2\alpha}(\mathbb{R}^3))} = 0. \tag{19}$$

[Proof of Lemma 3.1]
Using finit speed of propagation, we can see that v_n is supported in a fi ed compact set of $[0,T]\times\mathbb{R}^3$. Applying Cauchy-Schwarz, then, Hölder inequalities, we obtain for $\varepsilon>0$

$$\begin{aligned}\|v_n\|^\alpha_{L^\alpha([0,T],L^{2\alpha}(\mathbb{R}^3))} &\le T^{1/2}\Big\{\Big(\int_{\{|v_n|^\alpha<\varepsilon\}}|v_n|^{2\alpha}dxdt\Big)^{1/2} + \Big(\int_{\{|v_n|^\alpha\ge\varepsilon\}}|v_n|^{2\alpha}dxdt\Big)^{1/2}\Big\}\\ &\le C\varepsilon + \|v_n\|^\alpha_{L^5([0,T],L^{10}(\mathbb{R}^3))}\mu^{(\frac{1}{2})(1-\frac{\alpha}{5})}\{|v_n|^\alpha\ge\varepsilon\},\end{aligned}$$

where μ be the Lebesgue measure on $\mathbb{R}^4$. Noticing that C doesn't depend on n because of the finit speed of propagation. Applying Strichartz estimates, one can deduce that v_n is bounded in $L^5([0,T],L^{10}(\mathbb{R}^3))$. It suffice to prove that

$$\mu\{|v_n|^\alpha\ge\varepsilon\} \to 0 \quad \text{as} \quad n\to\infty.$$

Applying a result showed in [15], that is; under the hypothesis satisfie by the sequence (φ_n, ψ_n), and the metric $A(x)$, we have

$$v_n \rightharpoonup 0 \quad \text{in} \quad H^1(]0,T[\times\mathbb{R}^3),$$

we can deduce that up to extraction, and from Reillich Theorem, $v_n \longrightarrow 0$ in $L^2([0,T]\times\mathbb{R}^3)$. We conclude the proof by the following Bienaymé-Tchebychev inequality

$$\mu\{|v_n| \geq \varepsilon\} \leq \frac{1}{\varepsilon^2}\|v_n\|_{L^2}^2.$$

■

Now, we continue the proof of Theorem 2.1. Setting $h_n = u_n - v_n$. The remainder term h_n solves the following problem

$$\Box_A h_n = f_n, \qquad (h_n, \partial_t h_n)_{/t=0} = (0,0), \qquad \text{in} \quad [0,T]\times\mathbb{R}^3$$

where

$$f_n = |v_n|^4 v_n - |v_n + h_n|^4(v_n + h_n) - |v_n + h_n|^{\alpha-1}(v_n + h_n).$$

Applying Strichartz estimates (18) to the sequence h_n, we obtain

$$\|h_n\|_{L^5([0,T],L^{10}(\mathbb{R}^3))} \leq C\|f_n\|_{L^1([0,T],L^2(\mathbb{R}^3))}.$$

We have

$$\begin{aligned}\|f_n\|_{L^1([0,T],L^2(\mathbb{R}^3))} &\leq C\{\|v_n\|^5_{L^5([0,T],L^{10}(\mathbb{R}^3))} + \|(v_n+h_n)^5\|_{L^1([0,T],L^2(\mathbb{R}^3))}\} \\ &+ C\||v_n+h_n|^\alpha\|_{L^1([0,T],L^2(\mathbb{R}^3))},\end{aligned}$$

recalling that

$$\|(v_n+h_n)^5\|_{L^1([0,T],L^2(\mathbb{R}^3))} \leq \sum_0^5 \|(v_n)^i(h_n)^{(5-i)}\|_{L^1([0,T],L^2(\mathbb{R}^3))},$$

thus, using Hölder inequalities in the above sum and finit speed of propagation, we get

$$\begin{aligned}\|h_n\|_{L^5([0,T],L^{10}(\mathbb{R}^3))} &\leq C\{\|v_n\|^\alpha_{L^\alpha([0,T],L^{2\alpha}(\mathbb{R}^3))} + \|v_n^4 h_n\|_{L^1([0,T],L^2(\mathbb{R}^3))}\} \\ &+ C\{\|h_n\|^2_{L^5([0,T],L^{10}(\mathbb{R}^3))} + \|h_n\|^\alpha_{L^5([0,T],L^{10}(\mathbb{R}^3))}\} \\ &+ C\|h_n\|^5_{L^5([0,T],L^{10}(\mathbb{R}^3))}.\end{aligned}$$

C is a constant depending on T, R_0 and α. To conclude the proof, we need the following deformation Lemma.

Lemma 3.2 *For every $T, \varepsilon > 0$, for every sequence $(f_n) \in L^5([0,T])$ there exist a positive integer $m = m(\varepsilon)$ and a finite sequence of real numbers $0 < y_1^n < y_2^n < .... < y_m^n$ such that*

$$\overline{\lim_n}\ \|f_n\|_{L^5([y_{k-1}^n, y_k^n])} \leq \varepsilon, \qquad 1 \leq k \leq m, \tag{20}$$

where we have set, $y_0^n = 0$, $y_m^n = T$.

[Proof of Lemma3.2]
First, for a fi ed integer n, we shall prove the existence of a finit sequence of real numbers $0 < t_1^n < .... < t_{m(n)}^n$ satisfying

$$\|f_n\|_{L^5([t_{k-1}^n, t_k^n])} \leq \varepsilon, \qquad 1 \leq k \leq m(n),$$

where

$$\frac{T}{m(n)} = t_k^n - t_{k-1}^n, \quad t_0^n = 0 \quad \text{and} \quad t_{m(n)}^n = T.$$

By a density argument, we let $(\varphi_j^n)_j$ be a C_0^∞ approximation of (f_n) in $L^5([0,T])$. By Hölder inequalities, one can write

$$\|\varphi_j^n\|_{L^5([t_{k-1}^n, t_k^n])} \leq (t_k^n - t_{k-1}^n)^{1/10} \sup_{j\in\mathbb{N}} \|\varphi_j^n\|_{L^{10}([0,T])}.$$

It suffice to choose the integer $m(n)$ such that

$$\left(\frac{T}{m(n)}\right)^{1/10} \sup_{j\in\mathbb{N}} \|\varphi_j^n\|_{L^{10}([0,T])} \leq \varepsilon.$$

We have two possibilities about the sequence $(m(n))_n$. First, we may assume that $\sup\{m(n), n \in \mathbb{N}\}$ is finite thus, we can rearrange in order the terms t_k^n for all $0 \leq k \leq m := \max\{m(n), n \in \mathbb{N}\}$. Taking the necessary intersections, we construct a finit sequence of times $0 = y_0^n < y_1^n < y_2^n < .... < y_m^n = T$ such that for all integer n, we have

$$\|f_n\|_{L^5([y_{k-1}^n, y_k^n])} \leq \varepsilon, \qquad 1 \leq k \leq m.$$

Now, suppose that $\sup\{m(n), n \in \mathbb{N}\} = \infty$. Up to extraction, we let $(m_{\sigma(n)})_n$ be an increasing subsequence of $(m(n))_n$ going to the infinit . Set $x_k = t_{\sigma(k)}^{\sigma(k)}$ for all $k \in \mathbb{N}$, hence, we easily remark that $x_k - x_{k-1} < \frac{T}{m_{\sigma(k)}}$ for all integer $k \geq 1$. Consider $\eta_k > 0$ such that $x_k - x_{k-1} + \eta_k \leq \frac{T}{m_{\sigma(k)}}$ and fi two positive reals T' and l such that $l > T' > T$. Let (β_n) be an increasing sequence going to l, then, for all integer $n \geq N$ large enough, we have

$$\beta_n > x_{n+1} + \eta_{n+1} + \frac{T}{m_{\sigma(n+1)}} > x_n + \eta_n + \frac{T}{m_{\sigma(n)}}. \tag{21}$$

Using (21), one can write

$$[0,T'] \subset [0,x_1] \cup [x_1, x_1 + \frac{T}{m_{\sigma(1)}}] \cup \Big(\bigcup_{n\in\mathbb{N}, n\geq 2}]x_{n-1} + \frac{T}{m_{\sigma(n)}}, x_n + \eta_n + \frac{T}{m_{\sigma(n)}}[\Big)$$
$$\cup \Big(\bigcup_{n\in\mathbb{N}, n\geq N}]x_n + \eta_n + \frac{T}{m_{\sigma(n)}}, \beta_n[\Big).$$

Noticing that $\forall n \in \mathbb{N}^*$, we have

$$\|f_n\|_{L^5(]x_{n-1} + \frac{T}{m_{\sigma(n)}}, x_n + \eta_n + \frac{T}{m_{\sigma(n)}}[)} \leq \varepsilon.$$

Using a compactness argument, we can extract a finit decomposition of $[0,T']$ as following

$$[0,T'] \subset [0,x_1] \cup [x_1, x_1 + \frac{T}{m_{\sigma(1)}}] \cup \Big(\bigcup_{2\le n\le p}]x_{n-1} + \frac{T}{m_{\sigma(n)}}, x_n + \eta_n + \frac{T}{m_{\sigma(n)}}[\Big)$$
$$\cup \Big(\bigcup_{N\le n\le q}]x_n + \eta_n + \frac{T}{m_{\sigma(n)}}, \beta_n[\Big).$$

Replacing each interval of the above decomposition by its closure and taking the necessary intersections, then intersecting this decomposition with $[0,T]$, we obtain a finit decomposition of $[0,T]$ in closed intervals $[y^n_{k-1}, y^n_k]$, where $1 \le k \le m(\varepsilon)$. Moreover, we have

$$\sup_n \|f_n\|_{L^5([y^n_{k-1},y^n_k])} \le \varepsilon, \qquad 1 \le k \le m.$$

Setting

$$[0,T] = I^n_1 \cup I^n_2 \cup ..... \cup I^n_m, \qquad I^n_k = [y^n_{k-1}, y^n_k], \quad 1 \le k \le m,$$

we end the proof of Lemma 3.2. ∎

We continue the last part of the proof of Theorem 2.1. If we replace the sequence (f_n) in Lemma 3.2 by the sequence $(\|v_n(t,.)\|_{L^{10}(\mathbb{R}^3)})$, we obtain a decomposition of $[0,T]$ as define above and satisfying

$$\overline{\lim_n} \; \|v_n\|_{L^5([y^n_{k-1},y^n_k],L^{10}(\mathbb{R}^3))} \le \varepsilon, \qquad 1 \le k \le m,. \tag{22}$$

We iterate the Strichartz inequalities on each I^n_k by starting with I^n_1, we get

$$\begin{aligned} \|h_n\|_{L^5(I^n_1,L^{10}(\mathbb{R}^3))} &\le C\{\|v_n\|^{\alpha}_{L^{\alpha}(I^n_1,L^{2\alpha}(\mathbb{R}^3))} + \|v^4_n h_n\|_{L^1(I^n_1,L^2(\mathbb{R}^3))}\} \\ &+ C\{\|h_n\|^2_{L^5(I^n_1,L^{10}(\mathbb{R}^3))} + \|h_n\|^{\alpha}_{L^5(I^n_1,L^{10}(\mathbb{R}^3))}\} \\ &+ C\|h_n\|^5_{L^5(I^n_1,L^{10}(\mathbb{R}^3))}. \end{aligned}$$

From (22), one can assume that up to extraction,

$$\|v_n\|_{L^5(I^n_0,L^{10}(\mathbb{R}^3))} \le \frac{\varepsilon}{2}.$$

If the real ε is chosen such that $C\frac{\varepsilon^4}{16} < 1$, then, we have

$$\begin{aligned} \|h_n\|_{L^5(I^n_1,L^{10}(\mathbb{R}^3))} &\le C\{\|v_n\|^{\alpha}_{L^{\alpha}(I^n_1,L^{2\alpha}(\mathbb{R}^3))} + \|h_n\|^2_{L^5(I^n_1,L^{10}(\mathbb{R}^3))}\} \\ &+ C\{\|h_n\|^{\alpha}_{L^5(I^n_1,L^{10}(\mathbb{R}^3))} + \|h_n\|^5_{L^5(I^n_1,L^{10}(\mathbb{R}^3))}\}. \end{aligned}$$

By applying the absorption Lemma 2.14, there exists a constant $\tilde{C}$ such that

$$\|h_n\|_{L^5(I^n_1,L^{10}(\mathbb{R}^3))} \le \tilde{C}\|v_n\|^{\alpha}_{L^{\alpha}(I^n_1,L^{2\alpha}(\mathbb{R}^3))}.$$

Thus, in the light of Lemma 3.1, we get

$$\overline{\lim_n} \; \|h_n\|_{L^5(I^n_1,L^{10}(\mathbb{R}^3))} = 0.$$

Iterating on the intervals I_k^n, $1 \le k \le m$, the above result still holds on $[0,T]$ by induction. On the other hand, applying the energy estimate to h_n and arguing as before, we get

$$\overline{\lim_n} \sup_{t\in[0,T]} \|\nabla_{t,x} h_n\|_{L^2(\mathbb{R}^3)} = 0.$$

The proof of Theorem 2.1 is complete. ■

4. Study via Linear Concentrating Wave

In this section, we are invited to prove the description given by Theorem 2.10 of sequences of solutions of (1). The proof will be made in three steps according to the time t by comparison with the time of concentration t_∞. For some steps, we use the geometric-optics by "rescaling" the Cauchy problem into the microscopic variables given by

$$s = \frac{t - t_n}{\varepsilon_n} \quad \text{and} \quad y = \frac{x - x_n}{\varepsilon_n},$$

it follows that

$$w_n(t,x) = \frac{1}{\sqrt{\varepsilon_n}} \tilde{w}_n(s,y),$$

where $\tilde{w}_n$ is the solution of the system

$$\begin{cases} \partial_s^2 \tilde{w}_n - \operatorname{div}(A(\varepsilon_n . + x_n)\nabla_y \tilde{w}_n) & = 0 \\ (\tilde{w}_n, \partial_s \tilde{w}_n)_{/s=0} & = \quad (\varphi, \psi). \end{cases}$$

For the sake of simplicity, suppose that $t_n = t_\infty > 0$, $\lim_n x_n = x_\infty = 0$ and $A(0) = Id$. We fi a bounded interval of time I. Since $\lim_n(\varepsilon_n, x_n) = (0, x_\infty)$, one can notice the equivalence between the operators $\operatorname{div}(A(\varepsilon_n y + x_n)\nabla_y .)$ and $\operatorname{div}(A(x_\infty)\nabla_y .)$. On the other hand, by fixin the coefficient at the point of concentration x_∞, and in view of the conditions $x_\infty = 0$ and $A(0) = Id$, one can apply the scattering theory. This procedure enables one to guarantee the existence of the two linear concentrating waves $[(\varphi^\infty, \psi^\infty)_\pm, \underline{\varepsilon}, (\underline{x}, \underline{t})]$.

For the comfort of the reader, we recall that to every solution w^∞ of the equation

$$\Box_{A(x_\infty)} w^\infty = 0, \qquad (w^\infty, \partial_t w^\infty)_{/t=0} = (\varphi, \psi) \in \mathcal{E},$$

it is associated a unique function

$$v_\pm^\infty \in L^5(\mathbb{R}, L^{10}(\mathbb{R}^3)); \quad (v_\pm^\infty, \partial_t v_\pm^\infty) \in C(\mathbb{R}, \mathcal{E})$$

satisfying

$$\Box_{A(x_\infty)} v_\pm^\infty + |v_\pm^\infty|^4 v_\pm^\infty = 0; \qquad \lim_{s\to\pm\infty} E_0(v_\pm^\infty - w^\infty, s) = 0.$$

We recall that the wave-operators $\Omega_\pm^\infty$

$$\begin{aligned} \Omega_\pm^\infty : \mathcal{E} & \longrightarrow \mathcal{E} \\ (\varphi, \psi) & \longrightarrow (v_\pm^\infty, \partial_t v_\pm^\infty)_{/t=0} \end{aligned}$$

are homeomorphism from $\mathcal{E}$ onto itself, and the scattering operator is define by

$$S^{\infty} := (\Omega_{\pm}^{\infty})^{-1} o \Omega_{\pm}^{\infty}.$$

In this case, we set $\underline{w}^- = [(\varphi, \psi), \underline{\varepsilon}, (\underline{x}, t_\infty)]$ and $\underline{w}^+ = [S^\infty(\varphi, \psi), \underline{\varepsilon}, (\underline{x}, t_\infty)]$ the two linear concentrating waves in Theorem 2.8.
We suppose that the geometric condition $\mathbf{H}(I, x_\infty)$ is satisfied Before starting with the study of the nonlinear wave (u_n) described in the Theorem 2.10, we need the following Lemma.

Lemma 4.1 *The linear concentrating waves* $\underline{w}^\pm = [(\varphi^\infty, \psi^\infty)_\pm, \underline{\varepsilon}, (\underline{x}, \underline{t})]$ *satisfy*

$$\overline{\lim_n} \; \|w_n^\pm\|_{L^5(I_{n,\lambda}^\pm, L^{10}(\mathbb{R}^3))} \longrightarrow 0 \qquad as \quad \lambda \longrightarrow +\infty, \tag{23}$$

and

$$\overline{\lim_n} \; \|w_n^\pm\|_{L^\alpha(I_{n,\lambda}^\pm, L^{2\alpha}(\mathbb{R}^3))} \longrightarrow 0 \qquad as \quad \lambda \longrightarrow +\infty. \tag{24}$$

[Proof of Lemma 4.1]
We mention that (24) is a consequence of (23). We omit for the moment this fact and we start by proving (23) in the interval of time $I_{n,\lambda}^-$. For the sake of simplicity, set $I_{n,\lambda}^- = [0, t_\infty - \lambda\varepsilon_n]$. By interpolation, we get

$$\|w_n^-\|_{L^5(I_{n,\lambda}^-, L^{10}(\mathbb{R}^3))}^5 \leq \|w_n^-\|_{L^\infty(I_{n,\lambda}^-, L^6(\mathbb{R}^3))} \|w_n^-\|_{L^4(I_{n,\lambda}^-, L^{12}(\mathbb{R}^3))}^4.$$

Here, $(4, 12)$ is an admissible Strichartz couple. Since the initial data is supposed to be bounded in the energy space, then, applying the $L_t^4(L_x^{12})$ Strichartz estimate, implies that w_n^- is bounded in $L^4(I_{n,\lambda}^-, L^{12}(\mathbb{R}^3))$. On the other hand, taking into account the result of Lemma 2.9, we get

$$\overline{\lim_n} \; \|w_n^-\|_{L^5(I_{n,\lambda}^-, L^{10}(\mathbb{R}^3))} \longrightarrow 0 \qquad \text{as} \quad \lambda \longrightarrow +\infty.$$

Our result holds on the interval $I_{n,\lambda}^-$. It remains to prove (23) on $I_{n,\lambda}^+ := [t_\infty + \lambda\varepsilon_n, T]$. We proceed as previously by replacing w_n^- by w_n^+ and by taking into account the result

$$\overline{\lim_n} \; \|w_n^+\|_{L^\infty(I_{n,\lambda}^+, L^6(\mathbb{R}^3))} \longrightarrow 0 \qquad \text{as} \quad \lambda \longrightarrow +\infty.$$

We mention that the geometric condition $\mathbf{H}(I, x_\infty)$ enables one to avoid the lack of compactness of the sequence w_n^+ on intervals of type $[t_\infty + \lambda\varepsilon_n, T]$.
Now, in order to prove (24), we argue as in the proof of Lemma 3.1. For every $\varepsilon > 0$, by Hölder inequalities, we have

$$\begin{aligned}
\|w_n^-\|_{L^\alpha([0,t_\infty-\lambda\varepsilon_n], L^{2\alpha}(\mathbb{R}^3))}^\alpha &\leq (t_\infty - \lambda\varepsilon_n)^{1/2} \mu\{|w_n^-|^\alpha < \varepsilon\} \\
&+ \|w_n^-\|_{L^5([0,t_\infty-\lambda\varepsilon_n], L^{10}(\mathbb{R}^3))}^\alpha \mu^{(\frac{1}{2})(1-\frac{\alpha}{5})}\{|w_n^-|^\alpha \geq \varepsilon\} \\
&\leq C_{(t_\infty, r_0)}\varepsilon + \|w_n^-\|_{L^5([0,t_\infty-\lambda\varepsilon_n], L^{10}(\mathbb{R}^3))}^\alpha \mu^{(\frac{1}{2})(1-\frac{\alpha}{5})}\{|w_n^-|^\alpha \geq \varepsilon\}.
\end{aligned}$$

In view of (23), it suffice to show that the quantity $\mu^{(\frac{1}{2})(1-\frac{\alpha}{5})}\{|w_n^-|^\alpha \geq \varepsilon\}$ is at least bounded. Indeed, using the inequality

$$\mu\{|w_n^-| \geq \varepsilon\} \leq \frac{1}{\varepsilon^6} t_\infty \sup_{\{t\in[0,t_\infty-\lambda\varepsilon_n]\}} \|w_n^-(t,.)\|_{L^6(\mathbb{R}^3)}^6,$$

and taking in account the result of Lemma 2.9, we conclude the proof of (24) in $I_{n,\lambda}^-$. The same arguments can be used to prove (24) for times bigger than the time of concentration t_∞, which concludes the proof. ■

4.1. Study before the Time of Concentration

In this section, we are interested in studying the wave $\underline{u}$ for times smaller than the time of concentration t_∞, more precisely, on $I_{n,\lambda}^-$. We have to prove the firs assertion in Theorem 2.10, that is

$$\overline{\lim_n} \; |||\gamma_{n,\lambda}^-|||_{I_{n,\lambda}^-} \longrightarrow 0 \quad \text{as} \quad \lambda \longrightarrow +\infty. \tag{25}$$

The remainder term $\gamma_{n,\lambda}^- = u_n - w_n^-$ is solution to the following problem in $I_{n,\lambda}^- \times \mathbb{R}^3$,

$$\Box_A \gamma_{n,\lambda}^- = f_n, \qquad (\gamma_{n,\lambda}^-, \partial_t \gamma_{n,\lambda}^-)_{/t=0} = (0,0)$$

where

$$f_n = -|w_n^- + \gamma_{n,\lambda}^-|^4(w_n^- + \gamma_{n,\lambda}^-) - |w_n^- + \gamma_{n,\lambda}^-|^{\alpha-1}(w_n^- + \gamma_{n,\lambda}^-).$$

Applying Strichartz inequalities, we obtain

$$\|\gamma_{n,\lambda}^-\|_{L^5(I_{n,\lambda}^-, L^{10}(\mathbb{R}^3))} \leq C\|f_n\|_{L^1(I_{n,\lambda}^-, L^2(\mathbb{R}^3))}.$$

We have

$$\|f_n\|_{L^1(I_{n,\lambda}^-, L^2(\mathbb{R}^3))} \leq C\{\|(w_n^- + \gamma_{n,\lambda}^-)^5\|_{L^1(I_{n,\lambda}^-, L^2(\mathbb{R}^3))} + \||w_n^- + \gamma_{n,\lambda}^-|^\alpha\|_{L^1(I_{n,\lambda}^-, L^2(\mathbb{R}^3))}\}.$$

The rest of the proof is very similar to the proof of Theorem 2.1. We begin by applying the Strichartz estimates to w_n^- in $I_{n,\lambda}^-$. Then, we substitute in the Strichartz estimates applied to $\gamma_{n,\lambda}^-$ in $I_{n,\lambda}^-$, we get

$$\begin{aligned}
\|\gamma_{n,\lambda}^-\|_{L^5(I_{n,\lambda}^-, L^{10}(\mathbb{R}^3))} \leq\; & C_\alpha\{\|w_n^-\|_{L^\alpha(I_{n,\lambda}^-, L^{2\alpha}(\mathbb{R}^3))}^\alpha + \|\gamma_{n,\lambda}^-\|_{L^5(I_{n,\lambda}^-, L^{10}(\mathbb{R}^3))}^\alpha\} \\
+\; & C\{\|\gamma_{n,\lambda}^-\|_{L^5(I_{n,\lambda}^-, L^{10}(\mathbb{R}^3))}^5 + \|(w_n^-)^4\gamma_{n,\lambda}^-\|_{L^1(I_{n,\lambda}^-, L^2(\mathbb{R}^3))}\} \\
+\; & C\{E_0^{3/2}(w_n^-, 0)\|\gamma_{n,\lambda}^-\|_{L^5(I_{n,\lambda}^-, L^{10}(\mathbb{R}^3))}^2 + E_0(w_n^-, 0)\|\gamma_{n,\lambda}^-\|_{L^5(I_{n,\lambda}^-, L^{10}(\mathbb{R}^3))}^3\} \\
+\; & C\{E_0^{1/2}(w_n^-, 0)\|\gamma_{n,\lambda}^-\|_{L^5(I_{n,\lambda}^-, L^{10}(\mathbb{R}^3))}^4 + \|\gamma_{n,\lambda}^-\|_{L^5(I_{n,\lambda}^-, L^{10}(\mathbb{R}^3))}^5\} \\
+\; & C\|w_n^-\|_{L^5(I_{n,\lambda}^-, L^{10}(\mathbb{R}^3))}^5.
\end{aligned}$$

Since the initial data is bounded in the energy space, we obtain

$$\|\gamma_{n,\lambda}^-\|_{L^5(I_{n,\lambda}^-, L^{10}(\mathbb{R}^3))} \leq C_\alpha\{\|w_n^-\|_{L^\alpha(I_{n,\lambda}^-, L^{2\alpha}(\mathbb{R}^3))}^\alpha + \|(w_n^-)^4\gamma_{n,\lambda}^-\|_{L^1(I_{n,\lambda}^-, L^2(\mathbb{R}^3))}\}$$

$$\begin{aligned}
&+ \; C\{\|\gamma_{n,\lambda}^-\|^2_{L^5(I_{n,\lambda}^-,L^{10}(\mathbb{R}^3))} + \|\gamma_{n,\lambda}^-\|^3_{L^5(I_{n,\lambda}^-,L^{10}(\mathbb{R}^3))}\} \\
&+ \; C\{\|\gamma_{n,\lambda}^-\|^4_{L^5(I_{n,\lambda}^-,L^{10}(\mathbb{R}^3))} + \|\gamma_{n,\lambda}^-\|^\alpha_{L^5(I_{n,\lambda}^-,L^{10}(\mathbb{R}^3))}\} \\
&+ \; C\{\|\gamma_{n,\lambda}^-\|^5_{L^5(I_{n,\lambda}^-,L^{10}(\mathbb{R}^3))} + \|w_n^-\|^5_{L^5(I_{n,\lambda}^-,L^{10}(\mathbb{R}^3))}\}.
\end{aligned}$$

For all $t \in [0, t_\infty - \lambda\varepsilon_n]$, set $X_n(t) := \|\gamma_{n,\lambda}^-\|_{L^5([0,t],L^{10}(\mathbb{R}^3))}$. Using the result (23) on $I_{n,\lambda}^-$, it exists a real number λ_1 such that for all $\lambda \geq \lambda_1$ and for n greater than any $n(\lambda_1)$, $\|w_n^-\|_{L^5(I_{n,\lambda}^-,L^{10}(\mathbb{R}^3))} \leq \varepsilon/2$. Here, ε is chosen such that $C(\varepsilon/2)^4 < 1$, then, one can absorb the linear term in the above inequality and get

$$\begin{aligned}
X_n(t_\infty - \lambda\varepsilon_n) \; &\leq \; C\{\|w_n^-\|^\alpha_{L^\alpha([0,t_\infty-\lambda\varepsilon_n],L^{2\alpha}(\mathbb{R}^3))} + \|w_n^-\|^5_{L^5(I_{n,\lambda}^-,L^{10}(\mathbb{R}^3))})\} \\
&+ \; C\{X_n^2(t_\infty - \lambda\varepsilon_n) + X_n^3(t_\infty - \lambda\varepsilon_n) + X_n^4(t_\infty - \lambda\varepsilon_n)\} \\
&+ \; C\{X_n^\alpha(t_\infty - \lambda\varepsilon_n) + X_n^5(t_\infty - \lambda\varepsilon_n)\}.
\end{aligned}$$

Applying the absorption Lemma 2.14, a constant $\tilde{C}$ exists such that

$$X_n(t_\infty - \lambda\varepsilon_n) \leq \tilde{C}\{\|w_n^-\|^\alpha_{L^\alpha([0,t_\infty-\lambda\varepsilon_n],L^{2\alpha}(\mathbb{R}^3))} + \varepsilon/2\}.$$

Using (24), we obtain

$$\overline{\lim_n} \;\; \|\gamma_{n,\lambda}^-\|_{L^5(I_{n,\lambda}^-,L^{10}(\mathbb{R}^3))} \longrightarrow 0 \qquad \text{as} \qquad \lambda \longrightarrow +\infty.$$

Applying the energy estimate to $\gamma_{n,\lambda}^-$, we conclude the proof of (25). ■

4.2. Study for Times Close to the Time of Concentration

In this section, we study the nonlinear wave $\underline{u}$ for times close to the time of concentration t_∞, more precisely, we prove that for times belonging to $[t_\infty - \lambda\varepsilon_n, t_\infty + \lambda\varepsilon_n]$, the wave $\underline{u}$ has the same behavior according to the energy-Strichartz norm as in the constant coefficien case.

Let $\underline{u}$ be the sequence of solutions to (1) and let $\underline{\tilde{u}}$ the "rescaled" function associated to $\underline{u}$ into the microscopic variables that is $\underline{\tilde{u}}$ satisfie the following Cauchy problem

$$\begin{cases} \Box_{A(\varepsilon_n\cdot+x_n)}\tilde{u}_n \;\; +|\tilde{u}_n|^4\tilde{u}_n + \;\; |\tilde{u}_n|^{\alpha-1}\tilde{u}_n = 0, \\ (\tilde{u}_n, \partial_s\tilde{u}_n)_{/s=0} \;\; = (\varphi, \psi). \end{cases}$$

Consider the functions w^0 and v_0^- given by

$$\begin{cases} \Box w^0 = 0 \\ (w^0, \partial_s w^0)_{/s=0} = \;\; (\varphi, \psi) \end{cases} \qquad \text{and} \qquad \begin{cases} \Box v_0^- + |v_0^-|^4 v_0^- & = 0 \\ \lim\limits_{s\longrightarrow-\infty} E_0(v_0^- - w^0, s) & = 0. \end{cases}$$

We introduce the auxiliary family of smooth functions u_λ belonging to $L^5(\mathbb{R}, L^{10}(\mathbb{R}^3))$ and satisfying the following Cauchy problem

$$\begin{cases} \Box u_\lambda + |u_\lambda|^4 u_\lambda & = 0 \\ (u_\lambda, \partial_s u_\lambda)_{/s=-\lambda} & = \quad (w^0, \partial_s w^0)_{/s=-\lambda}. \end{cases}$$

We have then the result given by the following Lemma.

Lemma 4.2

$$\overline{\lim_n} \quad |||u_n - u_\lambda|||_{[t_\infty - \lambda\varepsilon_n, t_\infty + \lambda\varepsilon_n]} \longrightarrow 0 \quad as \quad \lambda \longrightarrow +\infty.$$

[Proof of Lemma 4.2]
It suffice to prove the following result related to the microscopic variables.

$$\overline{\lim_n} \quad |||\tilde{u}_n - u_\lambda|||_{[-\lambda,\lambda]} \longrightarrow 0 \quad \text{as} \quad \lambda \longrightarrow +\infty. \tag{26}$$

First, we need the following statements. Let $(\varphi_\lambda, \psi_\lambda)$ be a C_0^∞ approximation of (φ, ψ) in the energy space, that is

$$\|\nabla(\varphi - \varphi_\lambda)\|_{L^2(\mathbb{R}^3)} + \|\psi - \psi_\lambda\|_{L^2(\mathbb{R}^3)} \longrightarrow 0 \quad \text{as} \quad \lambda \longrightarrow +\infty. \tag{27}$$

Consider $\underline{w}^\lambda$ the smooth solutions to the following problem

$$\Box w_n^\lambda = 0, \qquad (w_n^\lambda, \partial_t w_n^\lambda)_{/t=t_\infty}(x) = \frac{1}{\sqrt{\varepsilon_n}}\Big(\varphi_\lambda, \frac{\psi_\lambda}{\varepsilon_n}\Big)\Big(\frac{x - x_n}{\varepsilon_n}\Big).$$

In the microscopic variables, we obtain,

$$\underline{w}^\lambda(t,x) = \frac{1}{\sqrt{\varepsilon_n}}\tilde{w}^\lambda\Big(\frac{t - t_\infty}{\varepsilon_n}, \frac{x - x_n}{\varepsilon_n}\Big) = \frac{1}{\sqrt{\varepsilon_n}}\tilde{w}^\lambda(s,y),$$

where $\tilde{w}^\lambda$ is the solution of the system

$$\Box\tilde{w}^\lambda = 0, \qquad (\tilde{w}^\lambda, \partial_s\tilde{w}^\lambda)_{/s=0} = (\varphi_\lambda, \psi_\lambda).$$

We introduce the auxiliary family of functions $\tilde{v}_{n,\lambda}$ solution of the following problem

$$\begin{cases} \Box_{A(\varepsilon_n\cdot + x_n)}\tilde{v}_{n,\lambda} & + \quad |\tilde{v}_{n,\lambda}|^4\tilde{v}_{n,\lambda} = 0, \\ (\tilde{v}_{n,\lambda}, \partial_s\tilde{v}_{n,\lambda})_{/s=-\lambda} & = \quad (\tilde{w}_n, \partial_s\tilde{w}_n)_{/s=-\lambda}, \end{cases}$$

where $\tilde{w}_n$ is the "rescaled" function into the microscopic variables associated to the linear concentrating wave w_n.

We recall a key Lemma introduced in [13] available for dimensions $d \geq 3$.

Lemma 4.3 *Let $f(s,y)$ be a regular function defined on $\mathbb{R} \times \mathbb{R}^3$ such that for all $\lambda \in \mathbb{R}$, its restriction on $[-\lambda, \lambda] \times \mathbb{R}^3$ is supported in some compact set K_λ, then, under the hypothesis satisfied by the matrix A and the sequences (ε_n) and (x_n), we have,*

$$\overline{\lim_n} \quad \|\text{div}_y[(A(\varepsilon_n\cdot + x_n) - A(0))\nabla_y f]\|_{L^1([-\lambda,\lambda],L^2(\mathbb{R}^3))} = 0. \tag{28}$$

[Proof of Lemma 4.3]
The regularity of the function f, enables us to write

$$\begin{aligned}
\|\mathrm{div}_y[(A(\varepsilon_n.+x_n) &- A(0))\nabla_y f]\|_{L^1([-\lambda,\lambda],L^2(\mathbb{R}^3))}\\
&\le \int_{[-\lambda,\lambda]}\Big(\int_{\mathbb{R}^d}|\mathrm{div}(A(\varepsilon_n y+x_n)-A(0))|^2|\sup_{\mathbb{R}^3}\nabla_y f|^2 dy\Big)^{\frac{1}{2}}dt\\
&+ \int_{[-\lambda,\lambda]}\Big(\int_{\mathbb{R}^3}\sup_{y\in K_\lambda}|(A(\varepsilon_n y+x_n)-A(0))|^2\sup_{\mathbb{R}^3}|\nabla_y\partial f|^2 dy\Big)^{\frac{1}{2}}dt\\
&\le C\{\sup_{y\in K_\lambda}|(A(\varepsilon_n y+x_n)-A(0))|\sup_{s\in\mathbb{R}}E_0^{\frac{1}{2}}(\partial f,s)\}\\
&+ C\{\sup_{y\in K_\lambda}\varepsilon_n|\partial A(y)|\sup_{s\in\mathbb{R}}E_0^{\frac{1}{2}}(f,s)\}.
\end{aligned}$$

■

We return to the proof of Lemma 4.2. We need to prove the following results.

$$\overline{\lim_n}\ |||\tilde{u}_n-\tilde{w}_\lambda|||_{[-\lambda,\lambda]}\to 0 \quad \text{as} \quad \lambda\to+\infty, \tag{29}$$

$$\overline{\lim_n}\ |||\tilde{w}_\lambda-\tilde{v}_{n,\lambda}|||_{[-\lambda,\lambda]}\to 0 \quad \text{as} \quad \lambda\to+\infty \tag{30}$$

and

$$\overline{\lim_n}\ |||\tilde{v}_{n,\lambda}-u_\lambda|||_{[-\lambda,\lambda]}\to 0 \quad \text{as} \quad \lambda\to+\infty. \tag{31}$$

We start by proving the assertion (30). Consider the remainder term $r_{n,\lambda}=\tilde{v}_{n,\lambda}-\tilde{w}_\lambda$, then

$$\begin{cases}\Box_{A(\varepsilon_n.+x_n)}r_{n,\lambda} &= \mathrm{div}_y[(A(\varepsilon_n.+x_n)-A(0))\nabla_y\tilde{w}_\lambda]-|\tilde{r}_{n,\lambda}+\tilde{w}_\lambda|^4(\tilde{r}_{n,\lambda}+\tilde{w}_\lambda)\\ (r_{n,\lambda},\partial_s r_{n,\lambda})_{/s=-\lambda} &= (\tilde{w}_n-\tilde{w}_\lambda,\partial_s(\tilde{w}_n-\tilde{w}_\lambda))_{/s=-\lambda}.\end{cases}$$

First, we begin by showing the following result related to the energy $E_0(\tilde{w}_n-\tilde{w}_\lambda)$. We have,

$$\overline{\lim_n}\ \sup_{s\in[-\lambda,\lambda]}E_0(\tilde{w}_n-\tilde{w}_\lambda,s)\longrightarrow 0 \quad \text{as} \quad \lambda\longrightarrow+\infty. \tag{32}$$

Let us prove (32). The difference $\tilde{w}_n-\tilde{w}_\lambda$ solves the following problem,

$$\begin{cases}\Box_{A(\varepsilon_n.+x_n)}(\tilde{w}_n-\tilde{w}_\lambda) &= \mathrm{div}_y[(A(\varepsilon_n.+x_n)-A(0))\nabla_y\tilde{w}_\lambda]\\ (\tilde{w}_n-\tilde{w}_\lambda,\partial_s(\tilde{w}_n-\tilde{w}_\lambda))_{/s=0} &= (\varphi-\varphi_\lambda,\psi-\psi_\lambda).\end{cases}$$

Since the wave equation is reversible in time, thus, applying the energy estimate to $\tilde{w}_n-\tilde{w}_\lambda$ on $[-\lambda,0]$, then on $[0,\lambda]$, we get

$$\begin{aligned}
\sup_{s\in[-\lambda,\lambda]}E_0^{1/2}(\tilde{w}_n-\tilde{w}_\lambda,s) &\le C\{\|\nabla(\varphi-\varphi_\lambda)\|_{L^2(\mathbb{R}^3)}+\|\psi-\psi_\lambda\|_{L^2(\mathbb{R}^3)}\}\\
&+ C\|\mathrm{div}_y[(A(\varepsilon_n.+x_n)-A(0))\nabla_y\tilde{w}_\lambda]\|_{L^1([-\lambda,\lambda],L^2(\mathbb{R}^3))}.
\end{aligned}$$

By finit speed of propagation, we may assume that the smooth function $\tilde{w}_\lambda$ is supported on a fi ed compact set, hence applying Lemma 4.3, we get

$$\overline{\lim_n}\ \|\mathrm{div}_y[(A(\varepsilon_n.+x_n)-A(0))\nabla_y\tilde{w}_\lambda]\|_{L^1([-\lambda,\lambda],L^2(\mathbb{R}^3))}=0. \tag{33}$$

Thus,

$$\overline{\lim_n} \sup_{s\in[-\lambda,\lambda]} E_0^{1/2}(\tilde{w}_n - \tilde{w}_\lambda, s) \leq C\{\|\nabla(\varphi - \varphi_\lambda)\|_{L^2(\mathbb{R}^3)} + \|\psi - \psi_\lambda\|_{L^2(\mathbb{R}^3)}\},$$

which clearly leads to the result (32) provided (27). Now applying Strichartz inequalities to $r_{n,\lambda}$ and using Hölder inequalities, we get

$$\begin{aligned}\|r_{n,\lambda}\|_{L^5([-\lambda,\lambda],L^{10}(\mathbb{R}^3))} &\leq CE_0^{1/2}(\tilde{w}_n - \tilde{w}_\lambda, s = -\lambda)\\ &+ C\|\mathrm{div}_y[(A(\varepsilon_n \cdot + x_n) - A(0))\nabla_y \tilde{w}_\lambda]\|_{L^1([-\lambda,\lambda],L^2(\mathbb{R}^3))}\\ &+ \|r_{n,\lambda}\|_{L^5([-\lambda,\lambda],L^{10}(\mathbb{R}^3))}\|\tilde{w}_\lambda\|^4_{L^5([-\lambda,\lambda],L^{10}(\mathbb{R}^3))}\\ &+ C\{\|r_{n,\lambda}\|^2_{L^5([-\lambda,\lambda],L^{10}(\mathbb{R}^3))} + \|r_{n,\lambda}\|^5_{L^5([-\lambda,\lambda],L^{10}(\mathbb{R}^3))}\}.\end{aligned}$$

In view of (32), for $\lambda \geq \lambda_a$ large, and $n \geq n(\lambda_a)$ large, the quantity $E_0^{1/2}(\tilde{w}_n - \tilde{w}_\lambda, s = -\lambda)$ is small. On the other hand, using the technique of deformation of time described by Bahouri-Gérard in [2]; we can choose a real number δ such that $C\delta^4 < 1$ and $\|\tilde{w}_\lambda\|_{L^5([-\lambda,\lambda],L^{10}(\mathbb{R}^3))} < \delta$, it follows that

$$\begin{aligned}\|r_{n,\lambda}\|_{L^5([-\lambda,\lambda],L^{10}(\mathbb{R}^3))} &\leq C\{o(1) + \|\mathrm{div}_y[(A(\varepsilon_n \cdot + x_n) - A(0))\nabla_y \tilde{w}_\lambda]\|_{L^1([-\lambda,\lambda],L^2(\mathbb{R}^3))}\}\\ &+ C\{\|r_{n,\lambda}\|^2_{L^5([-\lambda,\lambda],L^{10}(\mathbb{R}^3))} + \|r_{n,\lambda}\|^5_{L^5([-\lambda,\lambda],L^{10}(\mathbb{R}^3))}\}.\end{aligned}$$

Using the absorption Lemma 2.14 and the result (33) yields

$$\overline{\lim_n} \|r_{n,\lambda}\|_{L^5([-\lambda,\lambda],L^{10}(\mathbb{R}^3))} \longrightarrow 0 \quad \text{as} \quad \lambda \longrightarrow +\infty.$$

Applying the energy estimate on $r_{n,\lambda}$, and arguing in the same way, we conclude the proof of (30).

To prove (31), we consider $h_{n,\lambda} = \tilde{v}_{n,\lambda} - u_\lambda$, then,

$$\begin{cases} \Box_{A(\varepsilon_n \cdot + x_n)} h_{n,\lambda} &= \mathrm{div}_y[(A(\varepsilon_n \cdot + x_n) - A(0))\nabla_y u_\lambda] - |h_{n,\lambda} + u_\lambda|^4 + |u_\lambda|^4 u_\lambda\\ (h_{n,\lambda}, \partial_s h_{n,\lambda})_{/s=-\lambda} &= (\tilde{w}_n - v_0^-, \partial_s(\tilde{w}_n - v_0^-))_{/s=-\lambda}.\end{cases}$$

First, we mention that we can estimate the energy as following

$$\begin{aligned}E_0(\tilde{v}_{n,\lambda} - u_\lambda, s = -\lambda) &= E_0(\tilde{w}_n - w_0, s = -\lambda)\\ &\leq E_0(\tilde{w}_n - \tilde{w}_\lambda, s = -\lambda) + E_0(\tilde{w}_\lambda - w_0, s = -\lambda)\\ &:= (I) + (II).\end{aligned}$$

The result (32) implies that $(I) = o(1)$ as $\lambda \to +\infty$, and, in the same way such as the proof of (32), we show that $(II) = o(1)$ as $\lambda \to +\infty$.

Applying Strichartz inequalities to $h_{n,\lambda}$, we obtain

$$\begin{aligned}\|h_{n,\lambda}\|_{L^5([-\lambda,\lambda],L^{10}(\mathbb{R}^3))} &\leq CE_0^{\frac{1}{2}}(\tilde{w}_n - w_0^-, s = -\lambda)\\ &+ C\|\mathrm{div}_y[(A(\varepsilon_n \cdot + x_n) - A(0))\nabla_y u_\lambda]\|_{L^1([-\lambda,\lambda],L^2(\mathbb{R}^3))}\\ &+ C\|(h_{n,\lambda} + u_\lambda)^5 - u_\lambda^5\|_{L^1([-\lambda,\lambda],L^2(\mathbb{R}^3))}.\end{aligned}$$

Using the finit speed of propagation and Hölder inequalities, we get for λ large

$$\begin{aligned}\|h_{n,\lambda}\|_{L^5([-\lambda,\lambda],L^{10}(\mathbb{R}^3))} &\leq C\{o(1)+\|\mathrm{div}_y[(A(\varepsilon_n.+x_n)-A(0))\nabla_y u_\lambda]\|_{L^1([-\lambda,\lambda],L^2(\mathbb{R}^3))}\}\\ &+ C\|h_{n,\lambda}\|_{L^5([-\lambda,\lambda],L^{10}(\mathbb{R}^3))}\|u_\lambda\|^4_{L^5([-\lambda,\lambda],L^{10}(\mathbb{R}^3))}\\ &+ C\{\|h_{n,\lambda}\|^2_{L^5([-\lambda,\lambda],L^{10}(\mathbb{R}^3))}+\|h_{n,\lambda}\|^5_{L^5([-\lambda,\lambda],L^{10}(\mathbb{R}^3))}\}.\end{aligned}$$

We use the same technique such that in the proof of (29) to absorb the linear term in the previous inequality, then, we apply the absorption Lemma 2.14. Once again, we apply Lemma 4.3 to u_λ to obtain (30).

Now, let us prove (29). Setting $\tilde{r}_n=\tilde{u}_n-\tilde{w}_\lambda$, it follows that $\tilde{r}_n$ satisfie

$$\begin{cases}\Box_{A(\varepsilon_n.+x_n)}\tilde{r}_n &= f_n,\\ (\tilde{r}_n,\partial_s\tilde{r}_n)_{/s=0} &= ((\varphi-\varphi_\lambda),(\psi-\psi_\lambda)),\end{cases}$$

where

$$\begin{aligned}f_n: &= \mathrm{div}_y[(A(\varepsilon_n.+x_n)-A(0))\nabla_y\tilde{w}_\lambda]-|\tilde{r}_n+\tilde{w}_\lambda|^4(\tilde{r}_n+\tilde{w}_\lambda)\\ &- |\tilde{r}_n+w_\lambda|^{\alpha-1}(\tilde{r}_n+\tilde{w}_\lambda).\end{aligned}$$

Applying Strichartz inequalities to $\tilde{r}_n$, we get

$$\begin{aligned}\|\tilde{r}_n\|_{L^5([-\lambda,\lambda],L^{10}(\mathbb{R}^3))} &\leq C\{\|\nabla(\varphi-\varphi_\lambda)\|_{L^2(\mathbb{R}^3)}+\|\psi-\psi_\lambda\|_{L^2(\mathbb{R}^3)}\}\\ &+ C\{\|\mathrm{div}_y[(A(\varepsilon_n.+x_n)-A(0))\nabla_y\tilde{w}_\lambda]\|_{L^1([-\lambda,\lambda],L^2(\mathbb{R}^3))}\}\\ &+ C\{\|\tilde{r}_n+\tilde{w}_\lambda\|^5_{L^5([-\lambda,\lambda],L^{10}(\mathbb{R}^3))}+\|\tilde{r}_n+\tilde{w}_\lambda\|^\alpha_{L^\alpha([-\lambda,\lambda],L^{2\alpha}(\mathbb{R}^3))}\}.\end{aligned}$$

By finit speed of propagation, $\tilde{w}_\lambda$ is supported on a fi ed compact set of $[-\lambda,\lambda]\times\mathbb{R}^3$, and belongs to $L^5([-\lambda,\lambda],L^{10}(\mathbb{R}^3))$, then, by Hölder inequalities, we get

$$\begin{aligned}\|\tilde{r}_n\|_{L^5([-\lambda,\lambda],L^{10}(\mathbb{R}^3))} &\leq C\{\|\nabla(\varphi-\varphi_\lambda)\|_{L^2(\mathbb{R}^3)}+\|\psi-\psi_\lambda\|_{L^2(\mathbb{R}^3)}\}\\ &+ C\{\|\mathrm{div}_y[(A(\varepsilon_n.+x_n)-A(0))\nabla_y\tilde{w}_\lambda]\|_{L^1([-\lambda,\lambda],L^2(\mathbb{R}^3))}\}\\ &+ C\{\|\tilde{r}_n\|_{L^5([-\lambda,\lambda],L^{10}(\mathbb{R}^3))}\|\tilde{w}_\lambda\|^4_{L^5([-\lambda,\lambda],L^{10}(\mathbb{R}^3))}\}\\ &+ C\{\|\tilde{w}_\lambda\|^\alpha_{L^\alpha([-\lambda,\lambda],L^{2\alpha}(\mathbb{R}^3))}+\|\tilde{r}_n\|^2_{L^5([-\lambda,\lambda],L^{10}(\mathbb{R}^3))}\}\\ &+ C\{\|\tilde{r}_n\|^\alpha_{L^5([-\lambda,\lambda],L^{10}(\mathbb{R}^3))}+\|\tilde{r}_n\|^5_{L^5([-\lambda,\lambda],L^{10}(\mathbb{R}^3))}\}.\end{aligned}\tag{34}$$

On the other hand, for all $\lambda>0$, there exist a positive integer $m=m(\lambda)$ and a finit decomposition of $[-\lambda,\lambda]$ given by the following

$$-\lambda=t_0^\lambda<t_1^\lambda<t_2^\lambda<....<t_{m(\lambda)}^\lambda=\lambda,$$

such that

$$\|\tilde{w}_\lambda\|^\alpha_{L^\alpha([t_{k-1}^\lambda,t_k^\lambda],L^{2\alpha}(\mathbb{R}^3))}\leq 1/\lambda,\qquad 1\leq k\leq m(\lambda),\tag{35}$$

where we have set

$$\frac{2\lambda}{m(\lambda)}=t_k^\lambda-t_{k-1}^\lambda.$$

In fact, since $\tilde{w}_\lambda$ belongs to all the Lebesgue spaces, then, using Hölder inequalities, we obtain

$$\|\tilde{w}_\lambda\|^\alpha_{L^\alpha([t^\lambda_{k-1},t^\lambda_k],L^{2\alpha}(\mathbb{R}^3))} \leq (t^\lambda_k - t^\lambda_{k-1}) \sup_{t\in[-\lambda,\lambda]} \|\tilde{w}_\lambda(t,.)\|^\alpha_{L^{2\alpha}(\mathbb{R}^3)}.$$

It suffice to choose $m(\lambda)$ such that

$$m(\lambda) > 2\lambda^2 \sup_{t\in[-\lambda,\lambda]} \|\tilde{w}_\lambda(t,.)\|^\alpha_{L^{2\alpha}(\mathbb{R}^3)}.$$

Now, we continue the proof of (29). Setting $I^\lambda_k = [t^\lambda_{k-1}, t^\lambda_k]$, $1 \leq k \leq m(\lambda)$. If we iterate the inequality (34) on each I^λ_k taking into account (27) and Lemma 4.3, we get

$$\begin{aligned}
\|\tilde{r}_n\|_{L^5(I^\lambda_k,L^{10}(\mathbb{R}^3))} &\leq C\{o_\lambda(1) + \|\tilde{r}_n\|_{L^5(I^\lambda_k,L^{10}(\mathbb{R}^3))}\|\tilde{w}_\lambda\|^4_{L^5(I^\lambda_k,L^{10}(\mathbb{R}^3))}\} \\
&+ C\{\|\tilde{r}_n\|^2_{L^5(I^\lambda_k,L^{10}(\mathbb{R}^3))} + \|\tilde{r}_n\|^\alpha_{L^5(I^\lambda_k,L^{10}(\mathbb{R}^3))}\} \\
&+ C\|\tilde{r}_n\|^5_{L^5(I^\lambda_k,L^{10}(\mathbb{R}^3))}.
\end{aligned}$$

By the technique of deformation of time described by H.Bahouri and P.Gérard in [2], we can suppose that $C\|\tilde{w}_\lambda\|^4_{L^5(I^\lambda_k,L^{10}(\mathbb{R}^3))}$ is sufficientl small. Hence, using the absorption Lemma 2.14 we conclude that

$$\overline{\lim_n} \|\tilde{r}_n\|_{L^5(I^\lambda_k,L^{10}(\mathbb{R}^3))} \longrightarrow 0 \quad \text{as} \quad \lambda \longrightarrow +\infty.$$

By induction, the result holds on $[-\lambda, \lambda]$. The energy estimate applied to $\tilde{r}_n$ leads to

$$\overline{\lim_n} \sup_{s\in[-\lambda,\lambda]} E_0^{\frac{1}{2}}(\tilde{r}_n, s) \longrightarrow 0 \quad \text{as} \quad \lambda \longrightarrow +\infty,$$

which concludes the proof of (29). ■

4.3. Study after the Time of Concentration

We have to prove the second assertion in Theorem 2.10, that is

$$\overline{\lim_n} |||\gamma^+_{n,\lambda}|||_{I^+_{n,\lambda}} \longrightarrow 0 \quad \text{as} \quad \lambda \longrightarrow +\infty,$$

where we have set $I^+_{n,\lambda} = [t_\infty + \lambda\varepsilon_n, T]$. In the case of times bigger than the time of concentration t_∞, the linear concentrating wave $\underline{w}^+ = (w^+_n)_n$ define by the Theorem 2.1 is solution to the following problem

$$\begin{cases} \Box_A w^+_n = 0 & \text{in} \quad I^+_{n,\lambda} \times \mathbb{R}^3 \\ (w^+_n, \partial_t w^+_n)_{/t=t_\infty} = & S^\infty(\varphi,\psi) \end{cases}$$

Consider the "rescaled" functions into the microscopic variables

$$w^+_n(t,x) = \frac{1}{\sqrt{\varepsilon_n}}\tilde{w}^+_n(\frac{x-x_\infty}{\varepsilon_n}, \frac{t-t_\infty}{\varepsilon_n}), \qquad u_n(t,x) = \frac{1}{\sqrt{\varepsilon_n}}\tilde{u}_n(\frac{x-x_\infty}{\varepsilon_n}, \frac{t-t_\infty}{\varepsilon_n}).$$

In a similar way as in the proof of the firs assertion in Theorem 2.10, We begin by estimate the remainder term $\gamma^+_{n,\lambda}$ according to the Strichartz norm. then, taking into account the result (23) on $I^+_{n,\lambda}$, we apply the absorption Lemma 2.14 . Finally, using the result (24), we achieve the proof. We mention that using the microscopic variables, the results (23) and (24) are respectively given by

$$\overline{\lim_n} \quad \|\tilde{w}_n^+\|_{L^5([\lambda, \frac{T-t_\infty}{\varepsilon_n})], L^{10}(\mathbb{R}^3))} \to 0 \quad \text{as} \quad \lambda \to +\infty,$$

$$\overline{\lim_n} \quad \|\tilde{w}_n^+\|_{L^\alpha([\lambda, \frac{T-t_\infty}{\varepsilon_n})], L^{2\alpha}(\mathbb{R}^3))} \to 0 \quad \text{as} \quad \lambda \to +\infty.$$

We begin by proving a result related to the energy of the remainder term $\gamma^+_{n,\lambda}$ explained by the microscopic variables.

Lemma 4.4

$$\overline{\lim_n} \quad E_0(\tilde{u}_n - \tilde{w}_n^+, s = \lambda) \to 0 \quad \textit{as} \quad \lambda \to +\infty.$$

[Proof of Lemma 4.4]
The functions $\tilde{u}_n$ (respectively $\tilde{w}_n^+$) are solutions of the problems

$$\begin{cases} \Box_{A(\varepsilon_n.+x_n)}\tilde{u}_n \quad + \quad |\tilde{u}_n|^4\tilde{u}_n + |\tilde{u}_n|^{\alpha-1}\tilde{u}_n = 0 \\ (\tilde{u}_n, \partial_s\tilde{u}_n)_{/s=0} = (\varphi, \psi), \end{cases}$$

(respectively)

$$\begin{cases} \Box_{A(\varepsilon_n.+x_n)}\tilde{w}_n^+ = 0 \\ (\tilde{w}_n^+, \partial_t\tilde{w}_n^+)_{/s=0} = S^\infty(\varphi, \psi) \end{cases}$$

We introduce the smooth function w^∞ belonging to $L^5(\mathbb{R}, L^{10}(\mathbb{R}^3))$ solution of the following Cauchy problem

$$\begin{cases} \Box w^\infty = 0 \\ (w^\infty, \partial_s w^\infty)_{/s=0} = S^\infty(\varphi, \psi) \end{cases}$$

We observe that we can estimate the energy $E_0(\tilde{u}_n - \tilde{w}_n^+, s = \lambda)$ as follows

$$\begin{aligned} E_0(\tilde{u}_n - \tilde{w}_n^+, s = \lambda) &\le E_0(\tilde{u}_n - u_\lambda, s = \lambda) + E_0(u_\lambda - w^\infty, s = \lambda) \\ &+ E_0(w^\infty - \tilde{w}_n^+, s = \lambda). \end{aligned}$$

By scattering, we have

$$\lim_{\lambda \to +\infty} \quad E_0(u_\lambda - w^\infty, s = \lambda) = 0.$$

To estimate the term $E_0(w^\infty - \tilde{w}_n^+, s = \lambda)$, we set $\tilde{r}_{n,+} = \tilde{w}_n^+ - w^\infty$, then, $r_{n,+}$ satisfie

$$\begin{cases} \Box_{A(\varepsilon_n.+x_n)}\tilde{r}_{n,+} = \operatorname{div}[(A(\varepsilon_n. + x_n) - A(0))\nabla_y w^\infty] \\ (\tilde{r}_{n,+}, \partial_s\tilde{r}_{n,+})_{/s=0} = (0, 0). \end{cases}$$

The energy estimate implies that

$$\sup_{s\in[-\lambda,\lambda]} E_0(\tilde{r}_{n,+},s) \leq C\{\|\text{div}[(A(\varepsilon_n.+x_n)-A(0))\nabla_y w^\infty]\|_{L^1([-\lambda,\lambda],L^2(\mathbb{R}^3))}\}.$$

On the other hand, by finit speed of propagation, we may assume that w^∞ is supported on a fi ed compact set of $[-\lambda,\lambda]\times\mathbb{R}^3$, thus, using Lemma 4.3, we obtain

$$\overline{\lim_n} \|\text{div}[(A(\varepsilon_n.+x_n)-A(0))\nabla_y w^\infty]\|_{L^1([-\lambda,\lambda],L^2(\mathbb{R}^d))} = 0,$$

then,

$$\overline{\lim_n} E_0^{\frac{1}{2}}(\tilde{r}_{n,+},\lambda) \longrightarrow 0 \quad \text{as} \quad \lambda \longrightarrow +\infty.$$

The estimate of the term $E_0(\tilde{u}_n - u_\lambda,\lambda)$ can be seen as a consequence of the previous section, indeed,

$$\overline{\lim_n} \||\tilde{u}_n - u_\lambda\||_{[-\lambda,\lambda]} \longrightarrow 0 \quad \text{as} \quad \lambda \longrightarrow +\infty$$

then,

$$\overline{\lim_n} E_0^{\frac{1}{2}}(\tilde{u}_n - u_\lambda,\lambda) \longrightarrow 0 \quad \text{as} \quad \lambda \longrightarrow +\infty.$$

In conclusion, we obtain

$$\overline{\lim_n} E_0(\tilde{u}_n - \tilde{w}_n^+, s=\lambda) \longrightarrow 0 \quad \text{as} \quad \lambda \longrightarrow +\infty,$$

which concludes the proof of the second assertion in Theorem 2.10. The proof of Theorem 2.10 is completely achieved . ■

Acknowledgments

The author is deeply indebted to the referee for his suggestions which helped to improve this paper.

References

[1] S. Alinhac et P. Gérard: *Opérateurs pseudodifférentiels et théorème de Nash-Moser*, Savoirs Actuels, (1991).

[2] H. Bahouri and P. Gérard: *High frequency approximation of solutions to critical non-linear wave equation*, *Amer. Journal. Math* **121**. 131-175, (1999).

[3] H. Bahouri and J. Shatah: *Global estimate for the semilinear wave equation* , *Ann. Inst. H. Poincaré Anal. Non linéaire* **15**. 783-789, (1998).

[4] M. Bezard: Une version générale de l'inégalité de Strichartz, *C.R.A.S. Paris* **315**. 1241-1244, (1992).

[5] S. Gallot, D. Hulin and J. Lafontaine: *Riemannian geometry*, Second edition Springer verlag, (1993).

[6] P. Gérard: Oscilations and concentration effects in semilinear dispersive wave equations,*J .Funct. Anal.* **141**. 60-98, (1996).

[7] J. Ginibre, A. Soffer and G. Velo: The Global Cauchy problem for the critical nonlinear wave equation, *J. Funct. Anal.* **110**. 96-130, (1992).

[8] J. Ginibre and G. Velo: The Global Cauchy problem for nonlinear Klein-Gordon equation, *Math. Z* ,**189**. 487-505, (1985).

[9] J. Ginibre and G. Velo: Scattering theory in the energie space for a class of nonlinear wave equations, *Comm. Math. Phys. J. Functional Analysis,* **123**. 535-573, (1989).

[10] J. Ginibre and G. Velo: Generalized Strichartz inequalities for the wave equations, *J. Functional Analysis,* **133**. 50-68, (1995).

[11] M. Grillakis: Regularity and asymptotic behaviour of the wave equation with a critical nonlinearity, *Annal. of Math.* **132**. 485-509, (1990).

[12] M. Grillakis: Regularity for the wave equation with a critical nonlinearity, *Comm. Pure App. Math.* **XLVI**. 749-774, (1992).

[13] S. Ibrahim: Geometric-optics for nonlinear concentrating waves in focusing and non-focusing two geometries, *Communications in Contemporary Mathematics.* **1-23**, (2004).

[14] S. Ibrahim et M. Majdoub: Existence globale de solutions pour l'équation des ondes semi-linéaire critique à coefficient variables, *Comptes-Rendus de l'Académie des Sciences de Paris, Série I,* **328**. 579-584, (1999).

[15] S. Ibrahim et M. Majdoub: Comparaison des ondes linéaires et non-linéaires à coefficient variables, *Bull. Bel. Math. Soc*. **10**. 299-312, (2003).

[16] S. Ibrahim et M. Majdoub: Solutions globales de l'équation des ondes semi-lin ́eaire critique à coefficient variables, *Bull. Soc. math. France*, **131**. 1-22, (2003).

[17] L. V. Kapitanski: Some generalizations of the Strichartz-Brenner inequality, *Leningrad Math.*, **62**, Série I. 2746-2777, (1992).

[18] M. Majdoub: Qualitative study of the critical wave equation with a subcritical perturbation,*J. Math. Ann. Appl,* **301**. 354-365, (2005).

[19] J. Shatah and M. Struwe: Regularity results for nonlinear wave equations, *Ann. of Math,* **2**, **138**. 503-518, (1993).

[20] J. Shatah and M. Struwe: Well-Posedness in the energy space for semilinear wave equation with critical growth, *IMRN*,**7**. 303-309, (1994).

[21] H. Smith and C. Sogge: On the critical semilinear wave equation outside convex obstacles, *Journal of the American Mathematical Society*, **8,4**. 879-916, (1995).

[22] R. Strichartz: A priori estimates for the wave equation and some applications, *J.Funct. Analysis,* **5**. 218-235, (1970).

[23] M. Struwe: Semilinear wave equations, *Bull. Amer. Math. Soc*, **N.S**, **26**. 53-85, (1992).

In: Evolution Equations: Theories, Solutions and Functions ISBN 978-1-61324-090-8
Editor: Gaston M. N'Guerekata

Chapter 3

RENORMALIZED SOLUTION FOR NONLINEAR ANISOTROPIC DEGENERATE PARABOLIC EQUATIONS WITH NONLIPSCHITZ CONVECTION AND DIFFUSION FLUX FUNCTIONS*

***Adama Ouédraogo*[1,†] *and Mohamed Maliki*[2,‡]**
[1]Department of Mathematics 03 BP 7021, University of Ouagadougou 03, Burkina Faso
[2]Department of Mathematics BP 146, Hassan II University Mohammedia, Morocco

Abstract

We consider the anisotropic degenerate parabolic equation of the form:

$$\partial_t u + \nabla.F(u) - \sum_{i=1}^{N} \partial_{x_i} (a_{ii}(u)\partial_{x_i} u) = f(u) \text{ in } Q =]0, T[\times \mathbb{R}^N, \quad T > 0 \qquad (1)$$

where $a(u) = (a_{ij}(u))_{1 \le i,j \le N}$ is a nonnegative diagonal matrix such that each coefficien is locally integrable in $\mathbb{R}$. We prove the well-posedness (existence and uniqueness) of renormalized solution of the associated Cauchy problem for L^1 initial data and source term. The novelty in our paper is the fact that we work in an unbounded domain with u_0, f in L^1, the convection-diffusion flu functions are not necessarily Lipschitz on $\mathbb{R}$ and f depends on u that extends the results of [11] and [17]. We establish the comparison principle, the existence and uniqueness of renormalized solution under an assumption (H_2) on the modulus of continuity of F_i and $A_{ii}(r) = \int_0^r a_{ii}(\xi)d\xi$ for $i = 1, ..., N$. This assumption seems to be optimal for conservation laws included here and turns out to be the key-point of the uniqueness.

---

*Communicated by the firs author in the firs Conference of Research in Applied Mathematics (CRAM), Ouagadougou, November 12-17, 2007.

†E-mail address: adam_ouedraogo3@yahoo.fr
‡E-mail address: mohamedmaliki@yahoo.fr

Keywords: parabolic equation, hyperbolic equation, anisotropic equation, entropy solution, renormalized solution, Kato's inequality

AMS Subject Classification: Partial differential equations, 35 K 65, 35 L 65.

1. Introduction

Let $Q =]0, T[\times \mathbb{R}^N$ with $T > 0$, $N \geq 1$ and $A = (A_{ii})_{1\leq i\leq N}$.
We consider the Cauchy problem $(CP) = (CP)(A, F, f, u_0)$:

$$(CP) \qquad \begin{cases} \partial_t u + \nabla.F(u) - \sum_{i=1}^{N} \partial^2_{x_i x_i} A_{ii}(u) = f(u) & \text{in } Q \\ u(0, .) = u_0 & \text{on } \mathbb{R}^N, \end{cases}$$

where

$$F \in C(\mathbb{R}, \mathbb{R}^N), \quad F(0) = 0, \tag{2}$$

$$A_{ii}(r) = \int_0^r a_{ii}(\xi) d\xi \quad i = 1, ..., N \tag{3}$$

with $a = (a_{ij})_{1\leq i,j\leq N}$, a $N \times N$ nonnegative diagonal and locally integrable matrix such that

$$a_{ii}(u) = (\sigma_{ii}(u))^2, \; 0 \leq \sigma_{ii}(u) \in L^1_{loc}(\mathbb{R}) \quad i = 1, ..., N.$$

$(CP)(A, F, f, u_0)$ is a model of second order degenerate diffusion-convection motions of fluid It has important applications in two phase fl ws in porous media ($cf.$[30]) and sedimentation-consolidation processes ($cf.$[3]). It is well known that there is no classical solution and the weak solution is not unique: if $A(u) = 0$, (CP) degenerates into a scalar conservation law problem which possesses discontinuous solutions. Hence (CP) must be interpreted in the sense of entropy or renormalized solution according to the regularity conditions on the initial and source terms. Under various conditions, many results on (CP) are proved in [1], [9], [12], [13], [14], [15], [18], [23], [25] and recently in [24, 25, 26], [27], [17], [16], [11]. For $N = 1$, well-posedness results in the sense of entropy solution are given by Wu and Yin [20] and by Ph. Bénilan and H. Touré [13].
Later, the multidimensional isotropic case was developed by José Carrillo in a bounded domain [15] using Kru$\check{z}$khov doubling-of-variables method. Using the same method in [28], we obtain in the particular case ($g = Id_{\mathbb{R}}$ in [15]), existence and uniqueness results of the renormalized solution in the whole space $\mathbb{R}^N$.
The general anisotropic diffusion context is more delicate. It was recently solved by Gui-Qiang Chen and Benoît Perthame [17] in a sense of a new notion of entropy solution and by Mostafa Bendahmane and Kenneth H. Karlsen [11] in a sense of renormalized solution based on [17] and [23]. Note that in both works ([11], [17]), authors suppose that $F' \in L^\infty_{loc}(\mathbb{R}; \mathbb{R}^N)$, $\sigma_{ii} \in L^\infty_{loc}(\mathbb{R})$ and the source term depends only on (t, x). The contribution of the present paper is to extend in the case of diagonal diffusion matrix, the results of [11] and [17] to a set of convection-diffusion functions which are not necessarily Lipschitz continuous and a source term f depending on u. Under an assumption (H_2) on the

modulus of continuity of F and A, we prove the existence, the comparison principle and the uniqueness of renormalized solution. The condition (H_2) on the modulus of continuity has been introduced in [10] for conservation laws where Bénilan and Kru žkhov gave a counter-example to show its optimality. They have shown that if (H_2) does not hold then the entropy solution is not unique with $A \equiv 0$. They constructed infinitel many entropy solutions for $N = 2$, $F_i = r^{\alpha_i}$, $i = 1, 2$ with $\alpha_i \geq 0$ and $\alpha_1 + \alpha_2 < 1$. If $N = 1$ and $A(r) = r^m$, this condition is $m > \dfrac{N-1}{N}$.
The techniques, ideas and estimations developed in the present paper are based essentially on [15], [11], [17], [19], [28], [29], [24] and [27].
The present work falls into three sections. Section 1 is the introduction. Section 2 is intended to prove the existence and uniqueness of the weak (entropy) solution when the data u_0 is bounded and f a locally Lipschitz continuous function. In Section 3, we use the results of the previous section and prove the existence and uniqueness of the renormalized solution of (CP) in the case of L^1 initial data.

2. Entropy Solution of $(CP)(A, F, f, u_0)$.

In this section, we suppose that the initial data and the source term satisfy the following hypothesis:

$$(H_1) \qquad \begin{cases} 1)\ u_0 \in L^\infty(\mathbb{R}^N) \cap L^1(\mathbb{R}^N), \\ 2)\ f \in Lip(\mathbb{R}) \text{ and } f(0) = 0. \end{cases}$$

For all $\epsilon > 0$ and $k \geq 0$, we defin the truncation function T_k at the level k and the operators $H, H_0, H^1_\epsilon, H^2_\epsilon$ respectively by

$$T_k(s) = \begin{cases} k & \text{if } s > k \\ s & \text{if } |s| \leq k \\ -k & \text{if } s < -k, \end{cases} \quad H(s) = \begin{cases} 1 & \text{if } s > 0 \\ [0,1] & \text{if } s = 0 \\ 0 & \text{if } s < 0, \end{cases} \quad H_0(s) = \begin{cases} 1 & \text{if } s > 0 \\ 0 & \text{if } s \leq 0, \end{cases} \tag{4}$$

$$H^1_\epsilon(s) = \begin{cases} 1 & \text{if } s > \epsilon \\ \sin(\frac{\pi}{2\epsilon}s) & \text{if } 0 \leq s \leq \epsilon \\ 0 & \text{if } s < 0, \end{cases} \quad H^2_\epsilon(s) = \begin{cases} -1 & \text{if } s < -\epsilon \\ \sin(\frac{\pi}{2\epsilon}s) & \text{if } -\epsilon \leq s \leq 0 \\ 0 & \text{if } s > 0 \end{cases} \tag{5}$$

and for $k \in \mathbb{R}$, $1 \leq i \leq N$, $j = 1, 2$, the corresponding entropy functions by

$$\begin{cases} \gamma^j_\epsilon(r,k) = \displaystyle\int_k^r H^j_\epsilon(s-k)ds, \\ \theta^{ij}_\epsilon(r,k) = \displaystyle\int_k^r H^j_\epsilon(s-k)F'_i(s)ds, \\ \nu^{ij}_\epsilon(r,k) = \displaystyle\int_k^r H^j_\epsilon(s-k)a_{ii}(s)ds. \end{cases} \tag{6}$$

For $i = 1, ..., N$, we set $\rho = (\rho_i)$ and $\rho^\eta = (\rho^\eta_i)$ with

$$\rho_i(r) = \int_0^r \sigma_{ii}(\tau)d\tau, \quad \rho^\eta_i(r) = \int_0^r \eta(\tau)\sigma_{ii}(\tau)d\tau \ \ \forall \eta \in C(\mathbb{R}).$$

For any convex C^2 entropy function $\gamma : \mathbb{R} \to \mathbb{R}$, we defin the entropy flu $\theta = (\theta_i) : \mathbb{R} \to \mathbb{R}^N$, $\nu = (\nu_i) : \mathbb{R} \to \mathbb{R}^N$ by $\theta'(r) = \gamma'(r)F'(r)$, $\nu'(r) = \gamma'(r)a(r)$.

Remark 2.1
For all $r,\ k \in I\!\!R$ *and* $i = 1, ..., N$, *we have:*

$$H^1_\epsilon(r) = -H^2_\epsilon(-r),\ \gamma^1_\epsilon(r,k) = \gamma^2_\epsilon(k,r),\ \theta^{i1}_\epsilon(r,k) = \theta^{i2}_\epsilon(k,r),\ \nu^{i1}_\epsilon(r,k) = \nu^{i2}_\epsilon(k,r). \tag{7}$$

When ϵ *goes to* 0,

$$\begin{cases} H^1_\epsilon(r) \to H_0(r)\ a.e., \\ \gamma^1_\epsilon(r,k) \to \gamma(r,k) = H_0(r-k)(r-k)\ a.e., \\ \theta_\epsilon(r,k) \to \theta(r,k) = H_0(r-k)(F(r)-F(k))\ a.e, \\ \nu_\epsilon(r,k) \to \nu(r,k) = H_0(r-k)(A(r)-A(k))\ a.e. \end{cases} \tag{8}$$

Definition 2.1 *(Entropy solution of* $(CP)\ (A, F, f, u_0)$*).*
Let u_0 *and* f *be such that* (H_1) *is fulfilled. An entropy solution of* $(CP)\ (A, F, f, u_0)$ *is a measurable function* $u : Q \to I\!\!R$ *such that:*

(i) $u \in L^\infty\left(0, T; L^1(I\!\!R^N)\right) \cap L^\infty(Q)$ *and* $\nabla\rho(u) \in (L^2(Q))^N$.

(ii) *For any entropy flux triple* (γ, θ, ν),

$$\partial_t\gamma(u)+\nabla.\theta(u)-\sum_{i=1}^N \partial^2_{x_ix_i}\nu_i(u)-\gamma'(u)f(u) \le -\gamma''(u)\sum_{i=1}^N(\partial_{x_i}\rho_i(u))^2 \quad in\ \mathcal{D}'(Q). \tag{9}$$

(iii) *The initial condition is assumed in the following strong* L^1 *sense:*

$$ess \lim_{t\downarrow 0} \|u(t,.)-u_0\|_{L^1(I\!\!R^N)} = 0.$$

Remark 2.2

1. *A discussion about a weaker formulation of* (iii) *in the Definition* 2.1 *has been made in* [16].

2. *For* $i = 1, ..., N$, *arguing as in* [17], *we have always*

$$\partial_{x_i}\rho^\eta_i(u) = \eta(u)\partial_{x_i}\rho_i(u)a.e \texttt{ in } Q \texttt{ and in } L^2(Q),\ \forall\eta\in C(I\!\!R). \tag{10}$$

3. *By density,* (9) *is true with* $\gamma \in W^{2,\infty}(I\!\!R)$. *Since we are working with test functions which are compact supports, we just have to ask* γ *to be in* $W^{2,\infty}_{loc}(I\!\!R)$. *In particular,* (9) *still holds with* γ^1_ϵ *and* γ^2_ϵ *which are in* $W^{2,\infty}_{loc}(I\!\!R)$.

2.1. Uniqueness of Entropy Solution of $(CP)(A, F, f, u_0)$.

Let us recall the following lemma:

Lemma 2.1 *(cf. [10], [24])*
Let $\omega_1, \omega_2, ..., \omega_l, \alpha_1, \alpha_2, ..., \alpha_l$ *be nonnegative functions such that for* $i = 1, ..., l$ *with* $1 \le l \le N$:

$$\lim_{\epsilon\to 0}\frac{\omega_i(\epsilon)}{\epsilon} = \lim_{\epsilon\to 0}\frac{\alpha_i(\epsilon)}{\epsilon} = +\infty,\ \lim_{\epsilon\to 0}\omega_i(\epsilon) = \lim_{\epsilon\to 0}\alpha_i(\epsilon) = 0; \tag{11}$$

$$\liminf_{\epsilon\to 0} \epsilon^{1-l} \prod_{i=1}^{l} \left(\alpha_i(\epsilon) + \left(\alpha_i^2(\epsilon) + 2\epsilon\omega_i(\epsilon)\right)^{1/2}\right) < +\infty \;\; \textit{if}\, N > 2; \tag{12}$$

$$\liminf_{\epsilon\to 0} \epsilon^{1-l} \prod_{i=1}^{l} \left(\alpha_i(\epsilon) + \left(\alpha_i^2(\epsilon) + 2\epsilon\omega_i(\epsilon)\right)^{1/2}\right) = 0 \;\; \textit{if}\, N = 2; \tag{13}$$

$$\textit{for } N = 1, \textit{ there is no condition on } \alpha_1 \textit{ and } \omega_1. \tag{14}$$

Let $h \in L^1_{loc}(Q)$ such that $h^+ = \max(h,0) \in L^1(Q)$. Let $W_0 \in L^1(\mathbb{R}^N)$, $W \in L^\infty(Q)$, $W \geq 0$, and let $\lambda > 0$ so that:

$$\begin{cases} \displaystyle\int\!\!\int_Q \{W\partial_t\xi + \sum_{i=1}^{l}(W+\epsilon)\frac{\omega_i(\epsilon)}{\epsilon}|\partial^2_{x_ix_i}\xi| + \sum_{i=1}^{l}(W+\epsilon)\frac{\alpha_i(\epsilon)}{\epsilon}|\partial_{x_i}\xi| \\ \displaystyle\qquad + \sum_{i=l+1}^{N} \lambda W\left(|\partial^2_{x_ix_i}\xi| + |\partial_{x_i}\xi|\right) + h\xi\}dxdt \geq 0, \end{cases} \tag{15}$$

for any $\epsilon > 0$ and $\xi \in \mathcal{D}(Q)$, $\xi \geq 0$.
Suppose $(W(t,.) - W_0)^+ \to 0$ in $L^1_{loc}(\mathbb{R}^N)$, when $t \to 0$ essentially.
Then $h \in L^1(Q)$, $W \in L^\infty(0,T;L^1(\mathbb{R}^N))$ and

$$\int_{\mathbb{R}^N} W(\tau,x)\,dx \leq \int_{\mathbb{R}^N} W_0(x)\,dx + \int_{Q_\tau} h\,dxdt \tag{16}$$

for $\tau \in (0,T)$ a.e. with $Q_\tau = (0,\tau)\times\mathbb{R}^N$.

We give now the so called Kato's inequality for two entropy solutions of (CP).

THEOREM 2.1 *(Kato's inequality).*
Let (u_0,f), (v_0,g) satisfy $(H1)$ and u, v be entropy solutions of $(CP)(A,F,f,u_0)$, $(CP)(A,F,g,v_0)$ respectively. Then

$$\begin{cases} \displaystyle\int_Q H_0(u-v)\left(\sum_{i=1}^{N}(F_i(u)-F_i(v))\partial_{x_i}\phi - \sum_{i=1}^{N}(A_{ii}(u)-A_{ii}(v))\partial^2_{x_ix_i}\phi - (u-v)\partial_t\phi\right)dxdt \\ \displaystyle\qquad \leq \int_{\mathbb{R}^N}(u_0-v_0)^+\phi(0)dx + \int_Q \kappa(f(u)-g(v))\phi\,dxdt \end{cases} \tag{17}$$

for any $\kappa \in H(u-v)$ a.e. and $0 \leq \phi \in \mathcal{D}(Q)$.

The proof of Theorem 2.1 is obtained in the same way as in [15] and [16]; so we omitted it. Now in order to get the comparison principle, we denote by $\omega_{A_{ii}}$ and α_{F_i} the modulus of continuity of A_{ii} and F_i. Changing coordinates if necessary, we assume as in [24] that:

1. for $l \geq 1$ and $i = 1,...,l$, the functions A_{ii} and F_i are continuous but not Lipschitz continuous;

2. for $i = l+1,...,N$, the functions A_{ii} and F_i are Lipschitz continuous and denote their Lipschitz constants by $\lambda_{A_{ii}}$ and λ_{F_i};

3. $\lambda_{A_{ii}}$ and λ_{F_i} may be replaced respectively by $\tilde{\lambda_{A_{ii}}} = \lambda_{A_{ii}} + \epsilon^{-1/N+1}$ and $\tilde{\lambda_{F_i}} = \lambda_{F_i} + \epsilon^{-1/N+1}$ if necessary;

4. $\lambda = \max\left(\lambda_{A_{l+1}}, ..., \lambda_{A_{NN}}, \lambda_{F_{l+1}}, ..., \lambda_{F_N}\right);$

Now, we make the following assumption:

$$(H_2) \qquad (\omega_{A_{ii}})_{1\le i\le l},\ (\alpha_{F_i})_{1\le i\le l}\ \text{satisfy}\ (11)-(14).$$

THEOREM 2.2
Let (u_0, f), (v_0, g) satisfy $(H1)$ and u, v be entropy solutions of $(CP)(A, F, f, u_0)$, $(CP)(A, F, g, v_0)$ respectively. Suppose that (H_2) holds. Then

$$\int_{\mathbb{R}^N} (u(t) - v(t))^+ \, dx \le \int_{\mathbb{R}^N} (u_0 - v_0)^+ \, dx + \int_{Q_t} \kappa \left(f(u) - g(v)\right) dxds \tag{18}$$

for $\kappa \in H(u - v)$ a.e. so that:

$$\|u(t) - v(t)\|_{L^1(\mathbb{R}^N)} \le \|u_0 - v_0\|_{L^1(\mathbb{R}^N)} + \int_0^t \|f(u) - g(v)\|_{L^1(\mathbb{R}^N)} ds. \tag{19}$$

Proof
Set $W = (u - v)^+$, $W_0 = (u_0 - v_0)^+$, $h = (f(u) - g(v))^+$. Since u, v are entropy solutions of (CP), the relation (17) gives us:

$$0 \le \int_Q \left(\sum_{i=1}^N |F_i(u) - F_i(v)||\partial_{x_i}\phi| + \sum_{i=1}^N |A_{ii}(u) - A_{ii}(v)||\partial^2_{x_ix_i}\phi| + W\partial_t\phi + h\phi \right) dxdt \tag{20}$$

for any $\phi \in \mathcal{D}(Q)$, $\phi \ge 0$ and $\phi(0) = 0$. Combining the property of modulus of continuity and the points 1. and 2. above, we have

$$|F_i(u) - F_i(v)| \le (W + \epsilon)\frac{\alpha_{F_i}(\epsilon)}{\epsilon},\ i = 1, ..., l; \quad |F_i(u) - F_i(v)| \le \lambda|u - v|,\ \ i = l + 1, ..., N$$

and

$$|A_{ii}(u) - A_{ii}(v)| \le (W + \epsilon)\frac{\omega_{A_{ii}}(\epsilon)}{\epsilon},\ i = 1, ..., l; \quad |A_{ii}(u) - A_{ii}(v)| \le \lambda|u - v|,\ \ i = l + 1, ..., N.$$

Using theses inequalities in (20), we get (15) and (18) follows by application of Lemma 2.1.

We deduce from Theorem 2.2 the uniqueness of entropy solution of $(CP)(A, F, f, u_0)$.

COROLLARY 2.1
Let u and v be entropy solutions of $(CP)(A, F, f, u_0)$, $(CP)(A, F, f, v_0)$.

If $u_0 \le v_0$ a.e. in $\mathbb{R}^N$ then $u \le v$ a.e. on Q.

So if u is an entropy solution of $(CP)(A, F, f, u_0)$ then u is unique.

Proof
Suppose that $u_0 \le v_0$ a.e. in $\mathbb{R}^N$. Since $f \in Lip(\mathbb{R})$, there exists a positive constant $c(f)$ such that for any $\kappa \in H(u - v)$ a.e.,

$$\int_{Q_t} \kappa \left(f(u) - f(v)\right) dxds \le c(f) \int_0^t \int_{\mathbb{R}^N} (u - v)^+ \, dxds.$$

So from (18) we get

$$\int_{\mathbb{R}^N} (u(t) - v(t))^+ \, dx \leq c(f) \int_0^t \int_{\mathbb{R}^N} (u - v)^+ \, dx ds.$$

We deduce by Gronwall's inequality that $u \leq v$ a.e. on Q and the uniqueness of entropy solution of $(CP)(A, F, f, u_0)$ follows.

2.2. Existence of Entropy Solution of $(CP)(A, F, f, u_0)$.

To end this section, we deal with the existence of entropy solution of $(CP)(A, F, f, u_0)$. Here, the contraction principle (19) will be important for the convergence of approximate solutions used in the proof of the existence theorem below.

THEOREM 2.3
Suppose that (H_1), (H_2) hold. Then there exists an entropy solution $u \in C([0, T]; L^1(\mathbb{R}^N))$ of the Cauchy problem $(CP)(A, F, f, u_0)$.

Proof
This proof is divided in two steps; it is based on results of [17] and [18].
Step 1. We assume that $u_0 \in W^{2,1}(\mathbb{R}^N) \cap H^1(\mathbb{R}^N) \cap L^\infty(\mathbb{R}^N)$, $f \in W^{2,1}(\mathbb{R}) \cap H^1(\mathbb{R}) \cap L^\infty(\mathbb{R})$.
We consider the Cauchy problem $(CP)(A_\epsilon, F, f, u_0)$ where $(A_{ii}^\epsilon) = (A_{ii}) + \epsilon I$.
Thanks to results of [22], [18], for each $\epsilon > 0$, there exists a smooth solution $u_\epsilon(t, x)$ of $(CP)(A_\epsilon, F, f, u_0)$ satisfying

$$\|u_\epsilon(t, .)\|_{L^\infty(Q)} \leq \|u_0\|_{L^\infty(\mathbb{R}^N)} \tag{21}$$

and there exists some subsequence of $\{u_\epsilon(t, x)\}_{\epsilon > 0}$ still denoted $\{u_\epsilon(t, x)\}_{\epsilon > 0}$ such that when ϵ goes to 0,

$$u_\epsilon \to u \text{ in } \mathcal{C}\left([0, T]; L^1(\mathbb{R}^N)\right); \quad \text{so } u_\epsilon \to u \text{ a.e. in } Q. \tag{22}$$

It remains us to prove that the limit function u is an entropy solution of $(CP)(A, F, f, u_0)$ in the sense of Definitio 2.1.

If we integrate over Q the nondegenerate parabolic equation associate to $(CP)(A_\epsilon, F, f, u_0)$, we deduce by the Gronwall's inequality that $u_\epsilon(t, x)$ is bounded in $L^1(\mathbb{R}^N)$. More precisely, we have

$$\|u_\epsilon(t, .)\|_{L^1(\mathbb{R}^N)} \leq \|u_0\|_{L^1(\mathbb{R}^N)} \exp(c(f)T). \tag{23}$$

Since f is a Lipschitz continuous function on $\mathbb{R}$, then (22) implies that

$$f(u_\epsilon) \to f(u) \text{ in } \mathcal{C}\left([0, T]; L^1(\mathbb{R}^N)\right).$$

Furthermore, as in [17], we multiply the nondegenerate parabolic equation associate to $(CP)(A_\epsilon, F, f, u_0)$ by $\gamma'(u_\epsilon)$ with $\gamma \in \mathcal{C}^2(\mathbb{R})$, convex such that

$\gamma'(0) = \gamma(0) = 0$, and $|\gamma'| \le K$ (for some given positive constant K) and the following equality holds:

$$\begin{cases} \partial_t \gamma(u_\epsilon) + \nabla.\theta(u_\epsilon) - \sum_{i=1}^{N} \partial_{x_i} \left(\gamma'(u_\epsilon) a_{ii}(u_\epsilon)\partial_{x_i} u_\epsilon\right) - \epsilon\Delta\gamma(u_\epsilon) - \gamma'(u_\epsilon) f(u_\epsilon) \\ \qquad = -\gamma''(u_\epsilon)\left(\sum_{i=1}^{N} (\partial_{x_i}\rho_i(u_\epsilon))^2 + \epsilon|\nabla u_\epsilon|^2\right). \end{cases} \tag{24}$$

After an integration of (24) over Q, we use (23) to obtain

$$\int_Q \gamma''(u_\epsilon)\left(\sum_{i=1}^{N}(\partial_{x_i}\rho_i(u_\epsilon))^2 + \epsilon|\nabla u_\epsilon|^2\right) dxdt$$
$$\le \int_{\mathbb{R}^N} (\gamma(u_0) + KTc(f)u_0 \exp(c(f)T))\, dx \le C < \infty. \tag{25}$$

Choosing now the particular function $\gamma(r) = \dfrac{r^2}{2}$ in (25), we end up with

$$\int_Q \left(\sum_{i=1}^{N}(\partial_{x_i}\rho_i(u_\epsilon))^2 + \epsilon|\nabla u_\epsilon|^2\right) dxdt \le \frac{1}{2}\|u_0\|^2_{L^2(\mathbb{R}^N)}$$
$$+Kc(f)T\|u_0\|_{L^1(\mathbb{R}^N)} \exp(c(f)T) < \infty. \tag{26}$$

Using a subsequence, it follows that

$$\partial_{x_i}\rho_i(u_\epsilon) \rightharpoonup \partial_{x_i}\rho_i(u) \text{ in } L^2([0,T]\times\mathbb{R}^N). \tag{27}$$

So, (i) of Definitio 2.1 comes from (21), (23) and (27).
Using test function in (24) and letting ϵ goes to 0, it appears that u satisfie (9).
Since $u \in \mathcal{C}\left([0,T]; L^1(\mathbb{R}^N)\right)$, the part (iii) of Definitio 2.1 is immediate.
So u is an entropy solution of (CP).
Step 2: u_0 and f satisfy (H_1).
We approximate u_0, f by $u_0^\epsilon \in W^{2,1}(\mathbb{R}^N) \cap H^1(\mathbb{R}^N) \cap L^\infty(\mathbb{R}^N)$, $f^\epsilon \in W^{2,1}(\mathbb{R}) \cap H^1(\mathbb{R}) \cap L^\infty(\mathbb{R})$ such that

$$u_0^\epsilon \to u_0 \text{ in } L^1(\mathbb{R}^N), \quad f^\epsilon \to f \text{ in } L^1(\mathbb{R}) \text{ as } \epsilon \downarrow 0.$$

By Step 1, there exists an entropy solution $u^\epsilon \in \mathcal{C}([0,T]; L^1(\mathbb{R}^N))$ of $(CP)(A, F, f^\epsilon, u_0^\epsilon)$ for each $\epsilon > 0$. Using (H_1) and combining the contraction principle (19) with the Gronwall's inequality, one has

$$\|u^{\epsilon_1} - u^{\epsilon_2}\|_{\mathcal{C}([0,T];L^1(\mathbb{R}^N))} \le \|u_0^{\epsilon_1} - u_0^{\epsilon_2}\|_{L^1(\mathbb{R}^N)} \exp(c(f)T) \to 0, \quad \text{if } \epsilon_1,\ \epsilon_2 \to 0.$$

Thus, (u^ϵ) is a Cauchy sequence in $\mathcal{C}([0,T]; L^1(\mathbb{R}^N))$ and there exists $u \in \mathcal{C}([0,T]; L^1(\mathbb{R}^N))$ such that (22) is fulfilled As in Step 1, we conclude that u is an entropy solution of (CP).

3. Renormalized Solution of $(CP)(A, F, f, u_0)$.

We suppose that

$$(H_3) \qquad \begin{cases} 1) & u_0 \in L^1(\mathbb{R}^N), \\ 2) & f \in Lip(\mathbb{R}) \text{ and } f(0) = 0. \end{cases}$$

Following [11], we introduce herein the notion of renormalized solution of (CP).

DEFINITION 3.1 *(Renormalized solution of (CP))*
Let u_0, f satisfy (H_3). A renormalized solution of (CP) (A, F, f, u_0) is a measurable function $u : Q \to \mathbb{R}$ such that:

(i) $u \in L^\infty(0, T; L^1(\mathbb{R}^N))$ and $\rho(T_k(u)) \in (H^1(Q))^N$ for all $k > 0$.

(ii) For any $k > 0$ and any entropy flux triple (γ, θ, ν) with $|\gamma'| \leq K$ (for some given positive constant K), there exists a nonnegative bounded Radon measure $m_k^{u,K}$ on Q such that

$$\begin{cases} \partial_t\gamma(T_k(u)) + \nabla.\theta(T_k(u)) - \sum_{i=1}^{N} \partial^2_{x_i x_i}\nu_i(T_k(u)) - \gamma'(T_k(u))f(u) \\ \qquad \leq -\gamma''(T_k(u)) \sum_{i=1}^{N} (\partial_{x_i}\rho_i(T_k(u)))^2 + m_k^{u,K} \quad \text{in } \mathcal{D}'(Q). \end{cases} \tag{28}$$

(iii) The total mass of the renormalized measure $m_k^{u,K}$ vanishes as $k \uparrow \infty$:
$\lim_{k\uparrow\infty} m_k^{u,K}(Q) = 0$.
(iv) ess $\lim_{t\downarrow 0} \|u(t, .) - u_0\|_{L^1(\mathbb{R}^N)} = 0$.

REMARK 3.1

1. *For all $k > 0$, $T_k(u)$ belongs to $L^\infty(Q)$ and (28) is well defined. Moreover, any bounded renormalized solution of (CP) is also an entropy solution of the same problem in the sense of Defintion 2.1.*

2. *As in the previous section, for $i = 1, ..., N$, we have always*

$$\partial_{x_i}\rho_i^\eta(T_k(u)) = \eta(T_k(u))\partial_{x_i}\rho_i(T_k(u)) \text{ a.e in } Q \text{ and in } L^2(Q), \quad \forall k > 0 \text{ and } \eta \in C(\mathbb{R}). \tag{29}$$

3.1. Uniqueness of Renormalized Solution of (CP).

We start with the uniqueness theorem:

THEOREM 3.1
Suppose that (H_2), (H_3) are fulfilled and let (u_0, f), (v_0, f) satisfy (H_3). If u and v are renormalized solutions of $(CP)(A, F, f, u_0)$ and $(CP)(A, F, f, v_0)$ respectively, then for a.e. $t \in (0, T)$, we have:

$$\int_{\mathbb{R}^N} (u(t) - v(t))^+ \, dx \leq \int_{\mathbb{R}^N} (u_0 - v_0)^+ dx + \int_{Q_t} \kappa\,(f(u) - f(v))\, dxds \tag{30}$$

for $\kappa \in H(u-v)$ *a.e. so that:*

$$\|u(t)-v(t)\|_{L^1(\mathbb{R}^N)} \leq \|u_0-v_0\|_{L^1(\mathbb{R}^N)} + \int_0^t \|f(u)-f(v)\|_{L^1(\mathbb{R}^N)} ds. \tag{31}$$

In order to prove Theorem 3.1, let us give this proposition below.

Proposition 3.1
Suppose that conditions of Theorem 3.1 *are fulfilled. Then*

$$\begin{cases} \int_Q H_0\left(T_k(u)-T_k(v)\right)\{\sum_{i=1}^N \left(F_i(T_k(u))-F_i(T_k(v))\right)\partial_{x_i}\phi \\ \quad -\sum_{i=1}^N \left(A_{ii}(T_k(u))-A_{ii}(T_k(v))\right)\partial^2_{x_ix_i}\phi \\ \quad -\left(T_k(u)-T_k(v)\right)\partial_t\phi\}dxdt \leq \int_{\mathbb{R}^N}\left(T_k(u_0)-T_k(v_0)\right)^+\phi(0)dx \\ \quad +\int_Q \kappa\left(f(u)-f(v)\right)\phi\, dxdt + o(\frac{1}{k}) \end{cases} \tag{32}$$

for any $\kappa \in H\left(T_k(u)-T_k(v)\right)$ *a.e. and* $0 \leq \phi \in \mathcal{D}(Q)$.

For the proof of Proposition 3.1, we request the reader to see [11] and [16].

Proof of Theorem 3.1
Let $k > 0$ and u, v be two renormalized solutions of $(CP)(A,F,f,u_0)$ and $(CP)(A,F,f,v_0)$ respectively. Set $w_k = (T_k(u)-T_k(v))^+$, $w_{0,k} = (T_k(u_0)-T_k(v_0))^+$, $h = (f(u)-f(v))^+$. Using Proposition 3.1, we get as in the proof of Theorem 2.2, comparison and contraction principles with the truncation function T_k:

$$\begin{cases} \int_{\mathbb{R}^N}\left(T_k(u(t))-T_k(v(t))\right)^+ dx \leq \int_{\mathbb{R}^N}\left(T_k(u_0)-T_k(v_0)\right)^+ dx \\ \qquad\qquad + \int_{Q_t}\kappa\left(f(u)-f(v)\right)dxds + o(\frac{1}{k}) \end{cases} \tag{33}$$

for $\kappa \in H\left(T_k(u)-T_k(v)\right)$ a.e. so that:

$$\begin{cases} \|T_k(u(t))-T_k(v(t))\|_{L^1(\mathbb{R}^N)} \leq \|T_k(u_0)-T_k(v_0)\|_{L^1(\mathbb{R}^N)} \\ \qquad\qquad + \int_0^t \|f(u)-f(v)\|_{L^1(\mathbb{R}^N)} ds + o(\frac{1}{k}). \end{cases} \tag{34}$$

See that

$$T_k(u) \to u \text{ a.e. if } k \uparrow +\infty \text{ and } |T_k(u)-T_k(v)| \leq |u-v|.$$

As $u-v \in L^1$, we send $k \uparrow +\infty$ and by Lebesgue's convergence theorem, we get (30) and (31).
To end this last section, we deal now with the existence theorem of renormalized solution.

3.2. Existence of Renormalized Solution of (CP)

THEOREM 3.2 *Suppose that (H_2), (H_3) hold. Then there exists at least one renormalized solution of (CP).*

Proof.
As in the previous section, we divide the proof in two steps:
Step 1: u_0 and f satisfy (H_1).
Due to the second section, (CP) has a unique entropy solution u which is a limit of some sequence of solutions of the nondegenerate parabolic problem $(CP)(A_\epsilon, F, f, u_0)$.
Let us prove that this entropy solution is also a renormalized solution for $(CP)(A, F, f, u_0)$. By density, (24) is still true with $\gamma \in W^{2,\infty}(\mathbb{R})$. So as in [11], we make the below judicious choices of γ in (24) to obtain some useful a priori estimates.

- Choose $\gamma(u_\epsilon) = \dfrac{1}{k}\displaystyle\int_0^{u_\epsilon} T_k(s)s ds$ and sending $k \downarrow 0$, we get

$$\|u_\epsilon\|_{L^\infty(0,T;L^1(\mathbb{R}^N))} \leq (1 + \exp(c(f)T)) \|u_0\|_{L^1(\mathbb{R}^N)}. \tag{35}$$

  We deduce from (35) that $u \in L^\infty(0, T; L^1(\mathbb{R}^N))$.

- Choose $\gamma(u_\epsilon) = \displaystyle\int_0^{u_\epsilon} T_k(s)ds = \begin{cases} \dfrac{|u_\epsilon^2|}{2} & \text{if } |u_\epsilon| \leq k, \\ k|u_\epsilon| - \dfrac{k^2}{2} & \text{if } |u_\epsilon| > k, \end{cases}$

  we have

$$\int_Q \left(\sum_{i=1}^N (\partial_{x_i}\rho_i(T_k(u_\epsilon)))^2 + \epsilon|\nabla T_k(u_\epsilon)|^2 \right) dxdt \leq k\,(1 + \exp(c(f)T)) \|u_0\|_{L^1(\mathbb{R}^N)}. \tag{36}$$

  As in the second section, we check that $\partial_{x_i}\rho_i(T_k(u)) \in L^2(Q)$.

- Choose γ such that $\gamma(0) = 0$ and for any $k > 0$, $\delta > 0$,

$$\gamma'(u_\epsilon) = \frac{1}{\delta}(T_{k+\delta}(u_\epsilon) - T_k(u_\epsilon)) = \begin{cases} -1, & \text{if } u_\epsilon < -k-\delta, \\ \dfrac{-u_\epsilon - k}{\delta} & \text{if } -k-\delta < u_\epsilon < -k \\ 0 & \text{if } -k < u_\epsilon < k, \\ \dfrac{u_\epsilon - k}{\delta} & \text{if } k < u_\epsilon < k+\delta, \\ 1 & \text{if } u_\epsilon > k+\delta, \end{cases}$$

  we have

$$\begin{cases} \dfrac{1}{\delta}\displaystyle\int_{\{k<|u_\epsilon|<k+\delta\}} \left(\sum_{i=1}^N (\partial_{x_i}\rho_i(T_k(u_\epsilon)))^2 + \epsilon|\nabla T_k(u_\epsilon)|^2 \right) dxdt \\ \quad \leq \displaystyle\int_{\{|u_0|>k\}} |u_0|dx + \exp(c(f)T) \int_{\{|u_\epsilon|>k\}} |u_0|dx := \chi(k). \end{cases} \tag{37}$$

As u_0 is bounded, from (35) we deduce that $\chi(k) \to 0$ as $k \uparrow \infty$.

- Choose $\gamma = \gamma_{\eta,h_{k,\delta}}$ such as:
$\gamma_{\eta,h_{k,\delta}}(0) = 0$, $\gamma'_{\eta,h_{k,\delta}} = \eta' h'_{k,\delta}$, $\eta \in \mathcal{C}^2(\mathbb{R})$, convex, $|\eta'| \leq K$, $h_{k,\delta}(0) = 0$ and

$$h'_{k,\delta}(u_\epsilon) = \begin{cases} 1 & \text{if } |u_\epsilon| < k, \\ \dfrac{k+\delta-|u_\epsilon|}{\delta} & \text{if } k < |u_\epsilon| < k+\delta, \\ 0 & \text{if } |u_\epsilon| > k+\delta. \end{cases}$$

Remark that $h_{k,\delta} \in \mathcal{C}^2(\mathbb{R})$, $supp(h) \subset [-k,k]$. So we deduce from (24) that

$$\begin{cases} \partial_t \gamma_{\eta,h_{k,\delta}}(u_\epsilon) + \nabla.\theta^{\gamma_{\eta,h_{k,\delta}}}(u_\epsilon) - \displaystyle\sum_{i=1}^{N} \partial^2_{x_i x_i} \nu_i^{\gamma_{\eta,h_{k,\delta}}}(u_\epsilon) - \epsilon \triangle \gamma_{\eta,h_{k,\delta}}(u_\epsilon) \\ \qquad -\gamma'_{\eta,h_{k,\delta}}(u_\epsilon) f(u_\epsilon) = -\left(m_\epsilon^{\eta' h''} + m_\epsilon^{\eta'' h'}\right)(t,x), \end{cases} \tag{38}$$

where

$$\begin{cases} m_\epsilon^{\eta' h''}(t,x) = \eta'(u_\epsilon) h''(u_\epsilon) \left(\displaystyle\sum_{i=1}^{N} (\partial_{x_i} \rho_i(u_\epsilon))^2 + \epsilon |\nabla u_\epsilon|^2 \right), \\ m_\epsilon^{\eta'' h'}(t,x) = \eta''(u_\epsilon) h'(u_\epsilon) \left(\displaystyle\sum_{i=1}^{N} (\partial_{x_i} \rho_i(u_\epsilon))^2 + \epsilon |\nabla u_\epsilon|^2 \right). \end{cases}$$

Observe that for any u_ϵ, if $\delta \downarrow 0$ then

$$h'_{k,\delta}(u_\epsilon) \to 1_{\{|u_\epsilon|<k\}}, \quad h_{k,\delta}(u_\epsilon) \to T_k(u_\epsilon). \tag{39}$$

So for any u_ϵ and when $\delta \downarrow 0$, (39) gives us

$$\begin{cases} \gamma_{\eta,h_{k,\delta}}(u_\epsilon) \to \eta\left(T_k(u_\epsilon)\right), \quad \gamma'_{\eta,h_{k,\delta}}(u_\epsilon) \to \eta'\left(T_k(u_\epsilon)\right), \\ \theta^{\gamma_{\eta,h_{k,\delta}}}(u_\epsilon) \to \theta\left(T_k(u_\epsilon)\right), \quad \nu^{\gamma_{\eta,h_{k,\delta}}}(u_\epsilon) \to \nu\left(T_k(u_\epsilon)\right). \end{cases} \tag{40}$$

On the other hand, for any Borel set $B \subset Q$ we have clearly

$$\int_{B\cap\{k<|u_\epsilon|<k+\delta\}} |m_\epsilon^{\eta' h''}(t,x)| dx dt \leq \frac{K}{\delta} \int_{B\cap\{k<|u_\epsilon|<k+\delta\}} \left(\sum_{i=1}^{N} (\partial_{x_i} \rho_i(u_\epsilon))^2 + \epsilon |\nabla u_\epsilon|^2 \right) dx dt.$$

Let $m^K_{k,\epsilon,\delta}$ be the Radon measure on Q such that for any Borel set $B \subset Q$,

$$m^K_{k,\epsilon,\delta}(B) = \frac{K}{\delta} \int_{B\cap\{k<|u_\epsilon|<k+\delta\}} \left(\sum_{i=1}^{N} (\partial_{x_i} \rho_i(u_\epsilon))^2 + \epsilon |\nabla u_\epsilon|^2 \right) dx dt.$$

Then by (37), $m^K_{k,\epsilon,\delta}(Q) \leq \chi(k)$. Thus, there exists some nonnegative bounded Radon measures $m^K_{k,\epsilon}$, m^K_k such that the following convergences in the sense of measures on Q hold:

$$m^K_{k,\epsilon,\delta} \stackrel{\star}{\rightharpoonup} m^K_{k,\epsilon} \text{ as } \delta \downarrow 0, \quad m^K_{k,\epsilon} \stackrel{\star}{\rightharpoonup} m^K_k \text{ as } \epsilon \downarrow 0 \text{ with } m^K_k \leq \chi(k) \to 0 \text{ as } k \uparrow \infty. \tag{41}$$

Furthermore $m_\epsilon^{\eta''h'}(t,x) \geq \eta''(u_\epsilon)h'(u_\epsilon)\sum_{i=1}^{N}(\partial_{x_i}\rho_i(u_\epsilon))^2$. So for any $\xi \in \mathcal{D}(Q)$, $\xi \geq 0$,

$$\lim_{\delta \downarrow 0}\int_Q m_\epsilon^{\eta''h'}(t,x)\xi\ dxdt \geq \int_Q \eta''\left(T_k(u_\epsilon)\right)\sum_{i=1}^{N}(\partial_{x_i}\rho_i\left(T_k(u_\epsilon)\right))^2\,\xi\ dxdt. \quad (42)$$

Thanks to $(39)-(42)$, if $\delta \downarrow 0$ in (38) we obtain

$$\begin{cases} \partial_t\eta\left(T_k(u_\epsilon)\right) + \nabla.\theta\left(T_k(u_\epsilon)\right) - \sum_{i=1}^{N}\partial^2_{x_ix_i}\nu_i\left(T_k(u_\epsilon)\right) - \epsilon\Delta\eta\left(T_k(u_\epsilon)\right) - \eta'\left(T_k(u_\epsilon)\right)f(u_\epsilon) \\ \qquad \leq -\eta''\left(T_k(u_\epsilon)\right)\sum_{i=1}^{N}\left(\partial_{x_i}\rho_i(u_\epsilon)\right)^2 + m^K_{k,\epsilon} \text{ in } \mathcal{D}'(Q). \end{cases} \quad (43)$$

Clearly, when $\epsilon \downarrow 0$ in (43), the limit function u satisfie (28) and we conclude that u is a renormalized solution of $(CP)(A,F,f,u_0)$.

Step 2 : u_0 and f satisfy (H_3).
Note that for any $l > 1$, $u_{0,l} = T_l(u_0)$, $f_l = T_l(f)$ satisfy (H_1) and $u_{0,l} \to u_0$ in $L^1(\mathbb{R}^N)$, $f_l \to f$ in $L^1(\mathbb{R})$ as $l \uparrow \infty$.
From Step 1, we know that there exists a sequence $\{u_l\}$ of renormalized solutions of $(CP)(A,F,f_l,u_{0,l})$ and thanks to contraction principle (31), $\{u_l\}$ is a Cauchy sequence in $C(0,T;L^1(\mathbb{R}^N))$ and has a limit point u.
On the other hand, if $m^K_{k,l}$ is the corresponding Radon measure, we can prove that $\{u_l\}$ and $m^K_{k,l}$ verify the a priori estimates in Step 1 where bounds don't depend on l but only on k. So we conclude as in Step 1 that u is a renormalized solution of $(CP)(A,F,f,u_0)$.

REMARK 3.2

1. *If the source term f is not a function of u, we don't need to impose it to be Lipschitz continuous. If it depends only of (t,x) as in* [11]*, our well-posedness results in the sense of entropy solution and renormalized solution still hold with $f \in L^1(Q) \cap L^\infty(Q)$ and $f \in L^1(Q)$ respectively.*

2. *Notice that if f is continuous BV-function, its derivative f' is a measure and f can be decomposed in the form $f = f_+ + f_-$ where f_+ and f_- are precisely primitives of positive measures $f'^+, -f'^-$. We can check that the proof of Corollary* 2.1 *is still true with f_-; so in this work, our uniqueness results hold with a more general continuous source term $f \in BV(\mathbb{R})$ such that f_+ is Lipschitz continuous.*

3. *It is importent to see that if a is not a diagonal matrix, the relations* (10) *and* (29) *are not automatically true without more smoothness condition ($cf.$*[17]*for more details). This is one of the main difficulties to extend our results to more general non diagonal matrix. Now, we are working to handle this problem in our coming works* [31].

Acknowledgments

This work was carried out while the firs author visited Hassan II University of Mohammedia (Morocco). He thanks the University Agency of French-speaking World (UAFW) and the Reseau PDE, Modelisation and Control for their financia supports. He is grateful to the professor Hamidou

References

[1] Ammar (K.), Wittbold (P.), Existence of renormalized solution of degnerate elliptic-parbolic problems *Proc. Royal Soc. Edinb.* **113A**, 477-496,2003

[2] Arendt (W.), Bénilan (Ph.), Inégalités de Kato et semi-groupes sous-Markoviens, *Revista Mathematica Complutense Madrid* **5**,(1992) 281-308.

[3] Butos (M.C.), Concha (F.), Bürger (R.), Tory (E.M.), Sedimentation and thickenning: phenomenological foundation and mathematical theory, Kluwer Academic, Dordrecht, 1999.

[4] Alt (H.W.), Luckhauss (S.), Quasi-linear elliptic-parabolic differential equations, *Math.Z.,* **183**,(1983) 311-341.

[5] Bénilan (Ph.), Crandall (M.G.) , Pazy (A.), *Evolution Equation governed by Accretive operators* (book to appear).

[6] Bénilan (Ph.), Boccardo (L.), Gariepy (B.), Pierre (M.), Vazquez (J.L.), An L^1 theory of existence and uniqueness of solutions of nonlinear elliptic equations, *Ann. Scul. Norm. Sup.* **22**(2) 241-273, 1995

[7] Blanchart (D.), Murat (F.), Renormalized solutions of nonlinear parabolic problems with L^1 data: existence and uniqueness *Proc.royal Soc. Edinb A* **127** 1137-1152 1997.

[8] Blanchard (D.) Redouane (H.), Solutions renormalisées d'équations paraboliques à deux non linéaritées. *C.R.A.S. Paris* **319** 831-847 1994.

[9] Bénilan (Ph.), Gariepy (B.), Strong solution L^1 of degenerate parabolic equation. *J. of Diff. Equat.,* **119**, (1995) 473-502.

[10] Bénilan (Ph.), Kruskhov (S.N.), Quasilinear firs order equations with continuous non linearities. *Russian Acad. Sci. Dokl. Math.* Vol 50 $N°3$, (1995) 391-396.

[11] Bendahmane (M.), Karlsen (K.H.), Renormalized solutions for quasilinear anisotropic degenerate parabolic equations, *Siam J. Marth.Anal.* Vol.36, N0.2,PP.405-422(2004).

[12] Bénilan (Ph.), Touré (H.), Sur l'équation générale $u_t = \varphi(u)_{xx} - \psi(u)_x + v$, *C.R. Acad. Sc. Paris, serie 1,* **299**, 18 (1984).

[13] Bénilan (Ph.), Touré (H.), Sur l'équation générale $u_t = a(., u, \varphi(., u)_x)_x$ dans L^1 II. Le problème d'évolution, *Ann. Inst. Henri Poincaré,* vol. 12, 6, 1995, pp. 727-761.

[14] Bénilan (Ph.), Wittbold (P.), On mild and weak solution of elliptic-Parabolic *Problems. Adv. in Diff. Equat.* Vol. 1 (6) (1996) 1053-1072.

[15] Carrillo (J.), Entropy solutions for nonlinear degenerate problems. *Arch. Ratio. Mech. Anal.* **147** (1999) 269-361.

[16] Gui-Qiang Chen, Karlsen (H.K.), Quasilinear anisotropic degenerate parabolic equations with time-space dependant diffusion coefficients *Commun.Pure Appl.Anal.*, Vol.4,N0.2,(2005),pp.241-266.

[17] Gui-Qiang Chen, Perthame (B.), Well-posedeness for non-isotropic degenerate parabolic-hyperbolic equations, *Ann. I. H. Poincaré-An* **20**,4(2003)645-668.

[18] Vol'pert (A.I.), Hudjaev (S.N.), Cauchy's problem for degenerate second order quasilinear parabolic equation, Math. *USSR-Sbornik*, Vol 7, $N°3$, (1969) 365-387.

[19] Igbida (N.), Urbano (J.M.), Uniqueness for degenerate problems, NoDEA 10 (2003) 287-307.

[20] Yin (J.), On the uniqueness and stability of BV solutions for nonlinear diffusion equations, *Comm. Part. Diff. Equat.* **15**, 12, (1990) 1671-1683.

[21] Kruskhov (S.N.), Panov (E.Yu.), Conservative quasilinear firs order laws with an infinit domain of dependence on the initial data, *Soviet. Math. Dokl.* Vol. 42, 2, (1991) 316-321.

[22] Ladyzenskaja (O.A.), Solonnikov (V.A.), Ural'ceva (N.N.), Linear and quasilinear equations of parabolic type, *Transl. of Math. Monographs* **23** (1968).

[23] Kruskhov (S.N.), First order quasi-linear equations in several independent variables, *Mat. Sb. (S.N.),* **10** (1970), pp. 217-243.

[24] Maliki (M.), Touré (H.), Uniqueness entropy solutions for nonlinear degenerate parabolic problem *Journal of Evolution equation* **3** (2003), no. 4, 603–622. (Birkhauser).

[25] Maliki (M.), Touré (H.), Solution généralisée locale d'une équation parabolique quasi linéaire dégénérée du second ordre. *Ann. Fac. Sci. Toulouse.* Vol. VII 1, (1998) 113-133.

[26] Maliki (M.), Touré (H.), Dépendence continue de solutions généralisées locales, *Ann. Fac. Sci. Toulouse*. Vol. X 4, (2001) 701-711.

[27] Maliki (M.), Continuous dependence of the entropy solution of general parabolic equation, *Ann. Fac. Sci. Toulouse.* Vol.XV 3, (2006) 589-598.

[28] Maliki (M.), Ouédraogo (A.), Renormalized solution for nonlinear degenerate problems in the whole space, *Ann. Fac. Sci. Toulouse*. Vol. XVII 6, (2008) 159-173.

[29] Igbida (N.), Wittbold (P.), *Renormalized solution for stephan type problem: Existence and Uniqueness,* Preprint LAMFA, Université de Picadie Jules Vernes.

[30] Chavent (G.), Jaffre (J.), *Mathematical models and finite elements for reservoir simulation*, North Holland, Amsterdam, 1986.

[31] Ouédraogo (A.), Maliki (M.), Touré (H.), *Renormalized solution for a class of general nonlinear anisotropic degenerate parabolic problems without Lipschitz condition on the diffusion and convection flux* (Preprint).

In: Evolution Equations: Theories, Solutions and Functions ISBN 978-1-61324-090-8
Editor: Gaston M. N'Guerekata

Chapter 4

PERIODIC SOLUTIONS OF IMPULSIVE EVOLUTION EQUATIONS

***Khalil Ezzinbi*[1], *James H. Liu*[2] *and Nguyen Van Minh*[3]**
[1]Departement de Mathematiques, Faculte des Sciences Semlalia,
B.P. 2390, Marrakech, Morocco
[2]Department of Mathematics, James Madison University,
Harrisonburg, VA, USA
[3]Department of Mathematics, University of West Georgia,
Carrollton, GA, USA

Abstract

We study the following semilinear evolution equation

$$\begin{aligned} &u'(t) + A(t)u(t) = f(t, u(t)), \quad t > 0, \quad t \neq t_i, \\ &u(0) = u_0, \\ &\Delta u(t_i) = I_i(u(t_i)), \quad i = 1, 2, \cdots, \quad 0 < t_1 < t_2 < \cdots < \infty, \end{aligned}$$

in a Banach space $(X, \|\cdot\|)$ with unbounded operators $A(t)$. Here, $\Delta u(t_i) = u(t_i^+) - u(t_i^-)$ constitutes an impulsive condition, which can be used to model more physical phenomena than the traditional initial value problems $u(0) = u_0$. We assume that $f(t, u)$ and $A(t)$ are T–periodic in t and then prove with some compactness conditions that if solutions of the equation are ultimate bounded, then the equation has a T–periodic solution. The new results obtained here extend some earlier results in this area for equations without impulsive conditions.

Keywords and phrases: Impulsive evolution equations, periodic solutions

AMS Subject Classification: 34G, 35K.

1. Introduction

Let $(X, \|\cdot\|)$ be a Banach space and let $A(t)$ be unbounded operators. The study of periodic solutions for the following evolution equation without impulsive conditions,

$$\begin{aligned} &u'(t) + A(t)u(t) = f(t, u(t)), \quad t > 0, \\ &u(0) = u_0, \end{aligned}$$

was given in [9] using the boundedness of the solutions; see [1, 2, 8, 12, 15, 16, 17] for some related references.

The study of the impulsive evolution equation

$$\begin{aligned} &u'(t) = Au(t) + f(t, u(t)), \;\; t > 0, \;\; t \neq t_i, \\ &u(0) = u_0, \\ &\Delta u(t_i) = I_i(u(t_i)), \;\; i = 1, 2, \cdots, \;\; 0 < t_1 < t_2 < \cdots, \end{aligned}$$

was given in [10] using a semigroup approach, where $\Delta u(t_i) = u(t_i^+) - u(t_i^-)$ constitutes an impulsive condition. Note that impulsive conditions are the combinations of the traditional initial value problems and the short-term perturbations whose duration can be negligible in comparison with the duration of the process. They have advantage over the traditional initial value problems because they can be used to model other phenomena that cannot be modeled by the traditional initial value problems, such as the dynamics of populations subjected to abrupt changes (harvesting, diseases, etc.). See [4, 5, 7, 11, 14, 18] for some related references.

In this paper, we will combine these results and study periodic solutions for the following evolution equation with impulsive condition,

$$u'(t) + A(t)u(t) = f(t, u(t)), \;\; t > 0, \;\; t \neq t_i, \tag{1}$$
$$u(0) = u_0, \tag{2}$$
$$\Delta u(t_i) = I_i(u(t_i)), \;\; i = 1, 2, \cdots, \;\; 0 < t_1 < t_2 < \cdots < \infty, \tag{3}$$

by using the boundedness of the solutions, and obtain new results. That is, we will prove that if the solutions of Eq. (1)-(3) are ultimate bounded, then the equation has a periodic solution. The new results extend some earlier results in this area for equations without impulsive conditions. We will provide some details to show how the difficultie associated with the impulsive conditions are overcome.

We make the following assumption in this paper in order to study periodic solutions.

Assumption 1.1

1. $T > 0$ is a constant, $A(t + T) = A(t)$, $f(t + T, u) = f(t, u)$, $t \geq 0$, $u \in X$. *$f(t, u)$ is continuous in (t, u) and Lipschitzian in u; I_i, $i = 1, 2, \cdots$, are Lipschitzian.*

2. $0 < t_1 < t_2 < \cdots < t_p < T < t_{p+1}$ and $t_{p+k} = t_k + T$, $I_{p+k} = I_k$, $k \geq 1$.

2. When $A(t)$ Is Independent of t

In this section, we study periodic solutions for the following semilinear evolution equation with impulsive condition,

$$u'(t) + Au(t) = f(t, u(t)), \;\; t > 0, \;\; t \neq t_i, \tag{1}$$
$$u(0) = u_0, \tag{2}$$
$$\Delta u(t_i) = I_i(u(t_i)), \;\; i = 1, 2, \cdots, \;\; 0 < t_1 < t_2 < \cdots < \infty, \tag{3}$$

where the operator A generates a C_0 semigroup $S(\cdot)$.

Definition 2.1 *A mild solution $u(\cdot)$ of Eq. (1)-(3) with $u(0) = u_0$ is a piecewise continuous function with points of discontinuity t_i of the first kind, and satisfies*

$$u(t) = S(t)u_0 + \int_0^t S(t-s)f(s,u(s))ds + \sum_{0<t_i<t} S(t-t_i)I_i(u(t_i)), \ \ t \geq 0. \qquad (4)$$

It is easily seen that mild solutions satisfy the impulsive condition (3). To take care of the jumps, we need the following lemma.

Lemma 2.2 *[3] Assume that $M \geq 0, K > 0, \beta_i \geq 0$ are constants, and*

$$m(t) \leq M + \int_0^t p(s)m(s)ds + \sum_{0<t_i<t} \beta_i m(t_i), \ \ 0 \leq t \leq K, \qquad (5)$$

where $p(t) \geq 0$ is continuous, and $m(t) \geq 0$ is piecewise continuous with points of discontinuity t_i of the first kind and left continuous at t_i. Then,

$$m(t) \leq M \prod_{0<t_i<t} (1+\beta_i) \exp\Big(\int_0^t p(s)ds\Big), \ \ 0 \leq t \leq K.$$

To be able to study periodic solutions, we make the following assumption in this section.

Assumption 2.3 *For each $u_0 \in X$, Eq. (1)-(3) has a unique mild solution $u(\cdot, u_0)$ existing on $[0, \infty)$. Following some similar papers in this area, we will use "solutions" to mean "mild solutions".*

Accordingly, we can defin a mapping $P : X \to X$ as

$$Pu_0 = u(T) = S(T)u_0 + \int_0^T S(T-s)f(s,u(s))ds + \sum_{0<t_i<T} S(T-t_i)I_i(u(t_i)), \quad (6)$$

where $u(\cdot) = u(\cdot, u_0)$ is the unique solution of Eq. (1)-(3).

We firs prove the following basic results which are needed to derive periodic solutions.

Lemma 2.4 *Let Assumptions 1.1 and 2.3 be satisfied, and let $u(t, u_0)$ be the solution of Eq. (1)-(3) with $u(0) = u_0$. Then the following are true.*

1. *$y(t) = u(t+T, u_0)$ is the solution of Eq. (1)-(3) with $y(0) = u(T, u_0)$.*
2. *$u(t+nT, u_0) = u(t, P^n u_0)$, $t \geq 0$, $n \geq 1$. In particular, $u(nT, u_0) = P^n u_0$, $n \geq 1$.*
3. *P has a fixed point u_0 iff $u(t, u_0)$ is a periodic solution of Eq. (1)-(3).*

Proof. (1). Since $u(t, u_0)$ is the solution of Eq. (1)-(3) with $u(0) = u_0$, we have

$$\begin{aligned} y(t) &= u(t+T, u_0) = S(t+T)u_0 + \int_0^{t+T} S(t+T-s)f(s,u(s))ds \\ &+ \sum_{0<t_i<t+T} S(t+T-t_i)I_i(u(t_i)) \\ &= S(t)S(T)u_0 + S(t)\int_0^T S(T-s)f(s,u(s))ds \\ &+ \int_T^{t+T} S(t+T-s)f(s,u(s))ds + \sum_{0<t_i<t+T} S(t+T-t_i)I_i(u(t_i)). \quad (7) \end{aligned}$$

Note that

$$\int_T^{t+T} S(t+T-s)f(s,u(s))ds = \int_0^t S(t-h)f(h+T,u(h+T))dh = \int_0^t S(t-h)f(h,y(h))dh,$$

and (as $t_i \neq T$)

$$\sum_{0<t_i<t+T} S(t+T-t_i)I_i(u(t_i)) = S(t)\sum_{0<t_i<T} S(T-t_i)I_i(u(t_i)) + \sum_{T<t_i<t+T} S(t+T-t_i)I_i(u(t_i)).$$

From Assumption 1.1, for $t_i > T$, we have $i = p + k$ and $t_i = t_{p+k} = t_k + T$, thus we have

$$\begin{aligned} \sum_{T<t_i<t+T} S(t+T-t_i)I_i(u(t_i)) &= \sum_{0<t_k<t} S(t-t_k)I_{p+k}(u(t_{p+k})) \\ &= \sum_{0<t_k<t} S(t-t_k)I_k(u(t_k+T)) = \sum_{0<t_k<t} S(t-t_k)I_k(y(t_k)). \end{aligned}$$

Therefore, (7) becomes

$$\begin{aligned} y(t) &= S(t)S(T)u_0 + S(t)\int_0^T S(T-s)f(s,u(s))ds \\ &+ \int_0^t S(t-h)f(h,y(h))dh + S(t)\sum_{0<t_i<T} S(T-t_i)I_i(u(t_i)) \\ &+ \sum_{0<t_k<t} S(t-t_k)I_k(y(t_k)) \\ &= S(t)\Big[S(T)u_0 + \int_0^T S(T-s)f(s,u(s))ds + \sum_{0<t_i<T} S(T-t_i)I_i(u(t_i))\Big] \\ &+ \int_0^t S(t-h)f(h,y(h))dh + \sum_{0<t_k<t} S(t-t_k)I_k(y(t_k)) \\ &= S(t)u(T) + \int_0^t S(t-h)f(h,y(h))dh + \sum_{0<t_i<t} S(t-t_i)I_i(y(t_i)). \quad (8) \end{aligned}$$

(2). Let's use induction on n. From what we just proved in (1), we know that $y(t) = u(t+T, u_0)$ is the solution of Eq. (1)-(3) with $y(0) = u(T, u_0)$. By using the uniqueness, this implies $u(t+T, u_0) = u(t, Pu_0)$, so it is true for $n = 1$. Assume $u(t+nT, u_0) = u(t, P^n u_0)$. Repeat the result from (1), we get $u(t+T, P^n u_0) = u(t, P(P^n u_0))$. Thus, $u(t, P^{n+1} u_0) = u(t+T, P^n u_0) = u(t+T+nT, u_0) = u(t+(n+1)T, u_0)$.

(3). If P has a fi ed point u_0, then $Pu_0 = u_0$. Now, from (1) we just proved, $y(t) = u(t+T, u_0) = u(t, Pu_0) = u(t, u_0)$, so that $u(t, u_0)$ is a periodic solution of Eq. (1)-(3).

On the other hand, if $u(t, u_0)$ is a periodic solution of Eq. (1)-(3), then $Pu_0 = u(T, u_0) = u(0, u_0) = u_0$, so that P has a fi ed point u_0. This completes the proof.

From now on, we are going to derive fi ed points of the mapping P in order to obtain periodic solutions of Eq. (1)-(3). We follow our early approach of using boundedness of the solutions.

Definition 2.5 *[2] Solutions of Eq. (1)-(3) are bounded if for each $B_1 > 0$ there is $B_2 > 0$ such that $\{\|u_0\| \le B_1, t \ge 0\}$ implies $\|u(t, u_0)\| < B_2$.*

Definition 2.6 *[2] Solutions of Eq. (1)-(3) are ultimate bounded if there is a bound $B > 0$ such that for each $B_3 > 0$, there is $K > 0$ such that $\{\|u_0\| \le B_3, t \ge K\}$ implies $\|u(t, u_0)\| < B$.*

For the relationship between boundedness and ultimate boundedness, we have the following result.

Theorem 2.7 *Let Assumptions 1.1 and 2.3 be satisfied. If the solutions of Eq. (1)-(3) are ultimate bounded, then they are also bounded.*

Proof Let $B > 0$ be the bound in the definitio of ultimate boundedness, then for any $B_1 > 0$, there is a $K = K(B, B_1) > 0$ such that $\{\|u_0\| \le B_1,\ t \ge K\}$ imply $\|u(t, u_0)\| \le B$.

Next, for t in the interval $[0, K]$, we can let $w(t)$ be a fi ed solution of Eq. (1)-(3), so that we have

$$\begin{aligned}\|u(t, u_0) - w(t)\| &\le \|S(t)[u_0 - w(0)]\| + \int_0^t \|S(t-s)[f(s, u(s)) - f(s, w(s))\|ds \\ &\quad + \|\sum_{0<t_i<t} S(t-t_i)[I_i(u(t_i)) - I_i(w(t_i))]\| \\ &\le \|S(t)[u_0 - w(0)]\| + c\int_0^t |S(t-s)|\|u(s) - w(s)\|ds \\ &\quad + \sum_{0<t_i<t} |S(t-t_i)|c_i\|u(t_i) - w(t_i)\|,\end{aligned}$$

where c, c_i are Lipschitz constants and the norm $|S(t)|$ is bounded for $t \in [0, K]$. Now we can apply Lemma 2.2 to obtain a constant C which is dependent of $K, S(t), t_i, c, c_i$ but independent of $u(t, u_0) - w(t)$, such that

$$\|u(t, u_0) - w(t)\| \le C\|u_0 - w(0)\|, \quad t \in [0, K]. \tag{9}$$

This implies that for $\|u_0\| \le B_1$, the solutions $u(t, u_0)$ are bounded on $[0, K]$. That is, there is a $H = H(B_1, K) = H(B_1, B) > 0$ such that $\|u(t, u_0)\| \le H$ for $t \in [0, K]$ and $\|u_0\| \le B_1$. Now, we see that $\{\|u_0\| \le B_1,\ t \ge 0\}$ imply $\|u(t, u_0)\| \le B_2 = \max\{H, B\}$, which verifie the boundedness of the solutions. This completes the proof.

Remark 2.8 *If we assume*

$$\max_{u\in X} \|I_k(u)\| < \infty, \quad k \geq 1,$$

then the same result of Theorem 2.7 can be proved without Lipschitz conditions on I_k.

Assumption 2.9 *There is a sufficiently small* $\delta > 0$ *such that* $S(\delta)$ *is compact (i.e., continuous and maps a bounded set into a precompact set). (So that for the semigroup* $S(\cdot)$, $S(t)$ *is compact for* $t \geq \delta$.)

Lemma 2.10 *Let Assumptions 1.1, 2.3 and 2.9 be satisfied. Then* P *defined by (6) is a compact operator.*

Proof. First, note that (9) can be used to prove that the solution $u(\cdot, u_0)$ is continuously dependent on u_0 on finit intervals. Thus, P is continuous. Next, we write P as $Pu_0 = P_1u_0 + P_2u_0$, where

$$\begin{aligned} P_1u_0 &= S(T)u_0 + \int_0^T S(T-s)f(s,u(s))ds, \\ P_2u_0 &= \sum_{0<t_i<T} S(T-t_i)I_i(u(t_i)), \end{aligned}$$

and $u(\cdot) = u(\cdot, u_0)$ is the unique solution of Eq. (1)-(3).

Let E be a bounded set in X. Then (9) indicates that for $u_0 \in E$, the solutions $u(t, u_0)$ are bounded for $t \in [0,T]$. Thus P_2 is a compact operator as $S(t)$ is a compact semigroup.

For P_1, consider, for $0 < \epsilon < T$,

$$\begin{aligned} P_\epsilon u_0 &= S(T)u_0 + \int_0^{T-\epsilon} S(T-r)f(r,u(r,u_0))dr \\ &= S(T)u_0 + S(\epsilon)\int_0^{T-\epsilon} S(T-r-\epsilon)f(r,u(r,u_0))dr. \end{aligned}$$

We see now that for $u_0 \in E$, $f(t, u(t, u_0))$ are also bounded for $t \in [0,T]$. Thus we conclude that the set $\{P_\epsilon u_0 : u_0 \in E\}$ is precompact in X. Next, we have

$$\|P_1u_0 - P_\epsilon u_0\| \leq \int_{T-\epsilon}^T \|S(T-r)f(r,u(r,u_0))\|dr,$$

which implies that $\{P_1u_0 : u_0 \in E\}$ is also precompact in X. This completes the proof.

Theorem 2.11 *(Horn's fixed point theorem) Let* $E_0 \subset E_1 \subset E_2$ *be convex subsets of a Banach space* Z, *with* E_0 *and* E_2 *compact subsets and* E_1 *open relative to* E_2. *Let* $P : E_2 \to Z$ *be a continuous operator such that for some integer* m, *one has*

$$P^j(E_1) \subset E_2, \quad 1 \leq j \leq m-1, \tag{10}$$

$$P^j(E_1) \subset E_0, \quad m \leq j \leq 2m-1, \tag{11}$$

then P *has a fixed point in* E_2.

Theorem 2.12 *Let Assumptions 1.1, 2.3 and 2.9 be satisfied. If the solutions of Eq. (1)-(3) are ultimate bounded, then Eq. (1)-(3) has a T-periodic solution.*

Proof. From Theorem 2.7, the solutions of Eq. (1)-(3) are also bounded. Let $B > 0$ be the bound in the definitio of ultimate boundedness, then by boundedness, there is a $B_1 > B$ such that $\|u_0\| \leq B$ implies $\|u(t, u_0)\| < \frac{1}{2}B_1$ for $t \geq 0$. Furthermore, there is a $B_2 > B_1$ such that $\|u_0\| \leq B_1$ implies $\|u(t, u_0)\| < B_2$ for $t \geq 0$. Now, using ultimate boundedness, there is a positive integer m such that $\|u_0\| \leq B_1$ implies $\|u(t, u_0)\| < B$ for $t \geq (m-1)T$.

From Lemma 2.4, we know that $P^k(u_0) = u(kT, u_0)$, $k \geq 1$. Thus, we have

$$\|P^{j-1}(u_0)\| = \|u((j-1)T, u_0)\| < B_2 \text{ for } j = 1, 2, ..., m-1 \text{ and } \|u_0\| \leq B_1,$$

$$\|P^{j-1}(u_0)\| = \|u((j-1)T, u_0)\| < B \text{ for } j \geq m \text{ and } \|u_0\| \leq B_1.$$

Now let

$$\begin{cases} H = \{u_0 \in X : \|u_0\| < B_2\}, & E_2 = \overline{cov.(P(H))}, \\ K = \{u_0 \in X : \|u_0\| < B_1\}, & E_1 = K \cap E_2, \\ G = \{u_0 \in X : \|u_0\| < B\}, & E_0 = \overline{cov.(P(G))}, \end{cases} \tag{12}$$

where $cov.(F)$ is the *convex hull* of the set F define by $cov.(F) = \{\sum_{i=1}^{n} \lambda_i f_i : n \geq 1, f_i \in F, \lambda_i \geq 0, \sum_{i=1}^{n} \lambda_i = 1\}$. From Lemma 2.10, the operator P is compact. It is also known that a convex hull of a precompact set is also precompact, then we see that $E_0 \subset E_1 \subset E_2$ are all convex subsets of X with E_0 and E_2 compact subsets and E_1 open relative to E_2. Next, from (12), we have

$$P^j(E_1) \subset P^j(K) = PP^{j-1}(K) \subset P(H) \subset E_2, \quad 1 \leq j \leq m-1,$$

$$P^j(E_1) \subset P^j(K) = PP^{j-1}(K) \subset P(G) \subset E_0, \quad m \leq j \leq 2m-1.$$

Consequently, we know from Horn's fi ed point theorem that the operator P has a fi ed point $u_0 \in X$. This completes the proof by using Lemma 2.4.

3. Non-autonomous Equations

In this section, we study periodic solutions for the following semilinear non-autonomous evolution equation with impulsive condition,

$$u'(t) + A(t)u(t) = f(t, u(t)), \quad t > 0, \quad t \neq t_i, \tag{1}$$

$$u(0) = u_0, \tag{2}$$

$$\Delta u(t_i) = I_i(u(t_i)), \quad i = 1, 2, \cdots, \quad 0 < t_1 < t_2 < \cdots < \infty. \tag{3}$$

Assumption 3.1 *For $t \in [0, T]$ one has*

1. *The domain $D(A(t)) = D$ is independent of t and is dense in X.*

2. *For* $t \geq 0$*, the resolvent* $R(\lambda, A(t)) = (\lambda I - A(t))^{-1}$ *exists for all* λ *with* $Re\lambda \leq 0$ *and is compact, and there is a constant* M *independent of* λ *and* t *such that*

$$\|R(\lambda, A(t))\| \leq M(|\lambda| + 1)^{-1}, \quad Re\lambda \leq 0.$$

3. *There exist constants* L *and* $0 < a \leq 1$ *such that*

$$\|(A(t) - A(s))A(r)^{-1}\| \leq L|t - s|^a, \quad s, t, r \in [0, T].$$

Based on this assumption, the results from [1, 6, 13] indicate that Eq. (1)-(3) has a unique evolution system $U(t, s),\ 0 \leq s \leq t \leq T$, such that mild solutions of Eq. (1)-(3) are define using

$$u(t) = U(t, 0)u_0 + \int_0^t U(t, s)f(s, u(s))ds + \sum_{0<t_i<t} U(t, t_i)I_i(u(t_i)), \quad t \geq 0. \tag{4}$$

Assumption 3.2 *For each* $u_0 \in X$*, Eq. (1)-(3) has a unique mild solution* $u(\cdot, u_0)$ *existing on* $[0, \infty)$*. Again, we use "solutions" to mean "mild solutions".*

Based on Assumption 3.2, we can defin a mapping $Q : X \to X$ as

$$Qu_0 = u(T) = U(T, 0)u_0 + \int_0^T U(T, s)f(s, u(s))ds + \sum_{0<t_i<T} U(T, t_i)I_i(u(t_i)), \tag{5}$$

where $u(\cdot) = u(\cdot, u_0)$ is the unique solution of Eq. (1)-(3).

Then the treatment when $A(t) = A$ is a constant operator given in the previous section can be carried here. In particular, with the result of [9], we see that the operator Q is compact. Thus, similar to the previous section, we have

Theorem 3.3 *Let Assumptions 1.1, 3.1 and 3.2 be satisfied. If the solutions of Eq. (1)-(3) are ultimate bounded, then Eq. (1)-(3) has a* T*-periodic solution.*

Equations with delays will be treated in forth coming papers.

References

[1] H. Amann, Periodic solutions of semi-linear parabolic equations, *Nonlinear Analysis,* A Collection of Papers in Honor of Erich Roth, Academic Press, New York, 1978, 1-29.

[2] T. Burton, *Stability and periodic solutions of ordinary differential equations and functional differential equations* , Academic Press, Inc., 1985, 197-308.

[3] D. Bainov, V. Covachev and I. Stamova, Estimates of the solutions of impulsive quasilinear functional differential equations, *Ann. Fac. Sci. Toulouse Math.*, **5**(1991), 149-161.

[4] D. Bainov, Z. Kamont and E. Minchev, Comparison principles for impulsive hyperbolic equations of firs order, *J. Comput. Appl. Math.*, **60**(1995), 379-388.

[5] C. Chan, L. Ke and A. Vatsala, Impulsive quenching for reaction-diffusion equations, *Nonlinear Analysis,* **22**(1994), 1323-1328.

[6] A. Friedman, *Partial differential equations*, Krieger, Huntington, New York, 1976.

[7] D. Guo and X. Liu, Extremal solutions of nonlinear impulsive integrodifferential equations in Banach spaces, *J. Math. Anal. Appl.,* **177**(1993), 538-552.

[8] J. Hale, Asymptotic behavior of dissipative systems, *American Math.* Soc., Providence, Rhode Island, 1988.

[9] J. Liu, Bounded and periodic solutions of semi-linear evolution equations, *Dynamic Sys. & Appl.,* **4**(1995), 341-350.

[10] J. Liu, Nonlinear impulsive evolution equations, *Dynam. Conti. Discr. Impul. Sys.*, **6**(1999), 77-85.

[11] X. Liu and A. Willms, Stability analysis and applications to large scale impulsive systems: a new approach, *Canadian Appl. Math. Quart.,* **3**(1995), 419-444.

[12] J. Massera, The existence of periodic solutions of systems of differential equations, *Duke Math. J.,* **17**(1950), 457-475.

[13] A. Pazy, *Semigroups of linear operators and applications to partial differential equations*, Springer-Verlag, New York, 1983, 60-212.

[14] Y. Rogovchenko, Impulsive evolution systems: main results and new trends, *Dynamics Contin. Discr. Impulsive Sys.,* **3**(1997), 57-88.

[15] K. Schmitt and J. Ward, *Periodic and almost periodic solutions of nonlinear evolution equations*, Delay and Differential Equations, A. Fink, R. Miller and W. Kliemann (eds.), World Scientific London, 1992, 78-92.

[16] X. Xiang and N. Ahmed, Existence of periodic solutions of semilinear evolution equations with time lags, *Nonlinear Anal., T.M.A.,* **11**(1992), 1063-1070.

[17] T. Yoshizawa, Stability theory by Liapunov's second method, *Math. Soc.,* Japan, Tokyo, 1966.

[18] A. Zavalishchin, Impulse dynamic systems and applications to mathematical economics, *Dynam. Systems Appl.,* **3**(1994), 443-449.

In: Evolution Equations: Theories, Solutions and Functions ISBN 978-1-61324-090-8
Editor: Gaston M. N'Guerekata

Chapter 5

CRITERIA FOR BOUNDEDNESS OF SOLUTIONS FOR SET CONTROL DIFFERENTIAL EQUATIONS

***Nguyen Dinh Phu*[1,*]*, Le Thanh Quang*[2] *and Tran Thanh Tung*[3]**
[1]Faculty of Mathematics and Computer Sciences,
University of Natural Science, VNU, Vietnam
[2]306 Vo Van Hat, Phuoc Hiep, Long Truong,
District 9. Ho Chi Minh City, Vietnam
[3]Faculty of Natural Science and Technology, Tay Nguyen University,
Buon Ma Thuot City, Vietnam

Abstract

The paper introduces the criteria for boundedness of solutions of set control differential equations(SCDEs).

Keywords and phrases: Differential equations, Control theory, Set differential equations, Set control differential equations.

1. Introduction

Recently, set differential equations in the metric space have been studied. Some interesting results in this direction have been obtained in works of Prof. Lakshmikantham and other authors [1]- [4]. Based on these results, we introduced the concept of SCDEs and studied the existence and comparison of their solutions in [5]- [9]. In this paper, we investigate the boundedness of solutions for SCDEs. The paper is organized as follows: in Section 2, we recall some basic concepts and notations which are useful in next sections. In Section 3, we present the boundedness of solutions for SCDEs.

*E-mail address: ndphu@mathdep.hcmuns.edu.vn

2. Preliminaries

Some following notations and concepts are taken from [1]- [4]. Let $K_c(\mathbb{R}^n)$ denote the collection of all nonempty, compact, convex subsets of $\mathbb{R}^n$. Let A, B be two nonempty bounded subsets of $\mathbb{R}^n$. The Hausdorff distance between A and B is def ned as

$$D[A, B] = \max\{\sup_{a\in A}\inf_{b\in B} \| a - b \|, \sup_{b\in B}\inf_{a\in A} \| a - b \|\} \tag{2.1}$$

In particular $D[A, \theta] = \| A \| = \sup\{\| a \|: a \in A\}$, where θ is the zero element of $\mathbb{R}^n$ which is regarded as a one point set.

The Haussdorff metric (2.1) satisf es the following properties:

$$\begin{aligned} &D[A + C, B + C] = D[A, B] \text{ and } D[A, B] = D[B, A],\\ &D[\lambda A, \lambda B] = \lambda D[A, B],\\ &D[A, B] \leq D[A, C] + D[C, B]\\ &D[A + A', B + B'] \leq D[A, B] + D[A', B'] \end{aligned} \tag{2.2}$$

for all $A, B, C \in K_c(\mathbb{R}^n)$ and $\lambda \in \mathbb{R}_+$.

If $\alpha, \beta \in \mathbb{R}$ and $A, B \in K_c(\mathbb{R}^n)$, then

$$\alpha(A + B) = \alpha A + \alpha B, \quad \alpha(\beta A) = (\alpha\beta)A, \quad 1A = A$$

and if $\alpha, \beta \geq 0$, then $(\alpha + \beta)A = \alpha A + \beta A$.

And θ is zero element of $K_c(\mathbb{R}^n)$ which is regarded as a one point set.

The set $K_c(\mathbb{R}^n)$, with the metric D def ned above, is a complete metric space. It has been proven that $K_c(\mathbb{R}^n)$ becomes a semilinear metric space which can be embedded as a complete cone into a corresponding Banach space, if it is equipped with the natural algebraic operations of addition and nonnegative scalar multiplication.

Let $A, B \in K_c(\mathbb{R}^n)$. If there exists a set $C \in K_c(\mathbb{R}^n)$ such that $A = B + C$, then C is called the Hausdorff difference (the geometric difference) of the sets A and B and is denoted by the symbol $A - B$. The mapping $F : \mathbb{R}_+ \supset I = [t_0, T] \to K_c(\mathbb{R}^n)$ is said to have a Hukuhara derivative $D_H F(\tau)$ at a point $\tau \in I$, if

$$\lim_{h\to 0^+} \frac{F(\tau + h) - F(\tau)}{h} \quad \text{and} \quad \lim_{h\to 0^+} \frac{F(\tau) - F(\tau - h)}{h}$$

exist in the topology of $K_c(\mathbb{R}^n)$ and are equal to $D_H F(\tau)$.

By embedding $K_c(\mathbb{R}^n)$ as a complete cone in a corresponding Banach space and taking into account the result on the differentiation of Bochner integral, we f nd that if

$$F(t) = X_0 + \int_{t_0}^{t} \Phi(s)ds, X_0 \in K_c(\mathbb{R}^n),$$

where $\Phi : I \to K_c(\mathbb{R}^n)$ is integrable in the sence of Bochner, then $D_H F(\tau)$ exists and the equality

$$D_H F(\tau) = \Phi(\tau) \text{ a.e. on } I \text{ holds.}$$

The Hukuhara integral of F is given by

$$\int_{t_0}^{t} F(s)ds = \Big[\int_{t_0}^{t} f(s)ds : f \text{ is continuous selector of } F \Big]$$

for any compact set $I \subset \mathbb{R}_+$.

Some properties of Hukuhara integral are the following.

If $F, G : [t_0, T] \to K_c(\mathbb{R}^n)$ are integrable, we have

$$\int_{t_0}^{t_2} F(s)ds = \int_{t_0}^{t_1} F(s)ds + \int_{t_1}^{t_2} F(s)ds, \quad t_0 \le t_1 \le t_2 \le T \tag{2.3}$$

and

$$\int_{t_0}^{t} \lambda F(s)ds = \lambda \int_{t_0}^{t} F(s)ds, \quad \lambda \in \mathbb{R}, \ t_0 \le t \le T$$

and

$$D\Big[\int_{t_0}^{t} F(s)ds, \int_{t_0}^{t} G(s)ds\Big] \le \int_{t_0}^{t} D\Big[F(s), G(s)\Big]ds. \tag{2.4}$$

More details in the concepts and properties of continuity, Hukuhara derivative, Hukuhara integral, see [1]- [4].

3. Main Results

In [6, 7], authors introduced the initial valued problem for SCDEs as follows:

$$D_H X = F(t, X(t), U(t)), \quad X(t_0) = X_0, \tag{3.1}$$

where $F \in C[I \times K_C(\mathbb{R}^n) \times K_C(\mathbb{R}^p), K_C(\mathbb{R}^n)]$, $t \in I \subset \mathbb{R}_+$, state $X(t) \in K_C(\mathbb{R}^n)$ and the control $U(t) \in K_C(\mathbb{R}^p)$. If $U : I \longrightarrow K_C(\mathbb{R}^p)$ is integrable, then it is called an admissible control. Let $\boldsymbol{U}$ be a set of all admissible controls. The mapping $X \in C^1[I, K_C(\mathbb{R}^n)]$ is said to be a solution of (3.1) on I if it satisf es (3.1) on I. Since $X(t)$ is continuosly differentiable, we have

$$X(t) = X_0 + \int_{t_0}^{t} D_H X(s)ds, \quad t \in I.$$

Thus, we associate with the initial value problem the following:

$$X(t) = X_0 + \int_{t_0}^{t} F(s, X(s), U(s))ds, \quad t \in I \tag{3.2}$$

where the integral is the Hukuhara integeral. Observe that $X(t)$ is a solution of (3.1) iff it satisf es (3.2) on I.

The existence of solutions of SCDEs was introduced in [7] and comparison of solutions was studied in [6]. Some more results of SCDEs were presented in [8]. The stability of solutions of SCDEs was investigated in [5, 9]. The boundedness of solutions is a nice property of differential systems and useful in practice. In this paper, we introduce some criteria for boundedness of SCDEs.

Definition 3.1. The solution of SCDEs is said to be

(B1) equi - bounded, if for any $\alpha > 0$ and $t_0 \in \mathbb{R}_+$, there exists a $\beta = \beta(t_0, \alpha) > 0$ such that

$$\| X_0 \| < \alpha \quad \text{implies} \quad \| X(t) \| < \beta, \quad t \geq t_0 \ ;$$

(B2) uniform - bounded, if β in (B1) does not depend on t_0;

(B3) quasi - equi - ultimately bounded for a bound B if for any $\alpha \geq 0$, $t_0 \in \mathbb{R}_+$, there exists a $B > 0$ and a $T = T(t_0, \alpha) > 0$ such that

$$\| X_0 \| < \alpha \quad \text{implies} \quad \| X(t) \| < B, \quad t \geq t_0 + T \ ;$$

(B4) quasi-uniform ultimately bounded, if T in $(B3)$ does not depend on t_0;

(B5) equi- ultimately bounded, if B_1 and B_3 hold simultaneously;

(B6) uniform ultimately bounded, if B_2 and B_4 hold simultaneously;

Firstly, we need the following theorem.

Theorem 3.1 ([5]). *Assume that the Lyapunov-like function V satisfies the following*

(i) $V \in C[\mathbb{R}_+ \times K_C(\mathbb{R}^n) \times K_C(\mathbb{R}^p), \mathbb{R}_+]$ *and* $| V(t, A, A') - V(t, B, B') | \leq LD[A, B]$, *where L is bounded Lipschitz constant, for all* $A, \ B \in K_C(\mathbb{R}^n), \ A', \ B' \in K_C(\mathbb{R}^p)$ *and* $t \in \mathbb{R}_+$.

(ii) $g \in C[\mathbb{R}^2_+, \mathbb{R}]$ *and for* $t \in \mathbb{R}_+, \ A \in K_C(\mathbb{R}^n), \ B \in K_C(\mathbb{R}^p)$,

$$D^+V(t, A, B) \equiv \limsup_{h \to 0^+} \frac{1}{h}[V(t+h, A+hF(t, A, B), B) - V(t, A, B)] \leq g(t, V(t, A, B)) \tag{3.3}$$

If $X(t) = X(t_0, X_0, t, U(t))$ *is any solution of SCDEs* (3.1) *existing on* $[t_0, T)$ *such that* $V(t_0, X_0, U_0) \leq \omega_0$ *then we have*

$$V(t, X(t), U(t)) \leq r(t, t_0, \omega_0), \qquad t \in [t_0, T) \tag{3.4}$$

where $r(t, t_0, \omega_0)$ *is a maximal solution of original differential equation*

$$\frac{dw}{dt} = g(t, w), \qquad g(t_0) = \omega_0 \geq 0.$$

And its consequence is below.

Corollary 3.1 ([5]). *The function* $g(t, w) = 0$ *is admissible in the Theorem* 3.1 *to yield the estimate*

$$V(t, X(t), U(t)) \leq V(t_0, X(t_0), U(t_0)).$$

Now, we introduce some results on the boundedness of solutions for SCDEs.

Theorem 3.2. *Assume that*

(i) $V \in [\mathbb{R}_+ \times K_C(\mathbb{R}^n) \times K_C(\mathbb{R}^p), \mathbb{R}_+]$, $| V(t, X_1, U_1) - V(t, X_2, U_2) | \leq LD[X_1, X_2]$, $L > 0$ *and for* $(t, X, U) \in \mathbb{R}_+ \times K_C(\mathbb{R}^n) \times K_C(\mathbb{R}^p)$, $D^+V(t, X, U) \leq 0$

(ii) $b(\| X \|) \leq V(t, X, U) \leq a(t, \| X \|)$, *for* $(t, X, U)) \in \mathbb{R}_+ \times K_C(\mathbb{R}^n) \times K_C(\mathbb{R}^p)$, *where* $b, a(t, .) \in \mathbb{K} = \{\sigma \in C[\mathbb{R}_+, \mathbb{R}_+] : \rho(\omega)$ *is increasing in* ω *and* $\rho(\omega) \to \infty$ *as* $\omega \to \infty\}$.

Then, $(B1)$ *holds.*

Proof. Let $0 < \alpha$ and $t_0 \in \mathbb{R}_+$, be given. Choose $\beta = \beta(t_0, \alpha)$ such that

$$a(t_0, \alpha) < b(\beta). \tag{3.5}$$

with this β, $(B1)$ holds. If this is not true, there would exist a solution $X(t) = X(t, X_0, U_0)$ of (3.1) and a $t_1 > t_0$ such that

$$\| X(t_1) \| = \beta \quad \text{and} \quad \| X(t) \| \leq \beta, \quad t_0 \leq t \leq t_1.$$

Assumption (i) and Corollary 3.1 show that

$$V(t, X(t), U(t)) \leq V(t_0, X_0, U_0), \quad t_0 \leq t \leq t_1.$$

As result, condition (ii) and (3.5) yield

$$b(\beta) = b(\| X(t_1) \|) \leq V(t_1, X(t_1), U(t_1)) \leq V(t_0, X_0, U_0) \leq a(t_0, \| X(t_0) \|) < a(t_0, \alpha) < b(\beta)$$

This contradiction proves (B1) and we are done. □

Theorem 3.3. *Assume that*

(i) $V \in [\mathbb{R}_+ \times S^c(\rho) \times K_C(\mathbb{R}^p), \mathbb{R}_+]$, *where* ρ *may be large;* $| V(t, X_1, U_1) - V(t, X_2, U_2) | \leq LD[X_1, X_2]$, $L > 0$ *and for* $(t, X, U) \in \mathbb{R}_+ \times S^c(\rho) \times K_C(\mathbb{R}^p)$, $D^+V(t, X, U) \leq 0$;

(ii) $b(\| X \|) \leq V(t, X, U) \leq a(t, \| X \|)$, *for* $(t, X, U) \in \mathbb{R}_+ \times S^c(\rho) \times K_C(\mathbb{R}^p)$, *where* $a, b \in \mathbb{K}$, *which are defined only on* $[\rho, \infty)$.

Then, $(B2)$ *holds.*

The proof is similar to the proof of Theorem 3.2.

Theorem 3.4. *Let the assumptions of Theorem* 3.2 *hold except that we strengthen the estimate on* $D^+V(t, X, U)$ *as*

$$D^+V(t, X, U) \leq -\eta V(t, X, U), \eta > 0, (t, X, U) \in \mathbb{R}_+ \times S^c(\rho) \times K_C(\mathbb{R}^p), \tag{3.6}$$

and suppose that condition (ii) *holds for* $\| X \| \geq B$. *Then* $(B5)$ *holds.*

Proof. Clearly (B1) is obtained from Theorem 3.2. Hence

$$\| X_0 \| < \alpha \quad \text{implies} \quad \| X(t) \| < \beta,\ t \geq t_0.$$

Now (3.6) yields the estimate

$$V(t, X, U) \leq V(t_0, X_0, U_0)e^{-\eta(t-t_0)}, \quad t \geq t_0. \tag{3.7}$$

Let $T = \frac{1}{\eta} ln \frac{a(t_0,\alpha)}{b(B)}$ and suppose that for $t \geq t_0 + T, \| X(t) \| \geq B$. Then, we get from (3.7)

$$b(B) \leq b(\| X(t) \|) \leq V(t, X(t), U(t)) < a(t_0, \alpha)e^{-\eta T} = b(B).$$

This contradiction proves (B5) and the proof is complete. □

Theorem 3.5. *Assume that*

(i) $\rho > 0, V \in [\mathbb{R}_+ \times S(\rho) \times K_C(\mathbb{R}^p), \mathbb{R}_+]$, V_1*is bounnded for* $(t, X, U) \in \mathbb{R}_+ \times \partial S(\rho)$ *and* $| V_1(t, X_1, U_1) - V(t, X_2, U_2) | \leq L_1 D[X_1, X_2]$, $L_1 > 0$,

$$D^+V_1(t, X, U) \equiv \limsup_{h \to 0^+} \frac{1}{h}[V_1(t + h, X + hF(t, X, U), U) - V_1(t, X, U)] \\ \leq g_1(t, V_1), \quad (t, X, U) \in \mathbb{R}_+ \times S^c(\rho) \times K_C(\mathbb{R}^p)$$

where $g_1 \in C[\mathbb{R}^2_+, \mathbb{R}]$

(ii) $V_2 \in [\mathbb{R}_+ \times S^c(\rho) \times K_C(\mathbb{R}^p), \mathbb{R}_+]$,

$$b(\| X \|) \leq V_2(t, X, U) \leq a(t, \| X \|), \quad a, b \in \mathbb{K},$$

$$D^+V_1(t, X, U) + D^+V_2(t, X, U) \leq g_2(t, V_1(t, X, U) + V_2(t, X, U)), \quad g_2 \in C[\mathbb{R}^2_+, \mathbb{R}],$$

(iii) *the scalar differential equations*

$$w_1' = g_1(t, w_1), \quad w_1(t_0) = w_{10} \geq 0, \tag{3.8}$$

and

$$w_2' = g_2(t, w_2), \quad w_2(t_0) = w_{20} \geq 0, \tag{3.9}$$

are equi-bounded and uniformly bounded respectively.

Then the system (3.1) *is equi-bounded.*

Proof. Let $B_1 > \rho$ and $t_0 \in \mathbb{R}_+$, be given. Let

$$\alpha_1 = \alpha(t_0, B_1) = max\{\alpha_0, \alpha^*\}$$

where $\alpha_0 = max\{V_1(t_o, X_0, U_0) : X_0 \in cl\{S(B_1) \cap S^c(\rho)\}\}$
and $\alpha^* \geq V_1(t, X, U)\ for\ (t, X, U) \in R_+ \times \partial S(\rho)$.

Since equation (3.8) is equi-bounded, given $\alpha_1 > 0$, and $t_0 \in \mathbb{R}_+$, there exist a $\beta_0 = \beta(t_0, \alpha_1)$ such that

$$w_1(t, t_0, w_{10}) < \beta_0, \quad t \geq t_0, \tag{3.10}$$

provided $w_{10} < \alpha_1$, where $w_1(t, t_0, w_{10})$ is any solution (3.8). Let $\alpha_1 = a(B_1) + \beta_0$, the uniform boundedness of equation (3.9) yields that

$$w_2(t, t_0, w_{20}) < \beta(\alpha_2), \quad t \geq t_0, \tag{3.11}$$

provided $w_{20} < \alpha_2$, where $w_2(t, t_0, w_{20})$ is any solution of (3.9). Choose B_2 satisfying

$$b(B_2) > \beta_1(\alpha_2). \tag{3.12}$$

We know claim that $X_0 \in S(B_1)$ inplies that

$$X(t, X_0, U_0) \in S(B_2),$$

for $t \geq t_0$ where $X(t, X_0, U_0)$ is any solution of SCDEs (3.1).

If it is not true, there exists a solution $X(t, X_0, U_0)$ of (3.1) with $X_0 \in S(B_1)$ such that for some $t^* > t_0$, $\| X(t^*, X_0, U_0) \|) = B_2$. Since $B_1 > \rho$, there are two possibilities to consider:

(1) $X(t, X_0, U_0) \in S^c(\rho)$ for $t \in [t_0, t^*]$;

(2) there exist a $\overline{t} \geq t_0$ such that $X(\overline{t}, X_0, U_0) \in \partial S(\rho)$ and $X(t, X_0, U_0) \in S^c(\rho)$ for $t \in [\overline{t}, t^*]$.

If (1) holds, we can f nd $t_1 > t_0$ such that

$$\begin{gathered} X(t_1, X_0, U_0) \in \partial S(B_1), \\ X(t^*, X_0, U_0) \in \partial S(B_2), \\ X(t, X_0, U_0) \in S^c(B_1), \quad t \in [t_1, t^*]. \end{gathered} \tag{3.13}$$

Setting $m(t) = V_1(t, X(t, X_0, U_0), U(t)) + V_2(t, X(t, X_0, U_0), U(t))$ for $t \in [t_1, t^*]$, and then using Theorem 3.1, we can obtain the differential inequality

$$D^+ m(t) \leq g_2(t, m(t)), \quad t \in [t_1, t^*],$$

and so,

$$m(t) \leq \gamma_2(t, t_1, m(t)), \quad t \in [t_1, t^*],$$

where $\gamma_2(t, t_1, v_0)$ is the maximal solution of (3.8) with $\gamma_2(t_1, t_1, v_0) = v_0$.

Thus,

$$\begin{aligned} &V_1(t^*, X(t^*, X_0, U_0), U(t^*)) + V_2(t^*, X(t^*, X_0, U_0), U(t^*)) \\ &\qquad \leq \gamma_2(t^*, t_1, V_1(t_1, X(t_1, X_0, U_0), U(t_1))) + V_2(t_1, X(t_1, X_0, U_0), U(t_1)) \end{aligned} \tag{3.14}$$

Similarly, we also have

$$V_1(t_1, X(t_1, X_0, U_0), U(t_1)) \leq \gamma_1(t_1, t_0, V_1(t_0, X_0, U_0)) \tag{3.15}$$

where $\gamma_1(t_1, t_0, u_0)$ is the maximal solution of (3.8).

Set $w_{10} = V_1(t_0, X_0, U_0) < \alpha_1$. Then

$$V_1(t_1, X(t_1, X_0, U_0), U(t_1))) \leq \gamma_1(t_1, t_0, V_1(t_0, X_0, U_0)) \leq \beta_0,$$

since (3.10) holds.

Furthermore, $V_2(t_1, X(t_1, X_0, U_0), U(t))) \leq a(B1)$. Consequently, we have

$$w_{20} = V_1(t_1, X(t_1, X_0, U_0), U(t_1)) + V_2(t_1, X(t_1, X_0, U_0), U(t_1)) \leq \beta_0 + a(B1) = \alpha_2 \tag{3.16}$$

Combining (3.11), (3.12), (3.13) and (3.16), we obtain

$$b(B_2) \leq m(t^*) \leq \gamma(t^*) \leq \beta_1(\alpha_2) < b(B_2), \tag{3.17}$$

which is a contradiction

If case (2) holds, we also arrive at the inequality (3.14), where $t_1 > \bar{t}$ satisf es (3.13). We then have, in place of (3.15), the relation

$$V_1(t_1, X(t_1, X_0, U_0), U(t_1))) \leq \gamma_1(t_1, \bar{t}, V_1(\bar{t}, X(\bar{t}), U_0, U(\bar{t})).$$

Since $X(\bar{t}, t_0, X_0, U_0) \in \partial S(\rho)$ and $V_1(\bar{t}, X(\bar{t}, t_0, X_0, U_0), U(\bar{t}))) \leq \alpha^* \leq \alpha_1$, arguing as before, we get the contradiction (3.17). This proves that for any given $B_1 > \rho$, $t_0 > 0$, there exist a B_2 such that

$X_0 \in S(B_1)$ implies $X(t, t_0, X_0, U_0) \in S(B_2), t \geq t_0$.

For $B_1 < \rho$, we set $B_2(t_0, B_1) = B_2(t_0, \rho)$ and hence the proof is complete. □

Acknowledgments

The authors gratefully acknowledge the referees for their careful reading and many valuable remarks which improved the presentation of the paper.

References

[1] T. Gnana Bhaskar, J. Vasundhara Devi, Stability Criteria for Set Differential Equations, *Mathematical and Computer Modelling* **41** (2005), 1371-1378.

[2] T. Gnana Bhaskar, J. Vasundhara Devi, Nonuniform stability and boundedness criteria for set differential equations, *Applicable Analysis* **84 (2)** (2005), 131-142.

[3] V. Lakshmikantham, Set differential equations versus fuzzy differential equations, *J. Applied Mathematics and Computation* **164** (2005), 277-294.

[4] V. Lakshmikantham, T. Gnana Bhaskar, J. Vasundhara Devi, *Theory of set differential equations in metric spaces*, Cambridge Scientif c Publisher, UK, 2006.

[5] N.D. Phu, L.T. Quang, T.T. Tung, Stability criteria for set control differential equations, *Nonlinear Analysis TMA.* **69** (2008), 3715-3721.

[6] N.D. Phu, T.T. Tung, Some results on sheaf - solutions of sheaf set control problems, *Nonlinear Analysis* **67** (2007), 1309-1315.

[7] N.D. Phu, T.T. Tung, Existence of solutions of set control differential equations, *Journal of Science and Technology Development* **10(6)** (2007), 5-12.

[8] N.D. Phu, T.T. Tung, Some new results on fuzzy control differential equations, *Journal of Science and Technology Development* **11(1)** (2008), 5-20.

[9] N.N. Tu, T.T. Tung, Stability of set differential equations and applications, *Nonlinear Analysis TMA.* (In press DOI: 10.1016/j.na.2008.12.045).

In: Evolution Equations: Theories, Solutions and Functions
Editor: Gaston M. N'Guerekata
ISBN 978-1-61324-090-8

Chapter 6

Asymptotic Analysis of MHD Systems

Jamel Benameur*
Department of Mathematics College of Science,
King Saud University
Riyadh 11451, Kingdom of Saudi Arabia

Abstract

In this paper, we study the convergence of strong solutions of a Magneto-Hydro-Dynamic system. On the torus $\mathbb{T}^3$, the proof is based on Schochet's methods, whereas in the case of the whole space $\mathbb{R}^3$, we use Strichartz's type estimates and a product law's $2D \times 3D$.

2000 Mathematics Subject Classification: 35-xx, 35Bxx, 35Lxx.

Keywords and phrases: MHD system, Existence and Convergence, Filtered solutions, Strichartz's estimates.

1. Introduction

In this paper we study unique local existence and asymptotic behaviour of solutions for the MHD system, see [6],

$$\begin{cases} \partial_t u + u.\nabla u - \dfrac{E}{\varepsilon}\Delta u + \dfrac{e \times u}{\varepsilon} - \dfrac{\Lambda}{\varepsilon}\text{curl}(b) \times e' - \dfrac{\Lambda\theta}{\varepsilon}\text{curl}\, b \times b & = & -\dfrac{\nabla p}{\varepsilon}, \quad \text{in } \mathbb{R}_t^+ \times \Omega_x \\ \partial_t b - \dfrac{\Delta b}{\theta} + u.\nabla b - b.\nabla u - \text{curl}\, \dfrac{u \times e'}{\theta} & = & 0, \quad \text{in } \mathbb{R}_t^+ \times \Omega_x \\ \text{div}\, u = 0,\ \text{div}\, b & = & 0, \quad \text{in } \mathbb{R}_t^+ \times \Omega_x, \end{cases}$$

where Ω is the whole space $\mathbb{R}^3$ or the torus $\mathbb{T}^3 := \mathbb{R}^3/\mathbb{Z}^3$; u is the velocity field b is the magnetic fiel and e, e' are two fi ed vectors. The parameters E, ε, Λ and θ represent consecutively the Ekman number, the Rossby number, the Elsasser number and the magnetic Reynolds number. We notice that these parameters satisfy, according to [6],

$$\varepsilon \to 0, \quad \Lambda = \mathcal{O}(1), \quad \varepsilon\theta \to 0 \qquad \text{and} \qquad E \sim \varepsilon^2.$$

*E-mail address: jbenameur@ksu.edu.sa

These equations modelize the Magneto-Hydro-Dynamic fl w in the Earth's core which is believed to support a self-excited dynamo process generating the Earth's magnetic field Here, we present the analytical study of a simplifie problem where we choose $E = \varepsilon^2$, $\Lambda = \varepsilon^{3/2}$, $\theta = \varepsilon^{-1/2}$ and $e = e' = -e_3$. Then, we can see that our system takes the following form

$(\mathbf{MHD}^\varepsilon)$

$$\begin{cases} \partial_t u - \varepsilon\Delta u + u.\nabla u - \text{curl}\,(b) \times b + \sqrt{\varepsilon}\text{curl}\,(b) \times e_3 + \dfrac{u \times e_3}{\varepsilon} &= -\nabla p \\ \partial_t b - \sqrt{\varepsilon}\Delta b + u.\nabla b - b.\nabla u + \sqrt{\varepsilon}\text{curl}\,(u \times e_3) &= 0 \\ \text{div}\, u = 0,\ \text{div}\, b &= 0. \end{cases}$$

The goal of paper is to fin the "limit system" when ε goes to zero.

We denote by $\mathbb{P}$ the L^2 orthogonal projection on divergence-free vector fields Applying $\mathbb{P}$ to the firs equation of $(\mathbf{MHD}^\varepsilon)$, then one can see that $U = (u, b)$ is a solution of the following abstract form

$$(\mathcal{S}^\varepsilon) \qquad \begin{cases} \partial_t U + Q(U,U) + a_2^\varepsilon(D)U + L^\varepsilon(U) &= 0, \quad \text{in} \quad \mathbb{R}_t^+ \times \Omega_x \\ \text{div}\, u = 0,\ \text{div}\, b &= 0, \quad \text{in} \quad \mathbb{R}_t^+ \times \Omega_x, \end{cases}$$

where the quadratic term Q is define by

$$Q(U,U) = \Big(\mathbb{P}(u.\nabla u) - \mathbb{P}(b.\nabla b), u.\nabla b - b.\nabla u\Big),$$

the viscous term is

$$a_2^\varepsilon(D)U = \Big(-\varepsilon\Delta u, -\sqrt{\varepsilon}\Delta b\Big),$$

and the linear perturbation L^ε is given by

$$L^\varepsilon(U) := \Big(\mathbb{P}(\frac{u \times e_3}{\varepsilon}) + \sqrt{\varepsilon}\partial_3 b, \sqrt{\varepsilon}\partial_3 u\Big).$$

In the literature, $a_2^\varepsilon(D)$ is elliptic and, in the case of a rotating flui $L^\varepsilon = \frac{1}{\varepsilon}L$ with L a skew-symmetric linear operator.

Singular limits in system such as $(\mathcal{S}^\varepsilon)$ have been studied by several authors. In the hyperbolic case, namely $a_2^\varepsilon(D)U = 0$, A. Babin, A. Mahalov and B. Nicolaenko [1] studied the incompressible rotating Euler equation on the torus. Using the method introduced by S. Schochet (see [14] and [15]), I. Gallagher studied in [8] this problem in its abstract hyperbolic form. In the case of the incompressible rotating Navier-Stokes equation on the torus, it is shown (see [1] and [9]) that the solutions converge to a solution of a certain diffusion equation. Moreover, for a special initial condition, there exists a sequence of solutions convergent to a solution of a two dimensional Navier-Stokes equation. Motivated by this case, J. Y. Chemin, B. Desjardin, I. Gallagher and E. Grenier studied in [5] the incompressible fluid with anisotropic viscosity on the whole space, the key of their proof is an anisotropic version of Strichartz estimates. We refer to I. Gallagher [7] for the study of the abstract parabolic form. Among others, we also refer to the basic results of [5], [7], [9], [10], [11], [12], [13].

Notice that the existence results follow directly from the Friedrichs's method and the energy estimates. Following the approximation scheme of Friedrichs, we shall prove in section two global existence of "Leray's solutions" and local existence of strong solutions on uniform time; namely solutions define by the following results.

Theorem 1.1. *Let* $\varepsilon > 0$ *and* $U_0 = (u_0, b_0) \in L^2(\Omega)$ *such that*

$$\operatorname{div} u_0 = 0, \quad \operatorname{div} b_0 = 0.$$

There exists $U^\varepsilon := (u^\varepsilon, b^\varepsilon)$ *a solution of* $(\mathbf{MHD}^\varepsilon)$ *with* $U^\varepsilon \in L^\infty(\mathbb{R}^+, L^2) \cap L^2(\mathbb{R}^+, \dot{H}^1)$. *Moreover,* U^ε *satisfies the following energy estimate. For all* $t \geq 0$,

$$\|U^\varepsilon(t)\|_{L^2}^2 + 2\varepsilon \int_0^t \|\nabla u^\varepsilon\|_{L^2}^2 + 2\sqrt{\varepsilon} \int_0^t \|\nabla b^\varepsilon\|_{L^2}^2 \leq \|U_0\|_{L^2}^2. \tag{1.1}$$

Theorem 1.2. *Let* $U_0 = (u_0, b_0) \in H^s(\Omega)$ *with* $s > \frac{3}{2} + 2$ *an integer, such that*

$$\operatorname{div} u_0 = 0, \quad \operatorname{div} b_0 = 0.$$

Then there exists $T > 0$ *such that for all* $\varepsilon \in]0, 1[$, *there exists a unique solution* $U^\varepsilon := (u^\varepsilon, b^\varepsilon) \in C_T^0(H^s) \cap L_T^2(H^{s+1})$ *of system* $(\mathbf{MHD}^\varepsilon)$ *satisfying ; for all* $t \in [0, T]$

$$\|U^\varepsilon(t)\|_{H^s}^2 + 2\varepsilon \int_0^t \|\nabla u^\varepsilon\|_{H^s}^2 + 2\sqrt{\varepsilon} \int_0^t \|\nabla b^\varepsilon\|_{H^s}^2 \leq 2\|U_0\|_{H^s}^2. \tag{1.2}$$

Moreover, if $\|U_0\|_{H^s} \leq c\varepsilon$, *then the solution is global, where c's is constant with possible dependence on s only.*

We are interested now to seeing the limit of strong solutions of the $(\mathbf{MHD}^\varepsilon)$ system when ε goes to zero. Since $\partial_t u^\varepsilon$ is not bounded with respect to ε, one cannot takes the limit directly in the system, and then the classical proofs (see for example [16],[17]) no longer work.

• In the case $\Omega = \mathbb{T}^3$, a method to get round this difficult is to use the group $\mathcal{L}(t)$ associated with the operator $L(u) := \mathbb{P}(u \times e_3)$. We consider the filtere solution $v^\varepsilon := \mathcal{L}(-\frac{t}{\varepsilon})u^\varepsilon$ and we look for the limit system in $\mathcal{D}'$ satisfie by the eventual" limit of $(v^\varepsilon, b^\varepsilon)$,

$$(\mathbf{LS}) \quad \left\{ \begin{array}{rcl} \partial_t v + q^0(v, v) - \overline{\mathbb{P}(b.\nabla b)} & = & 0 \ \text{ in } \ \mathbb{R}_t^+ \times \mathbb{T}_x^3 \\ \partial_t b + \bar{v}.\nabla b - b.\nabla \bar{v} & = & 0 \ \text{ in } \ \mathbb{R}_t^+ \times \mathbb{T}_x^3 \\ \operatorname{div} v = 0, \quad \operatorname{div} b & = & 0 \ \text{ in } \ \mathbb{R}_t^+ \times \mathbb{T}_x^3 \\ (v, b)(0) & = & (u_0, b_0) \end{array} \right.$$

where

$$q^0(v, v) := \lim_{\varepsilon} \mathcal{L}(-t/\varepsilon)\Big[\mathbb{P}\Big((\mathcal{L}(-t/\varepsilon)v).\nabla(\mathcal{L}(-t/\varepsilon)v)\Big)\Big] \quad \text{in} \quad \mathcal{D}'$$

and here, we have denoted $\bar{f}(x_1, x_2) = \displaystyle\int f(x_1, x_2, x_3)dx_3$ and $f_{osc} = f - \bar{f}$.

Formally the $(\mathbf{MHD}^\varepsilon)$ system tends to the system $(\mathbf{LS})$. More precisely, the following results show that $(\mathbf{LS})$ is well posed in the appropriate space and that the family of solutions of $(\mathbf{MHD}^\varepsilon)$ converge strongly to those of $(\mathbf{LS})$.

Theorem 1.3. *Let* $s > \frac{3}{2} + 2$ *be an integer and* $U_0 = (u_0, b_0) \in H^s(\mathbb{T}^3)$ *such that* $\operatorname{div} u_0 = 0, \operatorname{div} b_0 = 0$. *Then there exists* $T > 0$, *and a unique solution* $(v, b) \in C_T^0(H^s)$ *of system* $(\mathbf{LS})$.

Theorem 1.4. *Let $s > \frac{3}{2} + 2$ be an integer and $U_0 = (u_0, b_0) \in H^s(\mathbb{T}^3)$ such that $\operatorname{div} u_0 = 0$, $\operatorname{div} b_0 = 0$. We denote by $U^\varepsilon = (u^\varepsilon, b^\varepsilon)$ the family of solutions of* $(\mathbf{MHD}^\varepsilon)$ *given by Theorem* 1.2. *Then, for all $s' < s$,*

$$\begin{array}{lcl} u^\varepsilon - \mathcal{L}(t/\varepsilon)v & = & o(1) \ \text{ in } \ L^\infty_T(H^{s'}(\mathbb{T}^3)) \\ b^\varepsilon - b & = & o(1) \ \text{ in } \ L^\infty_T(H^{s'}(\mathbb{T}^3)), \end{array}$$

where (v, b) is the solution of $(\mathbf{LS})$.

• In the case $\mathbf{\Omega} = \mathbb{R}^3$, we mention that the plane $\mathbb{R}^2 \times \{0\}$, where the resonance phenomenon is supposed to take place, is zero Lebesgue measure. The following theorem holds.

Theorem 1.5. *Let $s > \frac{3}{2} + 2$ be an integer and $U_0 = (u_0, b_0) \in H^s(\mathbb{R}^3)$ such that $\operatorname{div} u_0 = 0$, $\operatorname{div} b_0 = 0$. We denote by $U^\varepsilon = (u^\varepsilon, b^\varepsilon)$ the family of solutions of* $(\mathbf{MHD}^\varepsilon)$ *given by Theorem* 1.2. *Then, for all $s' < s$*

$$\begin{array}{lcl} u^\varepsilon & = & o(1) \ \text{ in } \ L^4_T(C^{s'-\frac{3}{2}}(\mathbb{R}^3)) \\ b^\varepsilon - b_0 & = & o(1) \ \text{ in } \ L^\infty_T(H^{s'}(\mathbb{R}^3)). \end{array}$$

The proof of the theorem above is based on Strichartz's inequality and of stationary phase's method. Since the limit of the system $(\mathbf{MHD}^\varepsilon)$ is the bidimensional Navier-Stokes equations, it is natural to consider initial data of the type $(u_0, b_0) = (\bar{u}_0 + w_0, b_0)$, where $\bar{u}_0 = \bar{u}_0(x_h)$, $w_0 = w_0(x_h, x_3)$ and $b_0 = b_0(x_h, x_3)$ (see [5]). Before stating the results, it will be useful to consider the following system

$$(\mathbf{NS2D}^\varepsilon) \qquad \left\{ \begin{array}{lcl} \partial_t \bar{u} - \varepsilon \Delta_h \bar{u} + \bar{u}.\nabla_h \bar{u} & = & -(\nabla_h p, 0) \ \text{ in } \ \mathbb{R}^+_t \times \mathbb{R}^2_h \\ \operatorname{div}_h \bar{u} & = & 0 \ \text{ in } \ \mathbb{R}^+_t \times \mathbb{R}^2_h \\ \bar{u}(0) & = & \bar{u}_0. \end{array} \right.$$

The presence of ε in this system is a technical task in order to study the asymptotic of $(\mathbf{MHD}^\varepsilon)$ system. For $(\mathbf{NS2D}^\varepsilon)$, using the classical Friedrich's scheme, we can prove the existence of strong solutions on uniform time. Precisely, we have the classical result.

Theorem 1.6. *Let $\bar{u}_0 \in \left(H^\sigma(\mathbb{R}^2)\right)^3$ be a divergence-free vector field with $\sigma > \frac{2}{2} + 2$ an integer. Then there exists $T_0^\sigma := \dfrac{1}{C(\sigma)\|\bar{u}_0\|_{H^\sigma}}$ such that for all $\varepsilon > 0$, there exists a unique solution $\bar{u}^\varepsilon \in C^0_{T_0^\sigma} H^\sigma(\mathbb{R}^2)) \cap L^2_{T_0^\sigma}(H^{\sigma+1}(\mathbb{R}^2))$ of system* $(\mathbf{NS2D}^\varepsilon)$ *satisfying; for all $t \in [0, T_0^\sigma]$*

$$\|\bar{u}^\varepsilon(t)\|^2_{H^\sigma} + 2\varepsilon \int_0^t \|\nabla \bar{u}^\varepsilon\|^2_{H^\sigma} \leq 2\|\bar{u}_0\|^2_{H^\sigma}.$$

In the following s denotes an integer greater than $\frac{3}{2} + 2$, and we defin the condition that helps us to analyze the $(\mathbf{MHD}^\varepsilon)$ systems and especially when ε goes to 0

$$(\mathbf{2D} \times \mathbf{3D})_s \qquad \left\{ \begin{array}{lcl} (u_0, b_0) & = & (\bar{u}_0 + w_0, b_0) \\ \operatorname{div}_h \bar{u}_0 & = & \operatorname{div} w_0 = \operatorname{div} b_0 = 0 \\ \bar{u}_0 & = & \bar{u}_0(x_h) \in H^{s+1}(\mathbb{R}^2) \\ w_0 & = & w_0(x_h, x_3) \in H^s(\mathbb{R}^3) \\ b_0 & = & b_0(x_h, x_3) \in H^s(\mathbb{R}^3). \end{array} \right.$$

Now we consider the following system which is, essentially, the $3D$ main part of $(\mathbf{MHD}^{\varepsilon})$ system when ε goes to zero.

$(\mathbf{MHD3D}^{\varepsilon})$

$$\begin{cases} \partial_t w - \varepsilon\Delta w + \dfrac{1}{\varepsilon} w \times e_3 + \sqrt{\varepsilon}\partial_3 B - B.\nabla B & = -\nabla p_L \;\text{ in }\; [0, T_0^{(s+1)}] \times \mathbb{R}^3 \\ \partial_t B - \sqrt{\varepsilon}\Delta B + \sqrt{\varepsilon}\partial_3 w + (\bar{u}^{\varepsilon} + w).\nabla B & \\ \qquad\qquad -B.\nabla(\bar{u}^{\varepsilon} + w) & = 0 \qquad \text{ in }\; [0, T_0^{(s+1)}] \times \mathbb{R}^3 \\ \qquad\qquad (\operatorname{div} w, \operatorname{div} B) & = (0,0) \;\text{ in }\; [0, T_0^{(s+1)}] \times \mathbb{R}^3 \\ \qquad\qquad (w, B)(0) & = (w_0, b_0). \end{cases}$$

Theorem 1.7. *Suppose that the condition* $(\mathbf{2D} \times \mathbf{3D})_s$ *is satisfied. There exists* $0 < T_1^s \leq T_0^{(s+1)}$ *such that for all* $\varepsilon > 0$*, there exists a unique solution* $(w^{\varepsilon}, B^{\varepsilon}) \in L^{\infty}_{T_1^s}(H^s(\mathbb{R}^3)) \cap L^2_{T_1^s}(H^{s+1}(\mathbb{R}^3))$ *of system* $(\mathbf{MHD3D}^{\varepsilon})$ *satisfying ; for all* $t \in [0, T_1^s]$

$$\|w^{\varepsilon}(t)\|^2_{H^s} + \|B^{\varepsilon}(t)\|^2_{H^s} + 2\varepsilon \int_0^t \|\nabla w^{\varepsilon}\|^2_{H^s} + 2\sqrt{\varepsilon} \int_0^t \|\nabla B^{\varepsilon}\|^2_{H^s} \leq 2\|(w_0, b_0)\|^2_{H^s}. \quad (1.3)$$

Moreover

$$w^{\varepsilon} \to 0 \quad \text{in} \quad L^4([0, T_1^s], C^{s'-\frac{3}{2}}(\mathbb{R}^3)) \; ; \; \forall s' < s.$$

Theorem 1.8. *Suppose that the condition* $(\mathbf{2D} \times \mathbf{3D})_s$ *is satisfied. Then, for all* $\varepsilon > 0$*, there exists a unique solution* $(u^{\varepsilon}, b^{\varepsilon})$ *of the system* $(\mathbf{MHD}^{\varepsilon})$ *such that*

$$u^{\varepsilon} - \bar{u}^{\varepsilon}, \quad b^{\varepsilon} \in L^{\infty}_{T_1^s}(H^s(\mathbb{R}^3)) \cap L^2_{T_1^s}(H^{s+1}(\mathbb{R}^3)).$$

Moreover, we have for all $t \in [0, T_1^s]$,

$$\begin{aligned} &\|u^{\varepsilon}(t) - \bar{u}^{\varepsilon}(t)\|^2_{H^s} + \|b^{\varepsilon}(t)\|^2_{H^s} \\ +2\varepsilon &\int_0^t \|\nabla(u^{\varepsilon} - \bar{u}^{\varepsilon})(\tau)\|^2_{H^s}\, d\tau + 2\sqrt{\varepsilon} \int_0^t \|\nabla b^{\varepsilon}\|^2_{H^s}\, d\tau \leq 2\|(w_0, b_0)\|^2_{H^s}. \end{aligned}$$

Now we are ready to state the main convergence result in the case of the whole space $\mathbb{R}^3$.

Theorem 1.9. *We keep the same hypothesis as in Theorem 1.8 above and we suppose* $s > \frac{3}{2} + 4$ *an integer. Then, for all* $s' > s$

$$u^{\varepsilon} - \bar{u}^{\varepsilon} - w^{\varepsilon} = o(1) \quad \text{in} \quad L^{\infty}_{T_1^s}(H^{s'}(\mathbb{R}^3))$$

$$b^{\varepsilon} - B^{\varepsilon} = o(1) \quad \text{in} \quad L^{\infty}_{T_1^s}(H^{s'}(\mathbb{R}^3)).$$

This paper is organized as follows. In the next section, we present the proofs of the existence theorems (Theorems 1.1 and 1.2). The third section is devoted to the proof of the convergence result in the case $\Omega = \mathbb{T}^3$ (Theorem 1.4) and the study of the system $(\mathbf{LS})$ (Theorem 1.3). In the fina section, we consider the case of the whole space $\mathbb{R}^3$; We give the proof of the Theorems 1.5, 1.7, 1.8, 1.9.

2. Existence Results

2.1. Proofs of Theorems 1.1 and 1.2

In this section we shall prove Theorems 1.1 and 1.2.

We begin by observing that using the energy methods, one can prove global existence of so-called "Leray's solutions" for the system $(\mathbf{MHD}^{\varepsilon})$. The crucial fact is the following L^2-energy estimate

$$\|U^{\varepsilon}(t)\|_{L^2}^2 + 2\varepsilon \int_0^t \|\nabla u^{\varepsilon}(\tau)\|_{L^2}^2\, d\tau + 2\sqrt{\varepsilon}\int_0^t \|\nabla b^{\varepsilon}(\tau)\|_{L^2}^2\, d\tau \leq \|U_0\|_{L^2}^2. \tag{2.4}$$

We now turn to the case of strong solutions. Let us introduce, for a strictly positive integer n, the Friedrich's operator J_n define by:

$$J_n\, u = \mathcal{F}^{-1}\Big(\mathbf{1}_{B(0,n)}\mathcal{F}(u)\Big).$$

After this definition we consider the following approximate Magneto-Hydro-Dynamic system $(\mathbf{MHD}_n)$

$$\begin{cases} \partial_t u_n - \varepsilon\Delta J_n u_n + J_n \text{div}\,(J_n u_n \otimes J_n u_n) - J_n\text{div}\,(J_n b_n \otimes J_n b_n) + \sqrt{\varepsilon}\partial_3(J_n b_n) + \\ \dfrac{J_n u_n \times e_3}{\varepsilon} = \nabla\Delta^{-1}\text{div}\,\Big(J_n\text{div}\,(J_n u_n \otimes J_n u_n) - J_n\text{div}\,(J_n b_n\otimes J_n b_n) + \dfrac{J_n u_n \times e_3}{\varepsilon}\Big) \\ \partial_t b_n - \sqrt{\varepsilon}\Delta J_n b_n + J_n\text{div}\,(J_n u_n \otimes J_n b_n) - J_n\text{div}\,(J_n b_n \otimes J_n u_n) + \sqrt{\varepsilon}\partial_3(J_n u_n) = 0 \\ (u_n, b_n)(0) = (J_n u_0, J_n b_0). \end{cases}$$

By the theory of ordinary differential equations in H^s we know that the system $(\mathbf{MHD}_n)$ has a unique maximal solution $U_n := (u_n, b_n)$ in the space $C^1([0, T_n^*(\varepsilon)), H^s)$. Using uniqueness and the fact that $\text{div}\, u_n = \text{div}\, b_n = 0$ and $J_n^2 = J_n$ we can re-write the system

$$(\mathbf{MHD}_n)\quad \begin{cases} \partial_t u_n - \varepsilon\Delta u_n + J_n(u_n.\nabla u_n) - J_n(b_n.\nabla b_n) + \sqrt{\varepsilon}\partial_3 b_n + \dfrac{u_n\times e_3}{\varepsilon} \\ \quad = \nabla\Delta^{-1}\text{div}\,\Big(J_n\text{div}\,(u_n\otimes u_n) - J_n\text{div}\,(b_n\otimes b_n) + \dfrac{u_n\times e_3}{\varepsilon}\Big) \\ \partial_t b_n - \sqrt{\varepsilon}\Delta b_n + J_n(u_n.\nabla b_n) - J_n(b_n.\nabla u_n) + \sqrt{\varepsilon}\partial_3 u_n = 0 \\ (u_n, b_n)(0) = (J_n u_0, J_n b_0). \end{cases}$$

To continue the proof, we recall without proof the following product law.

Lemma 2.1. *Let $\sigma > \frac{3}{2} + 2$ be an integer and a, b and c three vectors field in $H^{\sigma}(\Omega)$ such that* $\text{div}\, a = 0$. *Then, a constant C exists such that*

$$|\langle a.\nabla b, b\rangle_{H^\sigma}| \leq C\|\nabla a\|_{H^{\sigma-1}}\|\nabla b\|_{H^{\sigma-1}}^2, \tag{2.5}$$

$$|\langle a.\nabla b, c\rangle_{H^\sigma} + \langle a.\nabla c, b\rangle_{H^\sigma}| \leq C\|\nabla a\|_{H^{\sigma-1}}\|\nabla b\|_{H^{\sigma-1}}\|\nabla c\|_{H^{\sigma-1}}. \tag{2.6}$$

(See the Appendix for the proof of the lemma.)

We take the inner product in H^s and we use the last lemma, we obtain for all $t \in [0, T_n^*(\varepsilon))$,

$$\frac{1}{2}\frac{d}{dt}\|U_n(t)\|_{H^s}^2 + \varepsilon\|\nabla u_n(t)\|_{H^s}^2 + \sqrt{\varepsilon}\|\nabla b_n(t)\|_{H^s}^2 \leq C\|\nabla U_n(t)\|_{H^{s-1}}^3. \tag{2.7}$$

Then

$$\|U_n(t)\|_{H^s}^2 + 2\varepsilon\int_0^t \|\nabla u_n(\tau)\|_{H^s}^2 d\tau + 2\sqrt{\varepsilon}\int_0^t \|\nabla b_n(\tau)\|_{H^s}^2 d\tau$$
$$\leq \|U_0\|_{H^s}^2 + C\int_0^t \|\nabla U_n(\tau)\|_{H^{s-1}}^3 d\tau. \tag{2.8}$$

We set $T(n,\varepsilon) := \text{Sup}\{0 \leq t < T_n^*(\varepsilon);\quad \forall\tau \in [0,t], \|U_n(\tau)\|_{H^s} \leq 2\|U_0\|_{H^s}\}$. Using (2.1.) and Gronwall lemma we obtain, for all $t \in [0, T(n,\varepsilon)[$,

$$\|U_n(t)\|_{H^s}^2 \leq \|U_0\|_{H^s}^2 \exp(2Ct\|U_0\|_{H^s}).$$

Thus,

$$T(n,\varepsilon) > T := \frac{1}{C\|U_0\|_{H^s}} > 0.$$

Moreover, for all $t \in [0, T]$,

$$\|U_n(t)\|_{H^s}^2 + 2\varepsilon\int_0^t \|\nabla u_n(\tau)\|_{H^s}^2\, d\tau + 2\sqrt{\varepsilon}\int_0^t \|\nabla b_n(\tau)\|_{H^s}^2\, d\tau \leq 2\|U_0\|_{H^s}^2. \tag{2.9}$$

Now, the problem is to pass to the limit. Using Ascoli's theorem, the Cantor's diagonal process as in Navier-Stokes equations (see [4]) and the estimate (2.9), we obtain a solution satisfying, for all $t \in [0, T]$,

$$\|U^\varepsilon(t)\|_{H^s}^2 + 2\varepsilon\int_0^t \|\nabla u^\varepsilon(\tau)\|_{H^s}^2\, d\tau + 2\sqrt{\varepsilon}\int_0^t \|\nabla b^\varepsilon(\tau)\|_{H^s}^2\, d\tau \leq 2\|U_0\|_{H^s}^2. \tag{2.10}$$

This regularity implies in a standard way the uniqueness. It remains to prove the global existence when the initial data is small enough. We assume now that $\|U_0\|_{H^s} \leq c\varepsilon,\quad (c = \frac{1}{C})$, and we set

$$T_n(\varepsilon) := \text{Sup}\{0 \leq t < T_n^*(\varepsilon);\quad \forall\tau \in [0,t], \|U_n(\tau)\|_{H^s} \leq c\varepsilon\}.$$

Using (2.1.), it suffice to show that $T_n(\varepsilon) = T_n^*(\varepsilon)$. By (2.7) we have $\frac{d}{dt}\|U_n\|_{H^s}^2(0) < 0$, then there exists $t_n > 0$ such that $\|U_n(t_n)\|_{H^s} < c\varepsilon$. Since the quantity $\|U_n(t)\|_{H^s}$ is decreasing on $[t_n, T_n^*(\varepsilon))$, then $T_n(\varepsilon) = T_n^*(\varepsilon) = \infty$. This achieves the proof. □

3. The Case $\Omega = \mathbb{T}^3$

Let (U^ε) be a family of strong solutions of the system $(\mathcal{S}^\varepsilon)$ with initial data U_0. To take the limit when $\varepsilon \longrightarrow 0$, the classical proofs no longer work because $(\partial_t u^\varepsilon)$ is not uniformly bounded. An idea (as in [15] for instance) is to "filter the system by the group $\mathcal{L}(t)$ associated to L.

In what follows, we recall some properties of the Coriolis force $L(u)$. We consider, as in [9], the "wave equation"

$$\begin{cases} \partial_t u + L(u) = 0 \quad \text{in} \quad \mathbb{R}_t \times \mathbb{T}_x^3 \\ u(0) = u_0 \quad \text{with} \quad \text{div } u_0 = 0. \end{cases}$$

Lemma 3.1. *The above system has a global solution denoted by $u(t) = \mathcal{L}(t)u_0$, such that for all $s \in \mathbb{R}$ and for all $u_0 \in H^s(\mathbb{T}^3)$,*

$$\|\mathcal{L}(t)u_0\|_{H^s(\mathbb{T}^3)} = \|u_0\|_{H^s(\mathbb{T}^3)} \quad \textit{and} \quad \|{}^t\mathcal{L}(t)u_0\|_{H^s(\mathbb{T}^3)} = \|u_0\|_{H^s(\mathbb{T}^3)}.$$

Moreover, if we denote by $k = (k_1, k_2, k_3)$ the Fourier coordinates, then u is explicitly given by

$$\mathcal{F}u(t,k) = \exp(i\omega(k)t)(\mathcal{F}u(0,k), \nu_k^+)\nu_k^+ + \exp(-i\omega(k)t)(\mathcal{F}u(0,k), \nu_k^-)\nu_k^-,$$

where $\omega(k) = \frac{k_3}{|k|}$, $\nu_k^\pm$ are given unit vectors and $(.,.)$ denotes the usual scalar product.

Now we defin

$$v^\varepsilon(t) = \mathcal{L}(-t/\varepsilon)u^\varepsilon(t)$$

then, $V^\varepsilon := (v^\varepsilon, b^\varepsilon)$, satisfie the following system

$$\begin{cases} \partial_t v - \varepsilon\Delta v + \mathcal{L}(-t/\varepsilon)\mathbb{P}(\mathcal{L}(t/\varepsilon)v.\nabla\mathcal{L}(t/\varepsilon)v)\mathcal{L}(-t/\varepsilon)\mathbb{P}(\text{curl}b \times b) \\ \qquad\qquad + \sqrt{\varepsilon}\mathcal{L}(-t/\varepsilon)\mathbb{P}(\text{curl}b \times e_3) = 0 \\ \partial_t b - \sqrt{\varepsilon}\Delta b + (\mathcal{L}(t/\varepsilon)v).\nabla b - b.\nabla(\mathcal{L}(t/\varepsilon)v) + \sqrt{\varepsilon}\text{curl}((\mathcal{L}(t/\varepsilon)v) \times e_3) = 0 \\ \text{div}\, v = 0, \ \text{div}\, b = 0 \\ (v,b)(0) = (u_0, b_0). \end{cases}$$

This system can be re-written in the following way

$$(\tilde{\mathcal{S}}^\varepsilon) \qquad \begin{cases} \partial_t V + Q^\varepsilon(V,V) + a_2^\varepsilon(D)V + \tilde{L}^\varepsilon(V) = 0 \quad \text{in} \quad \mathbb{R}_t^+ \times \mathbb{T}_x^3 \\ \text{div}\, V = 0 \qquad\qquad \text{in} \quad \mathbb{R}_t^+ \times \mathbb{T}_x^3 \\ V(0) = U_0 = (u_0, b_0), \end{cases}$$

where

$$Q^\varepsilon(V,V) = \Big(q^\varepsilon(v,v) - \mathcal{L}(-t/\varepsilon)\mathbb{P}(b.\nabla b), (\mathcal{L}(t/\varepsilon)v).\nabla b - b.\nabla(\mathcal{L}(t/\varepsilon)v)\Big),$$

$$q^\varepsilon(v,v) = \mathcal{L}(-t/\varepsilon)\mathbb{P}(\mathcal{L}(t/\varepsilon)v.\nabla\mathcal{L}(t/\varepsilon)v)$$

$$a_2^\varepsilon(D)V = \Big(-\varepsilon\Delta v, -\sqrt{\varepsilon}\Delta b\Big),$$

and

$$\tilde{L}^\varepsilon(V) = \sqrt{\varepsilon}\Big(\mathcal{L}\Big(\frac{-t}{\varepsilon}\Big)\partial_3 b, \partial_3\mathcal{L}\Big(\frac{t}{\varepsilon}\Big)v\Big).$$

When ε goes to 0, we obtain formally the following limit system

$$(\mathbf{LS}) \qquad \begin{cases} \partial_t V + Q^0(V,V) = 0 & \text{in} \quad \mathbb{R}_t^+ \times \mathbb{T}_x^3 \\ \operatorname{div} v = 0, \ \operatorname{div} b = 0 & \text{in} \quad \mathbb{R}_t^+ \times \mathbb{T}_x^3 \\ V(0) = U_0 = (u_0, b_0), \end{cases}$$

where $Q^0(V,V)$ is the limit in $\mathcal{D}'$ of $Q^\varepsilon(V,V)$.

3.1. Proof of Theorem 1.3

The proof is similar to the one of Theorem 1.2. We have just to estimate the term

$$\left| \int_0^t \langle q^0(v_n, v_n), v_n \rangle_{H^s} \right|,$$

where (v_n, b_n) is the solution of the approximate limit system. Observe that

$$\lim_\varepsilon \int_0^t \langle q^\varepsilon(v_n, v_n), v_n \rangle_{H^s} = \int_0^t \langle q^0(v_n, v_n), v_n \rangle_{H^s}$$

and using the product law given by Lemma 2.1, we obtain

$$\left| \int_0^t \langle q^0(v_n, v_n), v_n \rangle_{H^s} \right| \leq C \int_0^t \|\nabla v_n(\tau)\|_{H^{s-1}}^3 \, d\tau$$

which completes the proof. □

3.2. Proof of Theorem 1.4

The proof of this theorem is based on a method used in [7], [9] for instance.

Let $W^\varepsilon = V^\varepsilon - V = (v^\varepsilon - v, b^\varepsilon - b) = (W_1^\varepsilon, W_2^\varepsilon)$, then W^ε satisfie

$$\begin{cases} \partial_t W^\varepsilon + Q^\varepsilon(W^\varepsilon, W^\varepsilon + 2V) + a_2^\varepsilon(D) W^\varepsilon + \tilde{L}^\varepsilon(W^\varepsilon) = F^\varepsilon + R_{osc}^\varepsilon \\ \operatorname{div} W_1^\varepsilon = 0, \ \operatorname{div} W_2^\varepsilon = 0 \\ W^\varepsilon(0) = (0,0), \end{cases}$$

where

$$\begin{aligned} F^\varepsilon &= -a_2^\varepsilon(D)V - \tilde{L}^\varepsilon V \\ &= (\varepsilon \Delta v - \sqrt{\varepsilon}\mathcal{L}(-t/\varepsilon) b; \sqrt{\varepsilon}\Delta b - \sqrt{\varepsilon}\mathcal{L}(t/\varepsilon) v) \end{aligned}$$

and

$$R_{osc}^\varepsilon = Q^0(V,V) - Q^\varepsilon(V,V).$$

We recall that v and b are in the space $C_T^0(H^s)$, then

$$F^\varepsilon \longrightarrow 0 \quad \text{in} \quad C_T^0(H^{s-2})$$

precisely we have

$$\|F^\varepsilon\|_{C_T^0(H^{s-2})} \leq C\sqrt{\varepsilon}. \tag{3.11}$$

On the other hand, the oscillating term R^{ε}_{osc} can be written as follows.

$$R^{\varepsilon}_{osc} = \sum_{k=0}^{3} A^{\varepsilon}_{k},$$

where

$$\begin{aligned} A^{\varepsilon}_{0} &= \mathcal{L}(-t/\varepsilon)\mathbb{P}(\mathcal{L}(t/\varepsilon)v.\nabla\mathcal{L}(t/\varepsilon)v) - q^{0}(v,v) \\ &= q^{\varepsilon}(v,v) - q^{0}(v,v), \\ A^{\varepsilon}_{1} &= (\mathcal{L}(-t/\varepsilon)\mathbb{P}(b.\nabla b) - \overline{\mathbb{P}(b.\nabla b)}, 0) \\ A^{\varepsilon}_{2} &= (0, -(\mathcal{L}(t/\varepsilon)v).\nabla b + \overline{v}.\nabla b) \\ A^{\varepsilon}_{3} &= \big(0, b.\nabla\overline{v} - b.\nabla(\mathcal{L}(t/\varepsilon)v)\big), \end{aligned}$$

and

$$q^{0}(v,v) = \lim_{\varepsilon} \mathcal{L}(-t/\varepsilon)\mathbb{P}(\mathcal{L}(t/\varepsilon)v.\nabla\mathcal{L}(t/\varepsilon)v) \quad \text{in} \quad \mathcal{D}'.$$

In the sequel, for any three-vector fiel X, we shall note,

$$X^{\pm}(n) = (X, \nu^{\pm}(n))\nu^{\pm}(n).$$

We begin by studying the term A^{ε}_{0}. By the non stationary phase's theorem we have

$$\mathcal{F}(A^{\varepsilon}_{0})(t,n) = \sum_{\sigma\in\{\pm\}^3} \sum_{\substack{k+m=n \\ \omega_\sigma(n,k,m)\neq 0}} e^{-i\frac{t}{\varepsilon}\omega_\sigma(n,k,m)}((\mathcal{F}v)^{\sigma_1}(t,k).(\mathcal{F}\nabla v)^{\sigma_2}(t,m))^{\sigma_3}(n),$$

where $\sigma = (\sigma_1,\sigma_2,\sigma_3)$ and $\omega_\sigma(n,k,m) = \sigma_1\frac{n_3}{|n|} - \sigma_2\frac{k_3}{|k|} - \sigma_3\frac{m_3}{|m|}$.

Hence we can write

$$A^{\varepsilon}_{0} = \mathcal{F}^{-1}\Big(\sum_{\sigma\in\{\pm\}^3} \sum_{\substack{k+m=n \\ \omega_\sigma(n,k,m)\neq 0}} e^{-i\frac{t}{\varepsilon}\omega_\sigma(n,k,m)} r_\sigma(n,k,m) f_\sigma(t,k) g_\sigma(t,m)\Big).$$

Similarly, we have

$$A^{\varepsilon}_{1} = \mathcal{F}^{-1}\Big(\mathbf{1}_{\{n_3\neq 0\}}(n)e^{-i\frac{t}{\varepsilon}\frac{n_3}{|n|}}(\mathcal{F}(b.\nabla b))^{+}(t,n) + \mathbf{1}_{\{n_3\neq 0\}}(n)e^{i\frac{t}{\varepsilon}\frac{n_3}{|n|}}(\mathcal{F}(b.\nabla b))^{-}(t,n)\Big),$$

$$A^{\varepsilon}_{2} = \mathcal{F}^{-1}\Big(\sum_{k_3\neq 0} e^{i\frac{t}{\varepsilon}\frac{k_3}{|k|}}(\mathcal{F}v)^{+}(t,k).(\mathcal{F}\nabla b)(t,n-k) + \sum_{k_3\neq 0} e^{-i\frac{t}{\varepsilon}\frac{k_3}{|k|}}(\mathcal{F}v)^{-}(t,k).(\mathcal{F}\nabla b)(t,n-k)\Big),$$

$$A^{\varepsilon}_{3} = \mathcal{F}^{-1}\Big(\sum_{k_3\neq 0} e^{i\frac{t}{\varepsilon}\frac{k_3}{|k|}}(\mathcal{F}b)(t,n-k).(\mathcal{F}\nabla v)^{+}(t,k) + \sum_{k_3\neq 0} e^{-i\frac{t}{\varepsilon}\frac{k_3}{|k|}}(\mathcal{F}b)(t,n-k).(\mathcal{F}\nabla v)^{-}(t,k)\Big).$$

Since R^{ε}_{osc} converges weakly to 0 (but not "strongly"), we shall divide it (as in [7] for instance) into high and low frequency terms. Precisely, for any integer $N > 1$, we defin

$$A^{\varepsilon}_{0,N} = \mathcal{F}^{-1}\Big(\mathbf{1}_{\{|n|\leq N\}} \sum_{\sigma\in\{\pm\}^3} \sum_{\substack{k+m=n \\ \omega_\sigma(n,k,m)\neq 0 \\ |k|,|m|\leq N}} e^{-i\frac{t}{\varepsilon}\omega_\sigma(n,k,m)} r_\sigma(n,k,m) f_\sigma(t,k) g_\sigma(t,m)\Big),$$

$$A^{\varepsilon}_{1,N} = \mathcal{F}^{-1}\Big(\mathbf{1}_{\{n_3\neq 0,|n|\leq N\}}(n)e^{-i\frac{t}{\varepsilon}\frac{n_3}{|n|}}(\mathcal{F}(b.\nabla b))^{+}(t,n)$$
$$+\mathbf{1}_{\{n_3\neq 0,|n|\leq N\}}(n)e^{i\frac{t}{\varepsilon}\frac{n_3}{|n|}}(\mathcal{F}(b.\nabla b))^{-}(t,n)\Big),$$
$$A^{\varepsilon}_{2,N} = \mathcal{F}^{-1}\Big(\mathbf{1}_{\{|n|\leq N\}} \sum_{\substack{k_3\neq 0\\ |n-k|,|k|\leq N}} e^{i\frac{t}{\varepsilon}\frac{k_3}{|k|}}(\mathcal{F}v)^{+}(t,k).(\mathcal{F}\nabla b)(t,n-k)$$
$$+e^{-i\frac{t}{\varepsilon}\frac{k_3}{|k|}}(\mathcal{F}v)^{-}(t,k).(\mathcal{F}\nabla b)(t,n-k)\Big)$$
$$A^{\varepsilon}_{3,N} = \mathcal{F}^{-1}\Big(\mathbf{1}_{\{|n|\leq N\}} \sum_{\substack{k_3\neq 0\\ |n-k|,|k|\leq N}} e^{i\frac{t}{\varepsilon}\frac{k_3}{|k|}}(\mathcal{F}b)(t,n-k).(\mathcal{F}\nabla v)^{+}(t,k)$$
$$+e^{-i\frac{t}{\varepsilon}\frac{k_3}{|k|}}(\mathcal{F}b)(t,n-k).(\mathcal{F}\nabla v)^{-}(t,k)\Big)$$

and

$$R^{\varepsilon}_{osc,N} = \sum_{k=0}^{3} A^{\varepsilon}_{k,N}.$$

Now, the idea is to absorb the low frequency terms. For that, we set

$$\tilde{A}^{\varepsilon}_{0,N} = \mathcal{F}^{-1}\Big(\mathbf{1}_{\{|n|\leq N\}} \sum_{\sigma\in\{\pm\}^3} \sum_{\substack{k+m=n\\ \omega_\sigma(n,k,m)\neq 0\\ |k|,|n|\leq N}} \frac{e^{-i\frac{t}{\varepsilon}\omega_\sigma(n,k,m)}}{i\omega_\sigma(n,k,m)} r_\sigma(n,k,m) f_\sigma(t,k) g_\sigma(t,m)\Big),$$

$$\tilde{A}^{\varepsilon}_{1,N} = \mathcal{F}^{-1}\Big(\mathbf{1}_{\{n_3\neq 0,|n|\leq N\}}(n)\frac{e^{-i\frac{t}{\varepsilon}\frac{n_3}{|n|}}}{i\frac{n_3}{|n|}}(\mathcal{F}(b.\nabla b))^{+}(t,n)$$
$$+\mathbf{1}_{\{n_3\neq 0,|n|\leq N\}}(n)\frac{e^{i\frac{t}{\varepsilon}\frac{n_3}{|n|}}}{-i\frac{n_3}{|n|}}(\mathcal{F}(b.\nabla b))^{-}(t,n)\Big),$$

$$\tilde{A}^{\varepsilon}_{2,N} = \mathcal{F}^{-1}\Big(\mathbf{1}_{\{|n|\leq N\}} \sum_{\substack{k_3\neq 0\\ |n-k|,|k|\leq N}} \frac{e^{i\frac{t}{\varepsilon}\frac{k_3}{|k|}}}{-i\frac{k_3}{|k|}}(\mathcal{F}v)^{+}(t,k).(\mathcal{F}\nabla b)(t,n-k)$$
$$+\frac{e^{-i\frac{t}{\varepsilon}\frac{k_3}{|k|}}}{i\frac{k_3}{|k|}}(\mathcal{F}v)^{-}(t,k).(\mathcal{F}\nabla b)(t,n-k)\Big)$$

$$\tilde{A}^{\varepsilon}_{3,N} = \mathcal{F}^{-1}\Big(\mathbf{1}_{\{|n|\leq N\}} \sum_{\substack{k_3\neq 0\\ |n-k|,|k|\leq N}} \frac{e^{i\frac{t}{\varepsilon}\frac{k_3}{|k|}}}{-i\frac{k_3}{|k|}}(\mathcal{F}b)(t,n-k).(\mathcal{F}\nabla v)^{+}(t,k)$$
$$+\frac{e^{-i\frac{t}{\varepsilon}\frac{k_3}{|k|}}}{i\frac{k_3}{|k|}}(\mathcal{F}b)(t,n-k).(\mathcal{F}\nabla v)^{-}(t,k)\Big),$$

and we defin

$$R^{\varepsilon,N}_{osc} = R^{\varepsilon}_{osc} - R^{\varepsilon}_{osc,N},$$

$$\tilde{R}^{\varepsilon}_{osc,N} = \sum_{k=0}^{3} \tilde{A}^{\varepsilon}_{k,N}.$$

Considering $\varphi^{\varepsilon}_N = W^{\varepsilon} + \varepsilon \tilde{R}^{\varepsilon}_{osc,N} = (\varphi^{\varepsilon}_{N,1}, \varphi^{\varepsilon}_{N,2})$, then φ^{ε}_N satisfie the following equation

$$\partial_t \varphi^{\varepsilon}_N + Q^{\varepsilon}(\varphi^{\varepsilon}_N, \varphi^{\varepsilon}_N + 2\varepsilon \tilde{R}^{\varepsilon}_{osc,N} + 2V) + a^{\varepsilon}_2(D)\varphi^{\varepsilon}_N + \tilde{L}^{\varepsilon}(\varphi^{\varepsilon}_N) = F^{\varepsilon} + R^{\varepsilon,N}_{osc} + \varepsilon r^{\varepsilon}_{osc,N},$$

where

$$\varepsilon r^{\varepsilon}_{osc,N} = \varepsilon\Big(Q^{\varepsilon}(\tilde{R}^{\varepsilon}_{osc,N}, \varepsilon \tilde{R}^{\varepsilon}_{osc,N} + 2V) + a^{\varepsilon}_2(D)\tilde{R}^{\varepsilon}_{osc,N} + \tilde{L}^{\varepsilon}(\tilde{R}^{\varepsilon}_{osc,N})\Big) + \Big(R^{\varepsilon}_{osc,N} + \varepsilon \partial_t \tilde{R}^{\varepsilon}_{osc,N}\Big).$$

Now, we have to study the low frequency terms. In fact, we have the following result.

Lemma 3.2. *A constant $C_N(T)$ exists, depending only T and N such that*

$$\|\tilde{R}^{\varepsilon}_{osc,N}\|_{C^0_T(H^{s-2}) \cap L^2_T(H^{s-1})} \leq C_N(T),$$

$$\|r^{\varepsilon}_{osc,N}\|_{L^2_T(H^{s-2})} \leq C_N(T).$$

Proof. Let us recall that all the functions considered here are truncated in low frequencies. Hence the result is simply due to the fact that $v, b \in C^0_T(H^s)$, $\partial_t v, \partial_t b \in C^0_T(H^{s-2})$ and the following product law. □

Proposition 3.1. *Let s be an integer, $\sigma > \frac{3}{2}$. A constant C exists such that for all $f \in H^{\sigma}(\mathbb{T}^3)$ and $g \in H^{\sigma+1}(\mathbb{T}^3)$ with $\operatorname{div} f = \operatorname{div} g = 0$, we have*

$$|\langle Q^{\varepsilon}(f, g), f\rangle_{H^{\sigma}}| \leq C\|f\|^2_{H^{\sigma}}\|g\|_{H^{\sigma+1}}$$

and

$$|\langle Q^{\varepsilon}(f, f), f\rangle_{H^{\sigma}}| \leq C\|f\|^3_{H^{\sigma}}.$$

Proof. Combining Lemma 2.1 with the properties of operator $\mathcal{L}(t)$ given by Lemma 3.1 we obtain the desired result. □

Lemma 3.3. *For any function $f \in C^0_T(H^s)$ with $s \in \mathbb{R}$, the high frequency term*

$$f^N = \mathcal{F}^{-1}\Big(\mathbf{1}_{[N,+\infty[}\mathcal{F}(f)\Big)$$

goes to zero when N goes to infinity in $C^0_T(H^s)$.

Proof. Let us recall that

$$\|f^N(t)\|^2_{H^s} = \sum_{|k| \geq N} |k|^{2s}|\mathcal{F}(f)(t,k)|^2.$$

So, the desired result is simply due to Dini's theorem which implies that $\|f^N(t)\|^2_{H^s}$ goes to zero uniformly in t. □

This lemma implies in a straightforward way the following result.

Lemma 3.4. *The high frequency term $R^{\varepsilon,N}_{osc}$ goes to zero in $C^0_T(H^{s-2}) \cap L^2_T(H^{s-1})$ when N goes to infinity, uniformly in ε; precisely $\|R^{\varepsilon,N}_{osc}\|_{C^0_T(H^{s-2})\cap L^2_T(H^{s-1})} \leq \eta_N$ with $\eta_N \to 0$.*

Now, we can end the proof of the theorem. By the energy estimate in $H^{s-2}(\mathbb{T}^3)$ we obtain

$$\begin{aligned}\frac{1}{2}\frac{d}{dt}\|\varphi^\varepsilon_N\|^2_{H^{s-2}} + \varepsilon\|\nabla\varphi^\varepsilon_{N,1}\|^2_{H^{s-2}} + \sqrt{\varepsilon}\|\nabla\varphi^\varepsilon_{N,2}\|^2_{H^{s-2}} \leq & |\langle F^\varepsilon + R^{\varepsilon,N}_{osc} + \varepsilon r^\varepsilon_{osc,N}, \varphi^\varepsilon_N\rangle_{H^{s-2}}| \\ & + 2|\langle Q^\varepsilon(\varphi^\varepsilon_N, \varepsilon\tilde{R}^\varepsilon_{osc,N} + V), \varphi^\varepsilon_N\rangle_{H^{s-2}}| \\ & + |\langle Q^\varepsilon(\varphi^\varepsilon_N, \varphi^\varepsilon_N), \varphi^\varepsilon_N\rangle_{H^{s-2}}|.\end{aligned}$$

This leads, using the product law (Proposition 3.1), to

$$\begin{aligned}\frac{1}{2}\frac{d}{dt}\|\varphi^\varepsilon_N\|^2_{H^{s-2}} \leq C\Big[&\|\varphi^\varepsilon_N\|^2_{H^{s-2}}(\|\varphi^\varepsilon_N\|_{H^{s-2}} + \|\varepsilon\tilde{R}^\varepsilon_{osc,N} + V\|_{H^{s-1}} + 1) \\ &+\|F^\varepsilon + R^{\varepsilon,N}_{osc} + \varepsilon r^\varepsilon_{osc,N}\|^2_{H^{s-2}}\Big].\end{aligned}$$

Integrating this inequality and using Lemma 3.2, we obtain

$$\begin{aligned}\|\varphi^\varepsilon_N(t)\|^2_{H^{s-2}} \leq & \Big(\varepsilon C_N(T)^2 + 4\varepsilon^2\|\Delta u\|^2_{L^2_T(H^{s-2})} + 4\|R^{\varepsilon,N}_{osc}\|^2_{L^2_T(H^{s-2})} \\ & +4\varepsilon^2 C_N(T)^2\Big) + C\int_0^t \|\varphi^\varepsilon_N(\tau)\|^2_{H^{s-2}}\Big(B(T) + \varepsilon C_N(T) + \|\varphi^\varepsilon_N(\tau)\|_{H^{s-2}}\Big)d\tau,\end{aligned}$$

where $B(T) = 1 + \|V\|_{L^\infty_T(H^{s-2})}$. We set

$$T^* = \sup\{0 \leq t < T; \quad \|\varphi^\varepsilon_N\|_{L^\infty_t(H^{s-2})} \leq B(T)\}.$$

Then, for all $0 \leq t < T^*$, we can write,

$$\begin{aligned}\|\varphi^\varepsilon_N(t)\|^2_{H^{s-2}} \leq & \ \varepsilon C_N(T)^2 + 4\varepsilon^2\|u\|^2_{L^2_{T_0}(H^s)} + 4\eta^2_N + 4\varepsilon^2 C_N(T)^2 \\ & +c(B(T) + \varepsilon C_N(T))\int_0^t \|\varphi^\varepsilon_N(\tau)\|^2_{H^{s-2}}d\tau.\end{aligned}$$

A classical Gronwall estimate gives

$$\begin{aligned}\|\varphi^\varepsilon_N(t)\|^2_{H^{s-2}} \leq & \Big(\varepsilon C_N(T)^2 + 4\varepsilon^2\|u\|^2_{L^2_T(H^s)} + 4\eta^2_N + 4\varepsilon^2 C_N(T)^2\Big) \\ & \exp\Big(cTB(T) + cT\varepsilon C_N(T)\Big).\end{aligned}$$

For N large and ε small enough we obtain, thanks to Lemma 3.4

$$\|\varphi^\varepsilon_N(t)\|_{H^{s-2}} \leq \frac{B(T)}{2},$$

which easily implies that $T^* = T$.

Using Lemma 3.2 and letting $\varepsilon \to 0$, $N \to +\infty$, we obtain

$$W^\varepsilon \to 0 \quad \text{in} \quad C^0_T(H^{s-2}).$$

An interpolation argument concludes the proof of Theorem 1.4. □

4. The Case $\Omega = \mathbb{R}^3$

This section is devoted to study of dispersion phenomenon in the $(\mathbf{MHD}^\varepsilon)$ system in the case of the space $\mathbb{R}^3$. Let us introduce the "linearized" equation in u^ε of the firs equation of $(\mathbf{MHD}^\varepsilon)$.

$$\begin{cases} \partial_t u + \dfrac{1}{\varepsilon} Lu = -\nabla p \quad \text{in} \quad \mathbb{R}_t \times \mathbb{R}^3_x \\ \operatorname{div} u = 0 \quad \text{in} \quad \mathbb{R}_t \times \mathbb{R}^3_x \\ u(0) = u_0. \end{cases}$$

In Fourier variables $\xi \in \mathbb{R}^3$, we obtain

$$\partial_t \mathcal{F}(u) + \frac{\xi_3}{\varepsilon |\xi|^2} \xi \times \mathcal{F}(u) = 0.$$

Hence, we are led to study the following family of operators

$$\begin{aligned} \mathcal{G}^\varepsilon : f \longmapsto & \int_{\mathbb{R}^3_\xi} \mathcal{F}(f)(\xi) \exp\Big(\mp it \frac{\xi_3}{\varepsilon|\xi|} + ix.\xi\Big) d\xi \\ & = \int_{\mathbb{R}^3_y \times \mathbb{R}^3_\xi} f(y) \exp\Big(\mp it \frac{\xi_3}{\varepsilon|\xi|} + i(x-y).\xi\Big) d\xi dy. \end{aligned}$$

We notice that the phase function $\frac{\xi_3}{|\xi|}$ is almost stationary when ξ_3 is almost equal to 0 as well as when $|\xi_3|$ is much larger then $|\xi_h|$. So, for some $0 < r < R$, let us defin the domain $\mathcal{C}_{r,R}$ by

$$\mathcal{C}_{r,R} = \{\xi \in \mathbb{R}^3; \quad |\xi_3| > r \quad \text{and} \quad |\xi| \leq R\}.$$

We consider ψ a cut-off function, which is radial with respect to horizontal variable $\xi_h = (\xi_1, \xi_2)$ and whose value is 1 near $\mathcal{C}_{r,R}$.

First, we study the case when $\mathcal{F}(f)$ is supported in $\mathcal{C}_{r,R}$. We can write

$$\mathcal{G}^\varepsilon f(t,x) = \Big(K\Big(\frac{t}{\varepsilon}, .\Big) * f \Big)(x),$$

where the kernel K is define by

$$K(t,z) = \int_{\mathbb{R}^3} \psi(\xi) e^{it\frac{\xi_3}{|\xi|} + iz.\xi} d\xi.$$

As in [5], we recall the following property of K.

Lemma 4.1. *For all r, R such that $0 < r < R$, there exists a constant $C_{r,R}$ such that*

$$\|K(t,.)\|_{L^\infty(\mathbb{R}^3)} \leq C_{r,R} \min\{1, t^{-\frac{1}{2}}\}.$$

Let us denote by w^ε the solution of

$$(PLF_\varepsilon) \qquad \begin{cases} \partial_t w + \dfrac{1}{\varepsilon} Lw = f \quad \text{in} \quad \mathbb{R}_t \times \mathbb{R}^3_x \\ w(0) = w_0. \end{cases}$$

Lemma 4.1 yields, in a standard way, the following Strichartz-estimate (see [5]).

Corollary 4.1. *For all constants r and R such that $0 < r < R$, let $\mathcal{C}_{r,R}$ be the domain defined above. Then a constant $C_{r,R}$ exists such that if*

$$supp\mathcal{F}(w_0) \cup supp\mathcal{F}(f) \subset \mathcal{C}_{r,R}$$

then the solution w^ε of (PLF_ε) with the forcing term f and initial data w_0 satisfies

$$\|w^\varepsilon\|_{L^4(\mathbb{R}_+,L^\infty)} \le C_{r,R}\varepsilon^{\frac{1}{4}}\Big(\|w_0\|_{L^2} + \|f\|_{L^1(\mathbb{R}^+,L^2)}\Big).$$

We notice that the constant $C_{r,R}$ does not depend on ε.

Using the above estimate, we are able to prove the convergence result.

4.1. Proof of Theorem 1.5

We start by proving a convergence result for the Leray's solutions. We shall apply the method used in [5] for the firs equation of our system.

Let $R > 0$ and χ be a cut-off function in $\mathcal{D}(\mathbb{R})$ taking the value 1 near the origin. We set $u_R^\varepsilon := \chi(|D|/R)u^\varepsilon = \mathcal{F}^{-1}\Big(\chi(|\xi|/R)\mathcal{F}(u^\varepsilon)(\xi)\Big)$. Then u_R^ε satisfie

$$\partial_t u_R^\varepsilon + \frac{1}{\varepsilon}\mathbb{P}(u_R^\varepsilon \times e_3) = f_R^\varepsilon,$$

with

$$f_R^\varepsilon = -\chi(|D|/R)\Big(\varepsilon\Delta u^\varepsilon + \mathbb{P}(u^\varepsilon\nabla u^\varepsilon) - \mathbb{P}(b^\varepsilon\nabla b^\varepsilon) - \sqrt{\varepsilon}\partial_3 b^\varepsilon\Big).$$

By Duhamel's formula, we can write

$$u_R^\varepsilon = e^{t\sigma^\varepsilon(D)}u_{0,R} + \int_0^t e^{(t-t')\sigma^\varepsilon(D)}f_R^\varepsilon(t')dt',$$

where $\sigma^\varepsilon(\xi)u = \frac{\xi_3}{\varepsilon|\xi|^2}u \times \xi$.

So we have

$$\|u_R^\varepsilon\|_{L^4_T(L^\infty(\mathbb{R}^3))} \le \|e^{t\sigma^\varepsilon(D)}u_{0,R}\|_{L^4_T(L^\infty(\mathbb{R}^3))} + \left\|\int_0^t e^{(t-t')\sigma^\varepsilon(D)}f_R^\varepsilon(t')dt'\right\|_{L^4_T(L^\infty(\mathbb{R}^3))}. \tag{4.12}$$

For the firs term of the right side of (4.12), we localize on $\mathcal{C}_{r,R}$. Hence, for all $0 < r < R$, we write

$$\begin{aligned}&\|e^{t\sigma^\varepsilon(D)}u_{0,R}\|_{L^4_T(L^\infty(\mathbb{R}^3))}\\ &\le \|e^{t\sigma^\varepsilon(D)}\chi(|D_3|/r)u_{0,R}\|_{L^4_T(L^\infty(\mathbb{R}^3))} + \|e^{t\sigma^\varepsilon(D)}(Id - \chi(|D_3|/r))u_{0,R}\|_{L^4_T(L^\infty(\mathbb{R}^3))}.\end{aligned}$$

Using Corollary 4.1, we obtain

$$\|e^{t\sigma^\varepsilon(D)}u_{0,R}\|_{L^4_T(L^\infty(\mathbb{R}^3))} \le CT^{\frac{1}{4}}(R^2 r)^{\frac{1}{2}}\|u_0\|_{L^2} + \varepsilon^{\frac{1}{4}}C_{r,R}\|u_0\|_{L^2}. \tag{4.13}$$

For the second term of (4.12),

$$\begin{aligned}\left\|\int_0^t e^{(t-t')\sigma^\varepsilon(D)} f_R^\varepsilon(t')dt'\right\|_{L_T^4(L^\infty)} &\leq \left\|\int_0^t e^{(t-t')\sigma^\varepsilon(D)}\chi(|D_3|/r) f_R^\varepsilon(t')dt'\right\|_{L_T^4(L^\infty)} \\ &+\left\|\int_0^t e^{(t-t')\sigma^\varepsilon(D)}(Id-\chi(|D_3|/r)) f_R^\varepsilon(t')dt'\right\|_{L_T^4(L^\infty)}.\end{aligned}$$

From Corollary 4.1 and Bernstein's lemma, we deduce that

$$\begin{aligned}\left\|\int_0^t e^{(t-t')\sigma^\varepsilon(D)} f_R^\varepsilon(t')dt'\right\|_{L_T^4(L^\infty)} &\leq CT^{\frac{5}{4}}R^{\frac{3}{2}}\|\chi(|D_3|/r)\chi(|D|/R)f^\varepsilon\|_{L_T^\infty(L^2)} \\ &+C_{r,R}\varepsilon^{\frac{1}{4}}\|(Id-\chi(|D_3|/r))\chi(|D|/R)f^\varepsilon\|_{L_T^1(L^2)} \\ &\leq CT^{\frac{5}{4}}R^{\frac{3}{2}}\|\chi(|D_3|/r)\chi(|D|/R)f^\varepsilon\|_{L_T^\infty(L^2)} \\ &+C_{r,R}\varepsilon^{\frac{1}{4}}\|\chi(|D|/R)f^\varepsilon\|_{L_T^\infty(L^2)}.\end{aligned}$$

Now, we focus our attention to the firs right term (4.12).

$$\begin{aligned}\chi(|D_3|/r)\chi(|D|/R)f^\varepsilon = &\ \varepsilon\chi(|D_3|/r)\chi(|D|/R)\Delta u^\varepsilon - \chi(|D_3|/r)\chi(|D|/R)\mathbb{P}(u^\varepsilon.\nabla u^\varepsilon) \\ + &\ \chi(|D_3|/r)\chi(|D|/R)\mathbb{P}(b^\varepsilon.\nabla b^\varepsilon) + \sqrt{\varepsilon}\chi(|D_3|/r)\chi(|D|/R)\partial_3 b^\varepsilon.\end{aligned}$$

First, the quadratic terms can be similarly bounded,

$$\begin{aligned}\|\chi(|D_3|/r)\chi(|D|/R)\mathbb{P}(u^\varepsilon.\nabla u^\varepsilon)\|_{L_T^\infty(L^2)} &\leq R\|\chi(|D_3|/r)\chi(|D|/R)(u^\varepsilon\otimes u^\varepsilon)\|_{L_T^\infty(L^2)} \\ &\leq R\|\mathcal{F}^{-1}\chi(|\xi_3|/r)\chi(|\xi|/R)*(u^\varepsilon\otimes u^\varepsilon)\|_{L_T^\infty(L^2)} \\ &\leq R\|\mathcal{F}^{-1}\chi(|\xi_3|/r)\chi(|\xi|/R)\|_{L_T^\infty(L^2)}\|u^\varepsilon\otimes u^\varepsilon\|_{L_T^\infty(L^1)} \\ &\leq R(R^2r)^{\frac{1}{2}}\|U_0\|_{H^s}^2. \\ \|\chi(|D_3|/r)\chi(|D|/R)\mathbb{P}(b^\varepsilon.\nabla b^\varepsilon)\|_{L_T^\infty(L^2)} &\leq R(R^2r)^{\frac{1}{2}}\|U_0\|_{H^s}^2,\end{aligned}$$

and finall

$$\|\chi(|D_3|/r)\chi(|D|/R)\partial_3 b^\varepsilon\|_{L_T^\infty(L^2)} \leq r\|\chi(|D_3|/r)\chi(|D|/R)b^\varepsilon\|_{L_T^\infty(L^2)} \leq r\|b^\varepsilon\|_{L_T^\infty(L^2)} \leq r\|U_0\|_{H^s}^2.$$

For the second right term we use

$$\begin{aligned}\|\chi(|D|/R)f^\varepsilon\|_{L_T^\infty(L^2)} &\leq \|\chi(|D|/R)\mathbb{P}(u^\varepsilon.\nabla u^\varepsilon)\|_{L_T^\infty(L^2)} + \|\chi(|D|/R)\mathbb{P}(b^\varepsilon.\nabla b^\varepsilon)\|_{L_T^\infty(L^2)} \\ &+\varepsilon\|\chi(|D|/R)\Delta u^\varepsilon\|_{L_T^\infty(L^2)} + \sqrt{\varepsilon}\|\chi(|D|/R)\partial_3 b^\varepsilon\|_{L_T^\infty(L^2)} \\ &\leq CR^{\frac{5}{2}}\|U_0\|_{H^s}^2 + CR^{\frac{5}{2}}\|U_0\|_{H^s}^2 + R\|b^\varepsilon\|_{L_T^\infty(L^2)} \\ &\leq CR^{\frac{5}{2}}\|U_0\|_{H^s}^2 + R\|U_0\|_{H^s}.\end{aligned}$$

Finally, we have

$$\begin{aligned}\left\|\int_0^t e^{(t-t')(\sigma^\varepsilon(D)} f_R^\varepsilon(t')dt'\right\|_{L_T^4(L^\infty)} &\leq CT^{\frac{5}{4}}R^{\frac{3}{2}}(\|U_0\|_{H^s}^2R^2r^{\frac{1}{2}} + r\|U_0\|_{H^s}) \\ &+C_{r,R}\varepsilon^{\frac{1}{4}}T(R^{\frac{5}{2}}\|U_0\|_{H^s}^2 + R\|U_0\|_{H^s}) \qquad (4.14)\end{aligned}$$

Combining (4.13) and (4.14) yields

$$u_R^\varepsilon \longrightarrow 0 \quad \text{as} \quad \varepsilon\to 0 \quad \text{in} \quad L_T^4(L^\infty).$$

Using the imbedding $L^\infty \hookrightarrow \mathcal{C}^{-1/4}$, we deduce that

$$u_R^\varepsilon \longrightarrow 0 \quad as \quad \varepsilon \to 0 \quad \text{in} \quad L_T^4(\mathcal{C}^{-1/4}). \tag{4.15}$$

On the other hand, we have

$$\begin{aligned}\|u^\varepsilon - u_R^\varepsilon\|_{L_T^2(\mathcal{C}^{-1/4})} &\le C\|u^\varepsilon - u_R^\varepsilon\|_{L_T^2(H^{5/4})}\\ &\le CR^{-s+\frac{5}{4}}\|u^\varepsilon\|_{L_T^2(H^s(\mathbb{R}^3))}\end{aligned}$$

and then by energy estimate (1.2), we conclude that

$$\|u^\varepsilon - u_R^\varepsilon\|_{L_T^2(\mathcal{C}^{-1/4})} \le CR^{-s+\frac{5}{4}}T^{1/2}\|U_0\|_{H^s(\mathbb{R}^3)}. \tag{4.16}$$

From (4.15), (4.16) we obtain

$$u^\varepsilon \longrightarrow 0 \quad \text{as} \quad \varepsilon \to 0 \quad \text{in} \quad L_T^4(\mathcal{C}^{-1/4}(\mathbb{R}^3)).$$

By an interpolation argument, we can deduce the firs part of Theorem 1.5.

To prove the second part of Theorem 1.5; let us take the inner product with $b^\varepsilon - b_0$ in H^{s-3},

$$\begin{aligned}\|b^\varepsilon(t) - b_0\|_{H^{s-3}}^2 \le &\int_0^t |\langle b^\varepsilon(\tau).\nabla u^\varepsilon(\tau), b^\varepsilon(\tau) - b_0\rangle_{H^{s-3}}|d\tau\\ &+\int_0^t |\langle u^\varepsilon(\tau).\nabla b^\varepsilon(\tau), b^\varepsilon(\tau) - b_0\rangle_{H^{s-3}}|d\tau\\ &+\varepsilon\int_0^t \|\Delta b^\varepsilon(\tau)\|_{H^{s-3}}\|b^\varepsilon(\tau) - b_0\|_{H^{s-3}}d\tau\\ &+\sqrt{\varepsilon}\int_0^t \|\partial_3 b^\varepsilon(\tau)\|_{H^{s-3}}\|b^\varepsilon(\tau) - b_0\|_{H^{s-3}}d\tau,\end{aligned}$$

using the energy estimate (1.2) and the following product law

Lemma 4.2. *Let $\sigma > 4$ be an integer and a, b and c three vectors field such that $a \in H^\sigma(\mathbb{R}^3) \cap \mathcal{C}^{\sigma+\frac{1}{4}}(\mathbb{R}^3)$, $b \in H^{\sigma+1}(\mathbb{R}^3) \cap \mathcal{C}^{\sigma+\frac{5}{4}}(\mathbb{R}^3)$, $c \in H^\sigma(\mathbb{R}^3)$ and $\operatorname{div} a = 0$. Then, there exists a constant $C(\sigma)$ such that*

$$|\langle a.\nabla b, c\rangle_{H^\sigma}| \le C(\sigma)\|c\|_{H^\sigma}\min\Big(\|a\|_{\mathcal{C}^{\sigma+\frac{1}{4}}}\|\nabla b\|_{H^\sigma}; \|a\|_{H^\sigma}\|b\|_{\mathcal{C}^{\sigma+\frac{5}{4}}}\Big).$$

We obtain

$$\begin{aligned}\|b^\varepsilon(t) - b_0\|_{H^{s-3}}^2 \le\ & C\sqrt{T}\|(w_0, b_0)\|_{H^s}^2\|u^\varepsilon\|_{L_T^2(\mathcal{C}^{s-\frac{7}{4}})}\\ &+C\sqrt{\varepsilon}\|(w_0, b_0)\|_{H^s}^2.\end{aligned}$$

As $u^\varepsilon \to 0$ in $L_T^4(\mathcal{C}^{s'-\frac{3}{2}})$, $(s' < s)$, decreasing inclusion of $\mathcal{C}^\sigma(\mathbb{R}^3)$ spaces and an interpolation argument, we can deduce the desired result. □

4.2. Proof of Theorem 1.7

We begin by recalling a fundamental lemma concerning some product laws in Sobolev spaces.

Lemma 4.3. *For* $\bar{a} \in \left(H^\sigma(\mathbb{R}^2)\right)^3$ *and* $b, c \in \left(H^\sigma(\mathbb{R}^3)\right)^3$ *($\sigma > 4$ an integer) such that* $\operatorname{div}_h \bar{a} = 0$*, then*

$$\begin{array}{rcl} |\langle \bar{a}.\nabla b, c\rangle_{H^\sigma} + \langle \bar{a}.\nabla c, b\rangle_{H^\sigma}| & \leq & C\|\bar{a}\|_{H^\sigma}\|b\|_{H^\sigma}\|c\|_{H^\sigma} \\ |\langle \bar{a}.\nabla b, b\rangle_{H^\sigma}| & \leq & C\|\bar{a}\|_{H^\sigma}\|b\|^2_{H^\sigma}. \end{array}$$

Moreover, if $\bar{a} \in \left(H^{\sigma+1}(\mathbb{R}^2)\right)^3$*, then*

$$|\langle b.\nabla \bar{a}, c\rangle_{H^\sigma}| \leq C\|\bar{a}\|_{H^{\sigma+1}}\|b\|_{H^\sigma}\|c\|_{H^\sigma}.$$

Now, we consider the following approximate system of $(\mathbf{MHD3D}^\varepsilon)$,
$(\mathbf{MHD3D_n})$

$$\begin{cases} \partial_t w - \varepsilon\Delta J_n w + \dfrac{1}{\varepsilon}J_n w \times e_3 + \sqrt{\varepsilon}\partial_3 J_n B - J_n(J_n B.\nabla J_n B) & \\ \qquad = \nabla\Delta^{-1}\operatorname{div}\left(J_n(J_n B.\nabla J_n B) - \dfrac{1}{\varepsilon}J_n w \times e_3\right) & \text{in } [0, T_0^{(s+1)}] \times \mathbb{R}^3 \\ \partial_t B - \sqrt{\varepsilon}\Delta J_n B + \sqrt{\varepsilon}\partial_3 J_n w + J_n((J_n\bar{u}^\varepsilon + J_n w).\nabla J_n B) & \\ \qquad - J_n(J_n B.\nabla(J_n\bar{u}^\varepsilon + J_n w)) = 0 & \text{in } [0, T_0^{(s+1)}] \times \mathbb{R}^3 \\ (\operatorname{div} w, \operatorname{div} B) = (0,0) & \text{in } [0, T_0^{(s+1)}] \times \mathbb{R}^3 \\ (w, B)(0) = (J_n w_0, J_n b_0). & \end{cases}$$

Thanks to Lemmas 2.1-3.1 and ordinary differential equations theory the system $(\mathbf{MHD3D}_n)$ has a unique maximal solution $W_n := (w_n, B_n)$ in the space $\mathcal{C}^1([0, T^*_{n,\varepsilon}), H^s(\mathbb{R}^3))$. The uniqueness and the fact $J_n^2 = J_n$ leads to $J_n W_n = W_n$.

Moreover $T^*_{n,\varepsilon} > T_1 := \dfrac{1}{C(s)(\|\bar{u}_0\|_{H^{s+1}} + \|(w_0, b_0)\|_{H^s})}$ and for all $t \in [0, T_1]$

$$\|W_n(t)\|^2_{H^s} + 2\varepsilon\int_0^t \|\nabla w_n\|^2_{H^s} + 2\sqrt{\varepsilon}\int_0^t \|\nabla B_n\|^2_{H^s} \leq 2\|(w_0, b_0)\|^2_{H^s}. \tag{4.17}$$

The end of the proof is the same as the proof for Theorem 1.5. The uniqueness is easy to prove. □

4.3. Proof of Theorem 1.8

Let $v^\varepsilon := u^\varepsilon - \bar{u}^\varepsilon$. $(v^\varepsilon, b^\varepsilon)$ satisfie the following system.
$(\mathbf{MHDF}^\varepsilon)$

$$\begin{cases} \partial_t v - \varepsilon\Delta v + v.\nabla v + v.\nabla\bar{u}^\varepsilon + \bar{u}^\varepsilon.\nabla v & & \\ \qquad -\text{curl} b \times b + \sqrt{\varepsilon}\text{curl} b \times e_3 + \dfrac{v \times e_3}{\varepsilon} & = -\nabla p & \text{in } [0, T_0] \times \mathbb{R}^3 \\ \partial_t b - \sqrt{\varepsilon}\Delta b + \bar{u}^\varepsilon.\nabla b + v.\nabla b - b.\nabla v & & \\ \qquad -b.\nabla\bar{u}^\varepsilon + \sqrt{\varepsilon}\text{curl}\,(v \times e_3) & = 0 & \text{in } [0, T_0] \times \mathbb{R}^3 \\ \operatorname{div} v = 0, \quad \operatorname{div} b & = 0 & \text{in } [0, T_0] \times \mathbb{R}^3 \\ (v, b)(0) & = (w_0, b_0) & \end{cases}$$

Now we consider the approximate system

$(\mathbf{MHDF}_n^\varepsilon)$

$$
\begin{cases}
\partial_t v & - \quad \varepsilon\Delta J_n v + J_n(J_n v.\nabla J_n v) + J_n(J_n v.\nabla J_n \bar{u}^\varepsilon) + J_n(J_n\bar{u}^\varepsilon.\nabla J_n v) \\
& + \quad J_n(J_n\bar{u}^\varepsilon.\nabla J_n v) - J_n(\mathrm{curl} J_n b \times J_n b) + \sqrt{\varepsilon}\mathrm{curl}\ J_n b \times e_3 + \\
\dfrac{J_n v \times e_3}{\varepsilon} & = \quad -\nabla\Delta^{-1}\mathrm{div}\Big(J_n(J_n v.\nabla J_n v) + J_n(J_n v.\nabla J_n\bar{u}^\varepsilon) + J_n(J_n\bar{u}^\varepsilon.\nabla J_n v) \\
+\dfrac{J_n v \times e_3}{\varepsilon} & + \quad J_n(J_n\bar{u}^\varepsilon.\nabla J_n v) - J_n(\mathrm{curl} J_n b \times J_n b) + \sqrt{\varepsilon}\mathrm{curl}\ J_n b \times e_3\Big) \\
\partial_t b & - \quad \sqrt{\varepsilon}\Delta J_n b + J_n(J_n\bar{u}^\varepsilon.\nabla J_n b) + J_n(J_n v.\nabla J_n b) \\
& - \quad J_n(J_n b.\nabla J_n v) - J_n(J_n b.\nabla J_n v) - J_n(J_n b.\nabla J_n \bar{u}^\varepsilon) \\
& + \quad \sqrt{\varepsilon}\mathrm{curl}\ (J_n v \times e_3) = 0 \\
& \qquad \mathrm{div}\, v = 0,\ \mathrm{div}\, b = 0 \\
& \qquad (v,b)(0) = (J_n w_0, J_n b_0).
\end{cases}
$$

Thanks to Lemmas 2.1 and 3.1, by ordinary differential equations theory the system $(\mathbf{MHDF}_n^\varepsilon)$ has a unique maximal solution $V_n := (v_n, b_n)$ in the space $C^1([0,T_n^*), H^s(\mathbb{R}^3))$. The uniqueness and the fact $J_n^2 = J_n$ leads to $J_n V_n = V_n$. Lemmas 2.1 and 3.1 give $T_n^* > T_1$ and for all $t \in [0, T_1]$

$$\|V_n(t)\|_{H^s}^2 + 2\varepsilon\int_0^t \|\nabla v_n\|_{H^s}^2 + 2\sqrt{\varepsilon}\int_0^t \|\nabla b_n\|_{H^s}^2 \le 2\|(w_0,b_0)\|_{H^s}^2. \tag{4.18}$$

The end is the same as the proof for Theorem 1.5, and we obtain existence of a $V^\varepsilon := (v^\varepsilon, b^\varepsilon) \in L^\infty_{T_1}(H^s) \cap L^2_{T_1}(H^{s+1})$ solution of $(\mathbf{MHDF}^\varepsilon)$, satisfy for all $t \in [0, T_1]$

$$\|V^\varepsilon(t)\|_{H^s}^2 + 2\varepsilon\int_0^t \|\nabla v^\varepsilon\|_{H^s}^2 + 2\sqrt{\varepsilon}\int_0^t \|\nabla b^\varepsilon\|_{H^s}^2 \le 2\|(w_0,b_0)\|_{H^s}^2. \tag{4.19}$$

The proof of the uniqueness is easy. □

4.4. Proof of Theorem 1.9

We fin that $\psi^\varepsilon := (u^\varepsilon - \bar{u}^\varepsilon - w^\varepsilon, b^\varepsilon - B^\varepsilon) = (\psi_1^\varepsilon, \psi_2^\varepsilon)$ obeys

$$\partial_t\psi_1^\varepsilon - \varepsilon\Delta\psi_1^\varepsilon + \sqrt{\varepsilon}\partial_3\psi_2^\varepsilon = f_1^\varepsilon \tag{4.20}$$

$$\partial_t\psi_2^\varepsilon - \sqrt{\varepsilon}\Delta\psi_2^\varepsilon + \sqrt{\varepsilon}\partial_3\psi_1^\varepsilon = f_2^\varepsilon, \tag{4.21}$$

where

$$
\begin{aligned}
f_1^\varepsilon &:= -\nabla p^\varepsilon - \psi_1^\varepsilon.\nabla\psi_1^\varepsilon - \psi_1^\varepsilon.\nabla\bar{u}^\varepsilon - \bar{u}^\varepsilon.\nabla\psi_1^\varepsilon + \psi_2^\varepsilon.\nabla\psi_2^\varepsilon + \psi_2^\varepsilon.\nabla B^\varepsilon + B^\varepsilon.\nabla\psi_1^\varepsilon - w^\varepsilon.\nabla\bar{u}^\varepsilon - \bar{u}^\varepsilon.\nabla w^\varepsilon \\
f_2^\varepsilon &:= -\psi_1^\varepsilon.\nabla\psi_2^\varepsilon - w^\varepsilon.\nabla\psi_2^\varepsilon - \bar{u}^\varepsilon.\nabla\psi_2^\varepsilon - \psi_1^\varepsilon.\nabla B^\varepsilon + B^\varepsilon.\nabla\psi_1^\varepsilon + \psi_2^\varepsilon.\nabla\bar{u}^\varepsilon + \psi_2^\varepsilon.\nabla\psi_1^\varepsilon + \psi_2^\varepsilon.\nabla w^\varepsilon.
\end{aligned}
$$

Taking the inner product in $H^{s-3}(\mathbb{R}^3)$ of (4.20)-(4.21) with ψ^ε yields

$$\frac{1}{2}\frac{d}{dt}\|\psi^\varepsilon\|_{H^{s-3}}^2 + \varepsilon\|\nabla\psi_1^\varepsilon\|_{H^{s-3}}^2 + \sqrt{\varepsilon}\|\nabla\psi_2^\varepsilon\|_{H^{s-3}}^2 \le \sum_{k=1}^{14} I_k, \tag{4.22}$$

where

$$I_1 := |\langle \psi_1^\varepsilon.\nabla\psi_1^\varepsilon, \psi_1^\varepsilon\rangle_{H^{s-3}}|, \quad I_2 := |\langle \bar{u}^\varepsilon.\nabla\psi_1^\varepsilon, \psi_1^\varepsilon\rangle_{H^{s-3}}|$$
$$I_3 := |\langle \psi_1^\varepsilon.\nabla\bar{u}^\varepsilon, \psi_1^\varepsilon\rangle_{H^{s-3}}|, \quad I_4 := |\langle \psi_2^\varepsilon.\nabla B^\varepsilon, \psi_1^\varepsilon\rangle_{H^{s-3}}|$$
$$I_5 := |\langle \psi_2^\varepsilon.\nabla\psi_2^\varepsilon, \psi_1^\varepsilon\rangle_{H^{s-3}} + \langle \psi_2^\varepsilon.\nabla\psi_1^\varepsilon, \psi_2^\varepsilon\rangle_{H^{s-3}}|$$
$$I_6 := |\langle B^\varepsilon.\nabla\psi_2^\varepsilon, \psi_1^\varepsilon\rangle_{H^{s-3}} + \langle B^\varepsilon.\nabla\psi_1^\varepsilon, \psi_2^\varepsilon\rangle_{H^{s-3}}|$$
$$I_7 := |\langle \psi_1^\varepsilon.\nabla\psi_2^\varepsilon, \psi_2^\varepsilon\rangle_{H^{s-3}}|, \quad I_8 := |\langle \psi_1^\varepsilon.\nabla B^\varepsilon, \psi_2^\varepsilon\rangle_{H^{s-3}}|$$
$$I_9 := |\langle \bar{u}^\varepsilon.\nabla\psi_2^\varepsilon, \psi_2^\varepsilon\rangle_{H^{s-3}}|, \quad I_{10} := |\langle w^\varepsilon.\nabla\psi_2^\varepsilon, \psi_2^\varepsilon\rangle_{H^{s-3}}|$$
$$I_{11} := |\langle \psi_2^\varepsilon.\nabla\bar{u}^\varepsilon, \psi_2^\varepsilon\rangle_{H^{s-3}}|, \quad I_{12} := |\langle \psi_2^\varepsilon.\nabla w^\varepsilon, \psi_1^\varepsilon\rangle_{H^{s-3}}|$$
$$I_{13} := |\langle \bar{u}^\varepsilon.\nabla w^\varepsilon, \psi_1^\varepsilon\rangle_{H^{s-3}}|, \quad I_{14} := |\langle w^\varepsilon.\nabla\bar{u}^\varepsilon, \psi_1^\varepsilon\rangle_{H^{s-3}}|.$$

Combining the Cauchy-Shwartz inequality and the fact that H^{s-3} is an algebra, we obtain

$$I_4 \le C\|B^\varepsilon\|_{H^{s-2}}\|\psi_2^\varepsilon\|_{H^{s-3}}\|\psi_1^\varepsilon\|_{H^{s-3}} \le C\|\psi^\varepsilon\|_{H^{s-3}}^2 \tag{4.23}$$
$$I_8 \le C\|B^\varepsilon\|_{H^{s-2}}\|\psi_1^\varepsilon\|_{H^{s-3}}\|\psi_2^\varepsilon\|_{H^{s-3}} \le C\|\psi^\varepsilon\|_{H^{s-3}}^2 \tag{4.24}$$
$$I_{12} \le C\|w^\varepsilon\|_{H^{s-2}}\|\psi_1^\varepsilon\|_{H^{s-3}}\|\psi_2^\varepsilon\|_{H^{s-3}} \le C\|\psi^\varepsilon\|_{H^{s-3}}^2. \tag{4.25}$$

By lemma 4.3,

$$I_1 \le C\|\psi_1^\varepsilon\|_{H^{s-3}}^3 \le C\|\psi^\varepsilon\|_{H^{s-3}}^2 \tag{4.26}$$
$$I_2 \le C\|\bar{u}^\varepsilon\|_{H^{s-2}}\|\psi_1^\varepsilon\|_{H^{s-3}}^2 \le C\|\psi^\varepsilon\|_{H^{s-3}}^2 \tag{4.27}$$
$$I_3 \le C\|\bar{u}^\varepsilon\|_{H^{s-2}}\|\psi_1^\varepsilon\|_{H^{s-3}}^2 \le C\|\psi^\varepsilon\|_{H^{s-3}}^2 \tag{4.28}$$
$$I_5 \le C\|\psi_1^\varepsilon\|_{H^{s-2}}\|\psi_2^\varepsilon\|_{H^{s-3}}^2 \le C\|\psi^\varepsilon\|_{H^{s-3}}^2 \tag{4.29}$$
$$I_6 \le C\|B^\varepsilon\|_{H^{s-3}}\|\psi_2^\varepsilon\|_{H^{s-3}}\|\psi_1^\varepsilon\|_{H^{s-3}} \le C\|\psi^\varepsilon\|_{H^{s-3}}^2 \tag{4.30}$$
$$I_7 \le C\|\psi_1^\varepsilon\|_{H^{s-3}}\|\psi_2^\varepsilon\|_{H^{s-3}}^2 \le C\|\psi^\varepsilon\|_{H^{s-3}}^2 \tag{4.31}$$
$$I_9 \le C\|\bar{u}^\varepsilon\|_{H^{s-3}}\|\psi_2^\varepsilon\|_{H^{s-3}}^2 \le C\|\psi^\varepsilon\|_{H^{s-3}}^2 \tag{4.32}$$
$$I_{10} \le C\|w^\varepsilon\|_{H^{s-2}}\|\psi_2^\varepsilon\|_{H^{s-3}}^2 \le C\|\psi^\varepsilon\|_{H^{s-3}}^2 \tag{4.33}$$
$$I_{11} \le C\|\bar{u}^\varepsilon\|_{H^{s-2}}\|\psi_2^\varepsilon\|_{H^{s-3}}^2 \le C\|\psi^\varepsilon\|_{H^{s-3}}^2. \tag{4.34}$$

The remaining terms I_{13} and I_{14} are treated in the same way, it is sufficien to consider I_{14}. Set $\chi \in \mathcal{D}(\mathbb{R})$ with value 1 near the origin. For $R > 0$ we get

$$w_R^\varepsilon := \mathcal{F}^{-1}(\chi(|\xi|/R)\mathcal{F}(w^\varepsilon)), \quad \tilde{w}_R^\varepsilon := w^\varepsilon - w_R^\varepsilon.$$

We have

$$I_{14} \le J + \tilde{J},$$

where

$$J := |\langle w_R^\varepsilon.\nabla\bar{u}^\varepsilon; \psi_1^\varepsilon\rangle_{H^{s-3}}|, \ \tilde{J} := |\langle \tilde{w}_R^\varepsilon.\nabla\bar{u}^\varepsilon; \psi_1^\varepsilon\rangle_{H^{s-3}}|$$

and these two terms can be bounded as follows.

To bound J, we write

$$\begin{aligned} J &= \Big| \sum_{|\alpha|\le s-3} \int \partial^\alpha (w_R^\varepsilon . \nabla \bar{u}^\varepsilon) \partial^\alpha \psi_1^\varepsilon \Big| \\ &= \Big| \sum_{|\alpha|\le s-3} \sum_{\beta\le\alpha} C_\alpha^\beta \int \partial^\beta w_R^\varepsilon . \nabla \partial^{\alpha-\beta} \bar{u}^\varepsilon \partial^\alpha \psi_1^\varepsilon \Big| \\ &\le \sum_{|\alpha|\le s-3} \sum_{\beta\le\alpha} C_\alpha^\beta \Big| \int \partial^\beta w_R^\varepsilon . \nabla \partial^{\alpha-\beta} \bar{u}^\varepsilon \partial^\alpha \psi_1^\varepsilon \Big|. \end{aligned}$$

For any α, β we have

$$\begin{aligned} \Big| \int \partial^\beta w_R^\varepsilon . \nabla \partial^{\alpha-\beta} \bar{u}^\varepsilon \partial^\alpha \psi_1^\varepsilon \Big| &= \Big| \int \mathcal{F}(\partial^\beta w_R^\varepsilon) \chi(\frac{|\xi|}{2R}) \overline{\mathcal{F}(\nabla \partial^{\alpha-\beta} \bar{u}^\varepsilon \partial^\alpha \psi_1^\varepsilon)} \Big| \\ &= \Big| \int (\partial^\beta w_R^\varepsilon) \mathcal{F}^{-1} \Big(\chi(\frac{|\xi|}{2R}) \overline{\mathcal{F}(\nabla \partial^{\alpha-\beta} \bar{u}^\varepsilon \partial^\alpha \psi_1^\varepsilon)} \Big) \Big| \\ &\le \|\partial^\beta w_R^\varepsilon\|_{L^\infty} \|\mathcal{F}^{-1} \Big(\chi\Big(\frac{|\xi|}{2R}\Big) \Big) * \mathcal{F}^{-1} \big(\overline{\mathcal{F}(\nabla \partial^{\alpha-\beta} \bar{u}^\varepsilon \partial^\alpha \psi_1^\varepsilon)} \big) \|_{L^1} \end{aligned}$$

By the Young inequality and the imbedding

$$C^{\frac{1}{4}} \hookrightarrow L^\infty \tag{4.35}$$

we have

$$\begin{aligned} \Big| \int \partial^\beta w_R^\varepsilon . \nabla \partial^{\alpha-\beta} \bar{u}^\varepsilon \partial^\alpha \psi_1^\varepsilon \Big| &\le C \|\partial^\beta w_R^\varepsilon\|_{C^{s-3+\frac{1}{4}}} \|\mathcal{F}^{-1} \Big(\chi\Big(\frac{|\xi|}{2R}\Big) \Big) \|_{L^1} \|\nabla \partial^{\alpha-\beta} \bar{u}^\varepsilon \partial^\alpha \psi_1^\varepsilon \|_{L^2} \\ &\le C_R \|w_R^\varepsilon\|_{C^{s-3+\frac{1}{4}}} \|\nabla \partial^{\alpha-\beta} \bar{u}^\varepsilon\|_{L^\infty} \|\partial^\alpha \psi_1^\varepsilon\|_{L^2} \end{aligned}$$

By (4.35) again,

$$\Big| \int \partial^\beta w_R^\varepsilon . \nabla \partial^{\alpha-\beta} \bar{u}^\varepsilon \partial^\alpha \psi_1^\varepsilon \Big| \le C_R \|w_R^\varepsilon\|_{C^{s-3+\frac{1}{4}}} \|\bar{u}^\varepsilon\|_{C^{s-2+\frac{1}{4}}} \|\psi_1^\varepsilon\|_{H^{s-3}}.$$

The Sobolev injection $H^\sigma(\mathbb{R}^2) \hookrightarrow C^{\sigma-1}(\mathbb{R}^2)$ gives

$$\begin{aligned} \Big| \int \partial^\beta w_R^\varepsilon . \nabla \partial^{\alpha-\beta} \bar{u}^\varepsilon \partial^\alpha \psi_1^\varepsilon \Big| &\le C_R \|w_R^\varepsilon\|_{C^{s-3+\frac{1}{4}}} \|\bar{u}^\varepsilon\|_{H^{s-\frac{3}{2}+\frac{1}{4}}} \|\psi_1^\varepsilon\|_{H^{s-3}} \\ &\le C_R \|w_R^\varepsilon\|_{C^{s-3+\frac{1}{4}}} \|\psi_1^\varepsilon\|_{H^{s-3}} \\ &\le C_R \|w_R^\varepsilon\|^2_{C^{s-3+\frac{1}{4}}} + \|\psi_1^\varepsilon\|^2_{H^{s-3}}. \end{aligned} \tag{4.36}$$

To bound $\tilde{J}$, we write

$$\begin{aligned} \tilde{J} &= |\langle \tilde{w}_R^\varepsilon . \nabla \bar{u}^\varepsilon ; \psi_1^\varepsilon \rangle_{H^{s-3}}| \\ &= \Big| \sum_{|\alpha|\le s-3} \int \partial^\alpha (\tilde{w}_R^\varepsilon . \nabla \bar{u}^\varepsilon) \partial^\alpha \psi_1^\varepsilon \Big| \\ &\le \sum_{|\alpha|\le s-3} \sum_{\beta\le\alpha} C_\alpha^\beta \Big| \int \partial^\beta \tilde{w}_R^\varepsilon . \nabla \partial^{\alpha-\beta} \bar{u}^\varepsilon \partial^\alpha \psi_1^\varepsilon \Big| \end{aligned}$$

$$
\begin{aligned}
&\le \sum_{|\alpha|\le s-3}\sum_{\beta\le\alpha} C_\alpha^\beta \|\nabla\partial^{\alpha-\beta}\bar{u}^\varepsilon\|_{L^\infty}\|\partial^\beta\tilde{w}_R^\varepsilon\|_{L^2}\|\partial^\alpha\psi_1^\varepsilon\|_{L^2}\\
&\le C\|\tilde{w}_R^\varepsilon\|_{H^{s-3}}\|\psi_1^\varepsilon\|_{H^{s-3}}\sum_{|\alpha|\le s-3}\sum_{\beta\le\alpha} C_\alpha^\beta\|\nabla\partial^{\alpha-\beta}\bar{u}^\varepsilon\|_{L^\infty}\\
&\le C\|\tilde{w}_R^\varepsilon\|_{H^{s-3}}\|\psi_1^\varepsilon\|_{H^{s-3}}\|\bar{u}^\varepsilon\|_{H^{s-2+\frac14}}\\
&\le C\|\tilde{w}_R^\varepsilon\|_{H^{s-3}}\|\psi_1^\varepsilon\|_{H^{s-3}}.
\end{aligned}
$$

Noticing that

$$\mathcal{F}(\tilde{w}_R^\varepsilon) \equiv 0 \quad \text{in} \quad B(0,R),$$

we deduce that

$$
\begin{aligned}
\tilde{J} &\le R^{-3}\|w^\varepsilon\|_{H^s}\|\psi_1^\varepsilon\|_{H^{s-3}}\\
&\le C.R^{-3}\|\psi_1^\varepsilon\|_{H^{s-3}}\\
&\le C.R^{-2}\|\psi_1^\varepsilon\|_{H^{s-3}}\\
&\le C.R^{-4}+\|\psi_1^\varepsilon\|_{H^{s-3}}^2.
\end{aligned}
$$

Putting together these estimates, we have

$$\|\psi^\varepsilon(t)\|_{H^{s-3}}^2 \le C.R^{-4} + C_R\int_0^{T_1}\|w_R^\varepsilon(\tau)\|_{C^{s-3+\frac14}}^2 d\tau + C\int_0^t \|\psi^\varepsilon(\tau)\|_{H^{s-3}}^2 d\tau.$$

By Gronwall's lemma

$$\|\psi^\varepsilon(t)\|_{H^{s-3}}^2 \le \Big(C.R^{-4} + C_R\int_0^{T_1}\|w_R^\varepsilon(\tau)\|_{C^{s-\frac{11}{4}}}^2 d\tau\Big)\exp(CT_1).$$

Proof is finishe if we show that

$$w_R^\varepsilon \to 0 \quad \text{in} \quad L^4([0,T_1], C^{s-\frac{11}{4}}).$$

It is exactly the firs step of the proof of Theorem 1.5: From the firs equation of $(\mathbf{MHD3D}^\varepsilon)$, w_R^ε satisfie

$$\partial_t w_R^\varepsilon + \frac{1}{\varepsilon}\mathbb{P}(w_R^\varepsilon \times e_3) = f_R^\varepsilon,$$

where

$$f_R^\varepsilon := \varepsilon\Delta w_R^\varepsilon - \sqrt{\varepsilon}\chi(|D|/R)\partial_3 B^\varepsilon + \chi(|D|/R)\mathbb{P}(B^\varepsilon.\nabla B^\varepsilon)$$

using energy estimate (1.3) and we follow closely the proof of Theorem 1.5. □

5. Appendix

In this section we give the proofs of Lemmas 2.1, 4.2 and 4.3.

5.1. Proof of Lemma 2.1

We suppose that a, b, c are three vectors field in $H^\infty(\mathbb{R}^3)$ such that div $a = 0$.

To prove (2.5) we write

$$\begin{aligned}
|\langle a.\nabla b, b\rangle_{H^\sigma(\mathbb{R}^3)}| &= \left|\sum_{|\alpha|\le\sigma}\int \partial^\alpha(a.\nabla b)\partial^\alpha b\right|,\\
&\le \sum_{|\alpha|\le\sigma}\left|\int \partial^\alpha(a.\nabla b)\partial^\alpha b\right|\\
&\le \sum_{|\alpha|\le\sigma}\sum_{\beta\le\alpha} C_\alpha^\beta A_{\alpha,\beta},
\end{aligned}$$

where

$$A_{\alpha,\beta} := \left|\int (\partial^\beta a.\nabla\partial^{\alpha-\beta} b)\partial^\alpha b\right|.$$

The most important term is for $|\alpha| = \sigma$.

- As div $a = 0$ then $A_{\alpha,\alpha} = 0$.
- For $0 < |\beta| < \sigma$, we have

$$\begin{aligned}
A_{\alpha,\beta} &\le \int |\partial^{\alpha-\beta} a||\nabla\partial^\beta b||\partial^\alpha b|\\
&\le \|\partial^{\alpha-\beta} a\|_{L^\infty}\|\nabla\partial^\beta b\|_{L^2}\|\partial^\alpha b\|_{L^2}\\
&\le \|\partial^{\alpha-\beta} a\|_{C^{\frac{1}{4}}}\|\nabla b\|^2_{H^{\sigma-1}}\\
&\le \|\nabla a\|_{C^{\sigma-\frac{11}{4}}}\|\nabla b\|^2_{H^{\sigma-1}}\\
&\le C\|\nabla a\|_{H^{\sigma-1}}\|\nabla b\|^2_{H^{\sigma-1}}.
\end{aligned}$$

- For $\beta = 0$,

$$\begin{aligned}
A_{\alpha,0} &\le \int |\partial^\alpha a||\nabla b||\partial^\alpha b|\\
&\le \|\partial^\alpha a\|_{L^2}\|\nabla b\|_{L^\infty}\|\partial^\alpha b\|_{L^2}\\
&\le \|\nabla a\|_{H^{\sigma-1}}\|\nabla b\|_{C^{\frac{1}{4}}}\|\nabla b\|_{H^{\sigma-1}}\\
&\le C\|\nabla a\|_{H^{\sigma-1}}\|\nabla b\|_{H^{\frac{7}{4}}}\|\nabla b\|_{H^{\sigma-1}}\\
&\le C\|\nabla a\|_{H^{\sigma-1}}\|\nabla b\|^2_{H^{\sigma-1}}.
\end{aligned}$$

We turn now to (2.6). As above, we write

$$\begin{aligned}
|\langle a\nabla b, c\rangle_{H^\sigma} + \langle a\nabla c, b\rangle_{H^\sigma}| &= \sum_{\substack{\alpha\in\mathbb{N}^3\\|\alpha|\le\sigma}}\left(\int \partial^\alpha(a\nabla b)\partial^\alpha c + \int \partial^\alpha(a\nabla c)\partial^\alpha b\right)\\
&\le \sum_{\substack{\alpha\in\mathbb{N}^3\\|\alpha|\le\sigma}}\sum_{\beta\le\alpha} C_\alpha^\beta B_{\alpha,\beta},
\end{aligned}$$

where

$$B_{\alpha,\beta} := \left|\int (\partial^{\alpha-\beta} a.\nabla\partial^\beta b)\partial^\alpha c + \int (\partial^{\alpha-\beta} a.\nabla\partial^\beta c)\partial^\alpha b\right|.$$

The most important term is for $|\alpha| = \sigma$.

- As div $a = 0$, then $B_{\alpha,\alpha} = 0$.
- For $\beta \neq \alpha$ we apply the firs step. □

5.2. Proof of Lemma 4.2

We write

$$\begin{aligned}
|\langle a.\nabla b, c\rangle_{H^\sigma(\mathbb{R}^3)}| &= \Big|\sum_{|\alpha|\le\sigma}\int \partial^\alpha(a.\nabla b)\partial^\alpha c\Big| \\
&= \Big|\sum_{|\alpha|\le\sigma}\int \partial^\alpha(a.\nabla b)\partial^\alpha c\Big| \\
&= \Big|\sum_{|\alpha|\le\sigma}\sum_{\beta\le\alpha} C_\alpha^\beta \int (\partial^{\alpha-\beta}a.\nabla\partial^\beta b)\partial^\alpha c\Big| \\
&\le \sum_{|\alpha|\le\sigma}\sum_{\beta\le\alpha} C_\alpha^\beta D_{\alpha,\beta},
\end{aligned}$$

where

$$D_{\alpha,\beta} := |\int (\partial^{\alpha-\beta}a.\nabla\partial^\beta b)\partial^\alpha c|.$$

We deduce from the Cauchy-Schwarz inequality that

$$\begin{aligned}
D_{\alpha,\beta} &\le \|\partial^{\alpha-\beta}a.\nabla\partial^\beta b\|_{L^2}\|\partial^\alpha c\|_{L^2} \\
&\le \|\partial^\alpha c\|_{L^2}\min\big(\|\partial^{\alpha-\beta}a\|_{L^2}\|\nabla\partial^\beta b\|_{L^\infty}, \|\partial^{\alpha-\beta}a\|_{L^\infty}\|\nabla\partial^\beta b\|_{L^2}\big) \\
&\le \|\partial^\alpha c\|_{L^2}\min\big(\|\partial^{\alpha-\beta}a\|_{L^2}\|\nabla\partial^\beta b\|_{C^{\frac14}}, \|\partial^{\alpha-\beta}a\|_{C^{\frac14}}\|\nabla\partial^\beta b\|_{L^2}\big) \\
&\le \|c\|_{H^\sigma}\min\big(\|a\|_{H^\sigma}\|\nabla b\|_{C^{\sigma+\frac14}}, \|a\|_{C^{\sigma+\frac14}}\|\nabla b\|_{H^\sigma}\big).
\end{aligned}$$

This achieves the proof. □

5.3. Proof of Lemma 4.3

We write

$$\begin{aligned}
|\langle \bar{a}.\nabla b, c\rangle_{H^\sigma(\mathbb{R}^3)} + \langle \bar{a}.\nabla c, b\rangle_{H^\sigma(\mathbb{R}^3)}| &= \Big|\sum_{|\alpha|\le\sigma}\int \partial^\alpha(\bar{a}.\nabla b)\partial^\alpha c + \int \partial^\alpha(\bar{a}.\nabla c)\partial^\alpha b\Big| \\
&= \Big|\sum_{|\alpha|\le\sigma}\sum_{\beta\le\alpha} C_\alpha^\beta\Big(\int (\partial^{\alpha-\beta}\bar{a}.\nabla\partial^\beta b)\partial^\alpha c \\
&\quad + \int (\partial^{\alpha-\beta}\bar{a}.\nabla\partial^\beta c)\partial^\alpha b\Big)\Big| \\
&\le \sum_{|\alpha|\le\sigma}\sum_{\beta\le\alpha} C_\alpha^\beta E_{\alpha,\beta},
\end{aligned}$$

where

$$E_{\alpha,\beta} := \Big|\int (\partial^{\alpha-\beta}\bar{a}.\nabla\partial^\beta b)\partial^\alpha c + \int (\partial^{\alpha-\beta}\bar{a}.\nabla\partial^\beta c)\partial^\alpha b\Big|.$$

The most important term is for $|\alpha| = \sigma$.

- As $\mathrm{div}_h\, \bar{a} = 0$, then $E_{\alpha,\alpha} = 0$.

For $\beta \neq \alpha$ we have $E_{\alpha,\beta} \leq E^{(1)}_{\alpha,\beta} + E^{(2)}_{\alpha,\beta}$,

where $E^{(1)}_{\alpha,\beta} = \left| \int (\partial^{\alpha-\beta}\bar{a}.\nabla\partial^\beta b)\partial^\alpha c \right|$ and $E^{(2)}_{\alpha,\beta} = \left| \int (\partial^{\alpha-\beta}\bar{a}.\nabla\partial^\beta c)\partial^\alpha b \right|$.

- $0 \leq |\beta| \leq \sigma - 2$. In this case we have

$$E^{(1)}_{\alpha,\beta} \leq \|\partial^\beta \bar{a}\|_{L^\infty(\mathbb{R}^2)} \|\nabla\partial^{\alpha-\beta} b\|_{L^2(\mathbb{R}^3)} \|\partial^\alpha c\|_{L^2(\mathbb{R}^3)}.$$

Using the injections

$$H^{1+\frac{1}{4}}(\mathbb{R}^2) \hookrightarrow C^{\frac{1}{4}}(\mathbb{R}^2) \hookrightarrow L^\infty(\mathbb{R}^2)$$

we deduce that

$$E^{(1)}_{\alpha,\beta} \leq C\|\bar{a}\|_{H^\sigma(\mathbb{R}^2)} \|\nabla b\|_{H^{\sigma-1}(\mathbb{R}^3)} \|\nabla c\|_{H^{\sigma-1}(\mathbb{R}^3)}.$$

- $\sigma - 1 \leq |\beta| \leq \sigma$. If we denote $x = (x_1, x_2, x_3) := (x_h, x_3)$, then

$$\begin{aligned} E^{(1)}_{\alpha,\beta} &\leq \int_{x_3} \Big(\int_{x_h} |\partial^\beta \bar{a}(x_h)| |\nabla\partial^{\alpha-\beta} b(x_h, x_3)| |\partial^\alpha c(x_h, x_3)| \Big) \\ &\leq \|\partial^\beta \bar{a}\|_{L^2(\mathbb{R}^2)} \Big(\int_{x_3} \|\nabla\partial^{\alpha-\beta} b(x_3)\|^2_{L^\infty(\mathbb{R}^2)} dx_3 \Big)^{\frac{1}{2}} \Big(\int_{x_3} \|\partial^\alpha c(x_3)\|^2_{L^2(\mathbb{R}^2)} dx_3 \Big)^{\frac{1}{2}}. \end{aligned}$$

But

$$\begin{aligned} \|\nabla\partial^{\alpha-\beta} b(x_3)\|_{L^\infty(\mathbb{R}^2)} &\leq C\|\nabla\partial^{\alpha-\beta} b(x_3)\|_{H^{\frac{3}{2}}(\mathbb{R}^2)} \\ &\leq C\|\nabla\partial^{\alpha-\beta} b(x_3)\|_{H^2(\mathbb{R}^2)}, \end{aligned}$$

$$\Big(\int_{x_3} \|\partial^\alpha c(x_3)\|^2_{L^2(\mathbb{R}^2)} dx_3 \Big)^{\frac{1}{2}} = \|\partial^\alpha c\|_{L^2(\mathbb{R}^3)},$$

and

$$\Big(\int_{x_3} \|\nabla\partial^{\alpha-\beta} b(x_3)\|^2_{H^2(\mathbb{R}^2)} \Big)^{\frac{1}{2}} \leq \|\nabla\partial^{\alpha-\beta} b\|_{H^2(\mathbb{R}^3)}$$

so

$$E_{\alpha,\beta} \leq \|\bar{a}\|_{H^\sigma(\mathbb{R}^2)} \|\nabla b\|_{H^{\sigma-1}(\mathbb{R}^3)} \|c\|_{H^\sigma(\mathbb{R}^3)}.$$

For the second estimate we remark that, if $c = b$,

$$\mathrm{div}_h\, \bar{a} = 0 \quad \text{imply} \quad E_{\alpha,\alpha} = 0 \quad \text{for all} \quad |\alpha| \leq \sigma.$$

For the third point we write

$$|\langle b.\nabla\bar{a}, c\rangle_{H^\sigma(\mathbb{R}^3)}| = \Big| \sum_{|\alpha|\leq\sigma} \int \partial^\alpha (b.\nabla\bar{a}) \partial^\alpha c \Big|,$$

then

$$|\langle b.\nabla\bar{a}, c\rangle_{H^\sigma(\mathbb{R}^3)}| \leq \sum_{|\alpha|\leq\sigma} \sum_{\beta\leq\alpha} C^\beta_\alpha F_{\alpha,\beta},$$

where

$$F_{\alpha,\beta} := \left| \int (\partial^\beta b.\nabla \partial^{\alpha-\beta}\overline{a})\partial^\alpha c \right|.$$

The most important case is when $|\alpha| = \sigma$.

• For $3 \leq |\beta| \leq \sigma$, we can write

$$F_{\alpha,\beta} \leq \|\nabla \partial^{\alpha-\beta}\overline{a}\|_{L^\infty(\mathbb{R}^2)} \|\partial^\beta b\|_{L^2(\mathbb{R}^3)} \|\partial^\alpha c\|_{L^2(\mathbb{R}^3)}.$$

But

$$\|\nabla \partial^{\alpha-\beta}\overline{a}\|_{L^\infty(\mathbb{R}^2)} \leq C\|\overline{a}\|_{H^\sigma(\mathbb{R}^2)}.$$

So

$$F_{\alpha,\beta} \leq C\|\overline{a}\|_{H^\sigma(\mathbb{R}^2)} \|b\|_{H^\sigma(\mathbb{R}^3)} \|c\|_{H^\sigma(\mathbb{R}^3)}.$$

• For $|\beta| \leq 2$, we have by the Cauchy-Schwarz inequality

$$\begin{aligned} F_{\alpha,\beta} &\leq \int_{x_3} \Big(\int_{x_h} |\nabla \partial^{\alpha-\beta}\overline{a}(x_h)||\partial^\beta b(x_h, x_3)||\partial^\alpha c(x_h, x_3)| \Big) \\ &\leq \|\nabla \partial^{\alpha-\beta}\overline{a}\|_{L^2(\mathbb{R}^2)} \Big(\int_{x_3} \|\partial^\beta b(x_3)\|^2_{L^\infty(\mathbb{R}^2)} dx_3 \Big)^{\frac{1}{2}} \Big(\int_{x_3} \|\partial^\alpha c(x_3)\|^2_{L^2(\mathbb{R}^2)} dx_3 \Big)^{\frac{1}{2}} \\ &\leq C\|\overline{a}\|_{H^{\sigma+1}(\mathbb{R}^2)} \|b\|_{H^\sigma(\mathbb{R}^3)} \|c\|_{H^\sigma(\mathbb{R}^3)}. \end{aligned}$$

This end of the proof of lemma 4.3. □

References

[1] A. Babin, A. Mahalov, B. Nicolaenko, Global Spiliting, Integrability and Regularity of 3D Euler and Navier-Stokes Equations for Uniformly Rotating Fluids, *European Journal of Mechanics,* **15** (1996), 3, 291-300.

[2] J. Benameur, M. Ghazel, M. Majdoub, About MHD System With Small Parameter, *Asymptotic Analysis,* **41** (2005), 653, 1-21.

[3] J. Benameur, S. Ibrahim, M. Majdoub, Asymptotic Study of Magneto-Hydro-Dynamic System, *Differential and Integral Equations,* Volume 18, Number 3, pages 299-324, 2004.

[4] J.-Y. Chemin, *About Navier-Stokes Equations*, Prépublication du Laboratoire d'Analyse Numérique de Paris VI, 1996.

[5] J.-Y. Chemin, B. Desjardins, I. Gallagher, E. Grenier, Anisotropy and dispersion in rotating fluids *Nonlinear Partial Differential Equations and their applications,* Collège de France Seminar, Vol XIV, Studies in Mathematics and its Applications, Vol. 31, 171-192, 2002.

[6] B. Desjardins, E. Dormy, E. Grenier, Stability of mixed Ekman-Hartmann boundary layers, *Nonlinearity* **12** (1999), 181-199.

[7] I. Gallagher, Applications of Schochet's Methods to Parabolic Equation, *Journal de Mathématiques Pures et Appliquées* **77** (1998), 989-1054.

[8] I. Gallagher, Asymptotics of the Solutions of Hyperbolic Equations With a Skew-Symmetric Perturbation , *Journal of Differential Equations,* **150** (1998), 363-384.

[9] E. Grenier, Oscillatory Perturbations of the Navier-Stokes Equations, *Journal de Mathématiques Pures et Appliquées* **76** (1997), 477-498.

[10] J. Joly, G.Métivier, J. Rauch Generic rigourous Asymptotic Expansions for Weakly Nonlinear Multidimensional Oscillatory Waves, *Duke Mathematical Journal,* **70** (1993), 373-404.

[11] J. Joly, G.Métivier, J. Rauch, Coherent and Focusing Multidimensional Oscillatory Nonlinear Geometric optics, *Annales Scientifiques de L'ENS,* **28** (1995), 51-113

[12] S. Klainerman, A. Majda, Singular Limits of Quasilinear Hyperbolic system with Large Parameters, and the Incompressible Limit of Compressible Fluids, *Communications on pure and applied Mathematics,* **34** (1981), 481-524.

[13] A. Majda, Compressible Fluid Flow and System of Conservation Laws in Several Space Variables, *Applied Mathematical Sciences,* **53**, Springer Verlag, New-York, 1984.

[14] S. Schochet, The Compressible Euler Equations in a Bounded Domain: Existence of Solutions and the Incompressible Limit, *Comm. Math. Physics,* **104** (1986), 49-75.

[15] S. Schochet, Fast Singular Limits of Hyperbolic PDEs, *Journal of Differential Equations,* **114** (1994), 476-512.

[16] J.-L. Lions, *Quelques méthodes de resolution des problèmes aux limites non linéaires* , Dunod.

[17] R. Temam, *Navier-Stokes equations*, North-Holland.

In: Evolution Equations: Theories, Solutions and Functions ISBN 978-1-61324-090-8
Editor: Gaston M. N'Guerekata

Chapter 7

NONLINEAR SPECTRAL THEORY AND CONTROLLABILITY OF SEMILINEAR EVOLUTION EQUATIONS

J. Appell[1,*]***, H. Leiva***[2] ***and N. Merentes***[3]
[1]Universität Würzburg, Institut für Mathematik,
D-97074 Würzburg, Germany
[2]Universidad de los Andes, Facultad de Ciencias,
Departamento de Matemática, Mérida 5101, Venezuela
[3]Universidad Central de Venezuela, Facultad de Ciencias,
Departamento de Matemática, Caracas 1020, Venezuela

Abstract

In this paper we introduce and apply a new nonlinear spectrum to characterize the exact controllability for a semilinear evolution equation. To this end, we firs recall a necessary and sufficien condition for the exact controllability of the corresponding linear system and prove that, under suitable hypotheses on the nonlinear right-hand side, the exact controllability of the linear system carries over to the perturbated semilinear system. Afterwards we apply this result to a nonlinear damped wave equation.

2000 Mathematics Subject Classification: 47J35, 93B05, 93C25, 47J10, 35J40, 35L35.

Keywords and phrases: Nonlinear spectral theory, Semilinear volution equation, Exact controllability.

1. Preliminaries

In view of the importance of spectral theory in both mathematics and physics, it is not surprising at all that various attempts have been made to introduce and study spectra also for some classes of nonlinear operators. As far as we know, the firs definition of nonlinear spectra seem to be Kachurovskij's spectrum [7] for Lipschitz continuous operators

$$\sigma_K(F) = \{\lambda \in \mathbb{K} : \lambda I - F \text{ is not a Lipschitz homeomorphism}\}, \tag{1}$$

*E-mail address: jurgen@dmuw.de

where $\mathbb{K} = \mathbb{R}$ or $\mathbb{K} = \mathbb{C}$, I denotes the identity operator, and $F : X \to X$ is a Lipschitz continuous nonlinear operator in a Banach space X, and Neuberger's spectrum [11] for Fréchet differentiable operators

$$\sigma_N(F) = \{\lambda \in \mathbb{K} : \lambda I - F \text{ is not a diffeomorphism}\}. \tag{2}$$

Both spectra reduce to the familiar spectrum in case of a linear operator F and have some pleasant properties in common with the linear spectrum. For instance, in [10] it is shown that the Kachurovskij spectrum (1) is always compact; however, it may be empty even in finit dimensional complex Banach spaces: For example, the map $F(z, w) = (\overline{w}, i\overline{z})$ in $X = \mathbb{C}^2$ has the property that $\lambda I - F$ is a Lipschitz homeomorphism for each $\lambda \in \mathbb{C}$ with Lipschitz continuous inverse

$$(\lambda I - F)^{-1}(\zeta, \omega) = \left(\frac{\overline{\lambda}\zeta + \overline{\omega}}{i + |\lambda|^2}, \frac{\overline{\lambda}\omega + i\overline{\zeta}}{i - |\lambda|^2}\right) \qquad ((\zeta, \omega) \in \mathbb{C}^2),$$

and so $\sigma_K(F) = \emptyset$. On the other hand, the Neuberger spectrum (2) may be unbounded or not closed, but is always nonempty in the complex case. So it is not accidental that the map $F(z, w) = (\overline{w}, i\overline{z})$ is not differentiable at any point $(z, w) \in \mathbb{C}^2$.

Subsequently, many other nonlinear spectra have been introduced in the literature which are more difficul to describe, but have nice properties and admit surprisingly many applications to nonlinear problems. Here we mention the asymptotic spectrum by Furi-Martelli-Vignoli [6], the global spectrum by Feng [5], and the local spectrum by Väth [12]. A detailed account of the properties and applications of these spectra may be found in the monograph [2] or the recent survey article [1].

Given two Hilbert spaces W and Z, a surjective bounded linear operator $L : W \to Z$, and a Lipschitz continuous nonlinear operator $H : W \to Z$, in this paper we propose a new spectrum $\sigma_L(H)$ consisting of all scalars $\lambda \in \mathbb{K}$ for which $\lambda L - H$ is not surjective. This spectrum is applied to characterize the exact controllability of the semilinear evolution equation

$$z'(t) = Az(t) + Bu(t) - f(t, z(t), u(t)) \qquad (t > 0, \;\; z(t) \in Z, \;\; u(t) \in U), \tag{3}$$

where U is another Hilbert space, $A : Z \supseteq \mathcal{D}(A) \to Z$ is the infinitesima generator of a strongly continuous semigroup $\{T(t) : t \geq 0\}$ in Z, $B : U \to Z$ is bounded and linear, the control function u belongs to $L^2(0, \tau; U)$, and $f : [0, \tau] \times Z \times U \to Z$ is a suitable nonlinearity. First, we give a necessary and sufficien condition for the exact controllability of the corresponding unperturbed linear system

$$z'(t) = Az(t) + Bu(t) \qquad (t > 0, \;\; z(t) \in Z, \;\; u(t) \in U). \tag{4}$$

Afterwards, under some suitable conditions on f, we prove that the exact controllability of the linear system (4) carries over to the semilinear system (3). The relation with the above mentioned spectrum consists then in the fact that the system (3) is exactly controllable if, and only if, $1 \notin \sigma_L(H)$, where $L : L^2(0, \tau; U) \to Z$ is the linear operator define by

$$Lu = \int_0^\tau T(\tau - s)Bu(s)\, ds, \tag{5}$$

while $H : L^2(0,\tau;U) \to Z$ is the nonlinear operator define by

$$Hu = \int_0^\tau T(\tau - s)f(s, z(s), u(s))\, ds. \tag{6}$$

Now, since the operator $LL^* : Z \to Z$ is always selfadjoint and positive, where $L^* : Z \to L^2(0,\tau;U)$ denotes the Hilbert adjoint of L, the operator $I - F$ is invertible on its range, where we have used the shortcut

$$F = HL^{\#}, \quad L^{\#} = L^*(LL^*)^{-1}. \tag{7}$$

To guarantee the spectral condition $1 \notin \sigma_L(H)$, it suffice in turn to fin upper estimates for the spectral radius of the operator $L - H$ in terms of the norm of the linear operator $L^{\#}$ in (7) and the minimal Lipschitz constant of the nonlinear operator H in (6). Moreover, we will show that, for exactly controllable systems (3), the control u steering an initial state z_0 to a fina state z_1 at time $\tau > 0$ is given by

$$u(t) = B^*T^*(\tau - t)(LL^*)^{-1}(I - F)^{-1}(z_1 - T(\tau)z_0) \qquad (0 \leq t \leq \tau). \tag{8}$$

Finally, in the last section we show how to apply our abstract results to a nonlinear controlled damped wave equation in N spatial dimensions.

2. A Nonlinear Defect Spectrum

Given two Hilbert spaces W and Z and a linear operator A between W and Z, throughout the paper we denote by $\mathcal{D}(A)$ the domain of definitio of A, by $\mathcal{N}(A) = \{w \in W : Aw = 0\}$ its nullspace (kernel), and by $\mathcal{R}(A) = \{Aw : w \in W\}$ its range.

If $L : W \to Z$ is bounded and linear with $\mathcal{R}(L) = Z$, and $H : W \to Z$ is Lipschitz continuous and (in general) nonlinear, we call the set

$$\sigma_L(H) = \{\lambda \in \mathbb{K} : \mathcal{R}(\lambda L - H) \neq Z\} \tag{9}$$

the *L-defect spectrum* of H. Of course, in case $W = Z$ and $L = I$ the spectrum $\sigma_I(H)$ contains the Kachurowski spectrum (1) of H. If, in addition, H is also linear, then $\sigma_I(H)$ reduces to the familiar *defect spectrum*

$$\sigma_d(L) = \{\lambda \in \mathbb{K} : \lambda I - L \text{ is not surjective}\} \tag{10}$$

of a bounded linear operator L which, together with the point spectrum (eigenvalues) $\sigma_p(L)$ constitutes the whole linear spectrum $\sigma(L)$. The above example in $W = Z = \mathbb{C}^2$ shows that the spectrum $\sigma_L(H)$ may be empty even in case $L = I$. The following elementary example shows how to calculate the spectrum $\sigma_L(H)$ in the case when $L \neq I$ but $LL^* = I$, and H is nonlinear, Lipschitz continuous, and compact.

Example 1. Let $W = Z = \ell^2(\mathbb{C})$, $L(\xi_1, \xi_2, \xi_3, \ldots) = (\xi_2, \xi_3, \xi_4, \ldots)$ the linear left shift operator, and $H(x) = \|x\|e_1 = (\|x\|, 0, 0, \ldots)$. Obviously, L is surjective and H is Lipschitz continuous. We claim that $\sigma_L(H) = \{\lambda \in \mathbb{C} : |\lambda| \leq 1\}$.

To see this, let firs $|\lambda| > 1$ and $y = (\eta_n)_n \in \ell^2(\mathbb{C})$. Then a straightforward calculation shows that the element $x = (\xi_n)_n$ define by

$$\xi_1 = 0, \quad \xi_2 = \frac{1}{\lambda}(\eta_1 + \|x\|), \quad \xi_3 = \frac{1}{\lambda}\eta_2, \quad \xi_4 = \frac{1}{\lambda}\eta_3, \ \ldots$$

belongs to $\ell^2(\mathbb{C})$ and satisfie $\lambda Lx - Hx = y$, so $\lambda L - H$ is onto and $\lambda \notin \sigma_L(H)$.

On the other hand, let $0 < |\lambda| \leq 1$ and suppose that there exists $x = (\xi_n)_n \in \ell^2(\mathbb{C})$ such that $\lambda Lx - Hx = e_1 = (1, 0, 0, \ldots)$. Then x must have the form $x = (\xi_1, \frac{1}{\lambda}(1 + \|x\|), 0, 0, \ldots)$, so taking norms we obtain

$$\|x\|^2 = |\xi_1|^2 + \frac{1}{|\lambda|^2} + \frac{2}{|\lambda|^2}\|x\| + \frac{\|x\|^2}{|\lambda|^2},$$

hence

$$1 \leq |\lambda|^2|\xi_1|^2 + 1 + 2\|x\| = (|\lambda|^2 - 1)\|x\|^2 \leq 0,$$

a contradiction. Finally, for $\lambda = 0$ the operator $\lambda L - H = -H$ is trivially not surjective. So we conclude that $\lambda \in \sigma_L(H)$ for $|\lambda| \leq 1$ as claimed. □

The following proposition shows that the spectrum (9) shares an important property with the familiar spectrum of a bounded linear operator.

Proposition 1. *The spectrum $\sigma_L(H)$ is a bounded subset of the real line (resp. the complex plane). More precisely, every $\lambda \in \sigma_L(H)$ satisfies the estimate*

$$|\lambda| \leq [H]_{Lip}\|L^{\#}\|, \tag{11}$$

where $L^{\#} : Z \to W$ is defined as in (7), and

$$[H]_{Lip} = \sup_{u \neq v} \frac{\|Hu - Hv\|}{\|u - v\|} \tag{12}$$

denotes the minimal Lipschitz constant of H.

Proof. Suppose that $|\lambda| > [H]_{Lip}\|L^{\#}\|$, and consider the operator $(\lambda L - H)L^{\#} = \lambda I - F$, with F given by (7). From

$$\left[\frac{F}{\lambda}\right]_{Lip} \leq \frac{1}{|\lambda|}[F]_{Lip} \leq \frac{1}{|\lambda|}[H]_{Lip}\|L^{\#}\| < 1$$

we conclude that the operator $I - F/\lambda$ is a Lipschitz homeomorphism (see [2, Prop. 5.1]), and so also $\lambda I - F$. But this implies that $\lambda L - H$ is surjective, and so $\lambda \notin \sigma_L(H)$. Indeed, given any $z \in Z$ we may choose $w = L^{\#}(\lambda I - F)^{-1}z$ and get $(\lambda L - H)w = z$. □

Observe that in Example 1 we have $LL^* = I$, hence $\|L^{\#}\| = \|L^*\| = 1$ and $[H]_{Lip} = 1$, so the estimate (11) for the spectral radius in Proposition 1 is sharp. In case $L = I$ the estimate (11) reduces to $|\lambda| \leq [H]_{Lip}$ for $\lambda \in \sigma_K(H)$, which is precisely the upper estimate for the Kachurovskij spectral radius proved in [10].

Recall that, as a consequence of the open mapping theorem, the defect spectrum (10) of a bounded linear operator L is always closed. We do not know whether or not the spectrum $\sigma_L(H)$ is also closed. In general, surjectivity of nonlinear operators is not stable with respect to small perturbations. For example, the map $F : \mathbb{C} \to \mathbb{C}$ define by $F(z) = \min\{|z|, 1\}e^z$ is certainly surjective, but for each $\varepsilon > 0$ one can fin another map $G : \mathbb{C} \to \mathbb{C}$ such that $|G(z)| \leq \varepsilon$ for all $z \in \mathbb{C}$ and $F + G$ is *not* surjective (e.g., $G(z) = \max\{\varepsilon(1 - |z|), 0\}$). Of course, this does not answer the question if this can also happen for perturbations of the very special form $G = \varepsilon I$.

3. Controllability of the Linear System

In this section we collect some results on the controllability of the linear system (4) that we will use in the following sections. To this end, we recall that, for all initial states $z_0 \in Z$ and controls $u \in L^2(0, \tau; U)$, the linear initial value problem

$$\begin{cases} z' = Az(t) + Bu(t) & \text{for } t > 0, \\ z(0) = z_0, \end{cases} \tag{13}$$

admits only one mild solution given by

$$z(t) = T(t)z_0 + \int_0^t T(t-s)Bu(s)\,ds \quad (0 \leq t \leq \tau). \tag{14}$$

Definition 1. The linear system (4) is said to be *exactly controllable* on $[0, \tau]$ if for all $z_0, z_1 \in Z$ there exists a control $u \in L^2(0, \tau; U)$ such that the solution $z = z(t)$ of (13) corresponding to u satisfie $z(\tau) = z_1$.

We point out that the Hilbert adjoint $L^* : Z \to L^2(0, \tau; U)$ of the operator (5) may be given explicitly in the form

$$L^*z(t) = B^*T^*(\tau - t)z \qquad (z \in Z,\ 0 \leq t \leq \tau), \tag{15}$$

and so the self-adjoint positive operator $LL^* : Z \to Z$ occurring in the definitio (7) of $L^\#$ is

$$LL^*z = \int_0^\tau T(\tau - s)BB^*T^*(\tau - s)z\,ds \qquad (z \in Z). \tag{16}$$

With this notation, the following theorem gives a precise characterization of the exact controllability of the linear system (4).

Theorem 1. *The linear system* (4) *is exactly controllable on* $[0, \tau]$ *if, and only if, any one of the following five conditions holds for some* $\gamma > 0$ *and all* $z \in Z$.

(a) $\mathcal{R}(L) = Z$;

(b) $\langle LL^*z, z\rangle \geq \gamma\|z\|^2$;

(c) $\|L^*z\|^2 = \int_0^\tau \|L^*z(s)\|^2\,ds \geq \gamma\|z\|^2$;

(d) $\int_0^\tau \|B^*T^*(\tau - s)z\|^2\,ds \geq \gamma\|z\|^2$;

(e) $\mathcal{N}(L^*) = \{0\}$ *and* $\mathcal{R}(L^*)$ *is closed.*

The proof of Theorem 1 is straightforward and can be found, e.g., in [3] or [4]. We point out that the equivalence of (a) and (e) follows immediately from the well-known equality $\sigma_d(L) = \sigma_{ap}(L^*)$ which relates the defect spectrum (10) of L with the *approximate point spectrum*

$$\sigma_{ap}(L^*) = \{\lambda \in \mathbb{K} : \inf_{\|x\|=1} \|\lambda x - L^* x\| = 0\}$$

of the adjoint operator L^*. Theorem 1 enables us to formulate and prove a new result on the exact controllability of the linear system (4).

Theorem 2. *The linear system* (4) *is exactly controllable on* $[0, \tau]$ *if, and only if, the operator* LL^* *is invertible. Moreover, in this case the control* $u \in L^2(0, \tau; U)$ *steering an initial state* z_0 *to a final state* z_1 *at time* $\tau > 0$ *is given by*

$$u(t) = B^* T^*(\tau - t)(LL^*)^{-1}(z_1 - T(\tau)z_0) \qquad (0 \le t \le \tau). \tag{17}$$

Proof. Suppose firs that the system (4) is exactly controllable on $[0, \tau]$. Then, from Theorem 1 (b) we conclude that $\langle LL^* u, u\rangle \ge \gamma\|z\|^2$ for all $z \in Z$, which shows that the operator LL^* is in fact invertible on its range $\mathcal{R}(LL^*)$. We claim that LL^* is also surjective, i.e., $\mathcal{R}(LL^*) = Z$.

Assume that $\mathcal{R}(LL^*)$ is strictly contained in Z. Applying the Cauchy-Schwarz inequality to (b) we get $\|LL^* z\| \ge \gamma\|z\|$, which immediately implies that $\mathcal{R}(LL^*)$ is closed. So we fin a nonzero element in the orthogonal complement $\mathcal{R}(LL^*)^\perp$ of $\mathcal{R}(LL^*)$, i.e., an element $z_0 \in Z \setminus \{0\}$ satisfying $\langle LL^* z, z_0\rangle = 0$ for all $z \in Z$. Choosing, in particular, $z = z_0$ we get $0 = \langle LL^* z_0, z_0\rangle \ge \gamma\|z_0\|^2$, a contradiction. We conclude that LL^* is a bijection, and so a linear isomorphism, by the open mapping theorem.

Conversely, suppose now that LL^* is a bijection. Then, given $z \in Z$ we fin a control $u \in L^2(0, \tau; U)$ such that $Lu = z$. This control u can be chosen in the form

$$u(t) = B^* T^*(\tau - t)(LL^*)^{-1} z;$$

in fact, the definitio (5) implies, together with (16), that

$$\begin{aligned} Lu = \int_0^\tau T(\tau - s)Bu(s)\,ds &= \int_0^\tau T(\tau - s)BB^*T^*(\tau - s)(LL^*)^{-1} z\,ds \\ &= (LL^*)(LL^*)^{-1} z = z. \end{aligned}$$

In the same way we can prove that the control u given by (17) steers the initial state z_0 to the fina state z_1 in time τ. □

Corollary 1. *Suppose that the system* (4) *is exactly controllable on* $[0, \tau]$. *Then the operator* $L^\# : Z \to L^2(0, \tau; U)$ *defined by* (7) *is a right inverse of L, i.e.,* $LL^\# = I$.

4. Controllability of the Nonlinear System

Consider now the nonlinear initial value problem

$$\begin{cases} z' = Az(t) + Bu(t) - f(t, z, u(t)) & \text{for } t > 0, \\ z(0) = z_0. \end{cases} \tag{18}$$

We assume that f is good enough such that the initial value problem (18) with control $u \in L^2(0,\tau,U)$ admits only one mild solution given by

$$z(t) = T(t)z_0 + \int_0^t T(t-s)Bu(s)\,ds - \int_0^t T(t-s)f(s,z(s),u(s))\,ds \quad (0 \le t \le \tau). \tag{19}$$

Definition 2. The nonlinear system (3) is said to be *exactly controllable* on $[0,\tau]$ if for all $z_0, z_1 \in Z$ there exists a control $u \in L^2(0,\tau;U)$ such that the solution $z = z(t)$ of (19) corresponding to u satisfie $z(\tau) = z_1$.

Together with the linear evolution operator (5) we consider now the nonlinear evolution operator (6) generated by the function f. The following trivial criterion for exact controllability of the system (3) is an immediate consequence of Definitio 2.

Proposition 2. *The nonlinear system* (3) *is exactly controllable on* $[0,\tau]$ *if, and only if,* $\mathcal{R}(L-H) = Z$.

In terms of the nonlinear defect spectrum (9) we may rephrase Proposition 2 in the following way: *The system* (3) *is exactly controllable on* $[0,\tau]$ *if, and only if,* $1 \notin \sigma_L(H)$, *with L given by* (5) *and H given by* (6). By Proposition 1, for this in turn a sufficien condition is

$$[F]_{Lip} \le [H]_{Lip}\|L^{\#}\| < 1 \tag{20}$$

with F given by (7), which requires upper estimates (or even explicit formulas) for the minimal Lipschitz constant of the operator (6) and for the norm of the operator (16). In order to get conditions for the exact controllability of the nonlinear system (3), we have to assume that the linear system (4) is exactly controllable. In this case we know from Corollary 1 that the operator $L^{\#}$ define by (7) is a right inverse of L. Now, observing that $(L-H)L^{\#} = I - HL^{\#} = I - F$, we obtain the operator equation for the exact controllability

$$(L-H)L^{\#}\xi = \xi - \int_0^{\tau} T(\tau-s)f(s,z_\xi(s),L^{\#}\xi(s))\,ds, \tag{21}$$

where $z_\xi = z_\xi(t)$ denotes the solution of (18) corresponding to the control function

$$u(t) = L^{\#}\xi(t) = B^*T^*(\tau-t)(LL^*)^{-1}\xi \qquad (0 \le t \le \tau).$$

Now, a comparison of (7) and (21) shows that the operator $F: Z \to Z$ is given by

$$F\xi = \int_0^{\tau} T(\tau-s)f(s,z_\xi(s),L^{\#}\xi(s))\,ds. \tag{22}$$

So, from condition (20) and Proposition 2 we immediately get the following

Theorem 3. *Suppose that the linear system* (4) *is exactly controllable on* $[0,\tau]$ *and the operator F given by* (22) *is globally Lipschitz with a Lipschitz constant* $[F]_{Lip} < 1$. *Then the nonlinear system* (3) *is exactly controllable on* $[0,\tau]$ *and the control steering the initial state* z_0 *to the final state* z_1 *is given by* (8).

Of course, Theorem 3 is only of limited use, inasmuch as the crucial condition (20) has to be *assumed*. In Theorem 5 below we give a condition on the nonlinear function f which *implies* (20). We start with a simple remark on the linear case which we state as

Theorem 4. *Suppose that the linear system* (4) *is exactly controllable on* $[0, \tau]$ *and the operator* F *given by* (22) *is linear and negative semidefinite, i.e.,* $\langle Fz, z\rangle \leq 0$ *for all* $z \in Z$. *Then the assertion of Theorem* 3 *holds true.*

Proof. The proof is similar to that of Theorem 2. Under the hypotheses of Theorem 4, the operator $I - F$ is clearly injective and satisfie

$$\|z - Fz\| \geq \|z\| \quad (z \in Z), \tag{23}$$

which implies that $\mathcal{R}(I - F)$ is closed. Now we prove with a similar reasoning as in Theorem 2 that $\mathcal{R}(I-F)$ is dense in Z, and so $I-F$ is surjective. Assume that $\mathcal{R}(I-F)$ is strictly contained in Z. Then we fin a nonzero element z_0 in $\mathcal{R}(I - F)^{\perp}$, which means that $\langle z - Fz, z_0\rangle = 0$ for all $z \in Z$. Choosing, in particular, $z = z_0$ yields

$$0 = \langle z_0 - Fz_0, z_0\rangle = \|z_0\|^2 - \langle Fz_0, z_0\rangle \geq \|z_0\|^2,$$

hence $z_0 = 0$, a contradiction. We conclude that $I - F$ is a bijection, and so a linear isomorphism, by the open mapping theorem. □

The following lemma may be proved in exactly the same way as Lemma 5.1 from [8].

Lemma 1. *Suppose that the function* f *satisfies a Lipschitz condition of the form*

$$\|f(t, z_1, u_1) - f(t, z_2, u_2)\| \leq c\,\{\|z_1 - z_2\| + \|u_1 - u_2\|\} \tag{24}$$

for all $(t, z_1, u_1), (t, z_2, u_2) \in [0, \tau] \times Z \times U$ *and some* $c > 0$. *Then the minimal Lipschitz constant* (12) *of the operator* F *given by* (22) *satisfies the upper estimate* $[F]_{Lip} \leq cc_1c_2$, *where we have used the shortcut*

$$c_1 = M[\|B\| + c]e^{Mc\tau}\tau + 1, \quad c_2 = M^2\|B\|\|(LL^*)^{-1}\|\tau, \quad M = \sup_{0\leq s\leq t\leq \tau}\|T(t-s)\|.$$

As an immediate consequence of Lemma 1 we may now deduce the following

Theorem 5. *Suppose that* f *satisfies the Lipschitz condition* (24), *the linear system* (4) *is exactly controllable on* $[0, \tau]$, *and*

$$c(M[\|B\| + c]e^{Mc\tau}\tau + 1)M^2\|B\|\|(LL^*)^{-1}\|\tau < 1. \tag{25}$$

Then the assertion of Theorem 3 *holds true.*

Proof. In the notation of Lemma 1, the condition (25) simply means that $cc_1c_2 < 1$, and so the assertion follows from Theorem 3. □

5. Application to the Damped Wave Equation

As an application of our abstract results we consider the controlled damped wave equation

$$\begin{cases} w_{tt} + \alpha w_t - \beta w_{xx} = u(t,x) - g(t, u(t,x), w, w_t) & \text{for } 0 < x < 1, \\ w(t,0) = w(t,1) = 0 & \text{for } -\infty < t < \infty \end{cases} \tag{26}$$

on the strip $\mathbb{R} \times (-1,1)$. Here we assume that $\alpha \geq 0$, $\beta > 0$, the distributed control function u belongs to $L^2(0,\tau; L^2(0,1))$, and the nonlinear term $g : [0,\tau] \times \mathbb{R}^3 \to \mathbb{R}$ satisfie a Lipschitz condition

$$|g(t, w_1, v_1, u_1) - g(t, w_2, v_2, u_2)| \leq c\{|w_1 - w_2| + |v_1 - v_2| + |u_1 - u_2|\} \tag{27}$$

for all $(t, w_1, v_1, u_1), (t, w_2, v_2, u_2) \in [0,\tau] \times \mathbb{R}^3$ and some $c > 0$. To treat this problem in the abstract framework of the preceding section, we consider in $X = L^2(0,1)$ the unbounded linear differential operator $D\phi = -\phi_{xx}$ with domain of definitio

$$\mathcal{D}(D) = \{\phi \in X : \phi, \phi_x \text{ are absolutely continuous}, \ \phi_{xx} \in X, \ \phi(0) = \phi(1) = 0\}.$$

The operator $D : \mathcal{D}(D) \to X$ frequently occurs in the theory of partial differential equations and has the following well-known properties:

- The spectrum of D is purely discrete and consists of a sequence of simple eigenvalues λ_n satisfying
$$0 < \lambda_1 < \lambda_2 < \cdots < \lambda_n \to \infty \quad (n \to \infty), \tag{28}$$
namely $\lambda_n = n^2\pi^2$.

- There exists a complete orthonormal set $\{\phi_n : n = 1, 2, 3, \ldots\}$ of corresponding eigenvectors of D, namely $\phi_n(x) = \sqrt{2}\sin n\pi x$.

- For all $x \in \mathcal{D}(D)$, the Fourier expansion
$$Dx = \sum_{n=1}^{\infty} \lambda_n \langle x, \phi_n \rangle \phi_n = \sum_{n=1}^{\infty} \lambda_n E_n x$$
holds, where $\langle \cdot, \cdot \rangle$ denotes the inner product in X and $E_n x = \langle x, \phi_n \rangle \phi_n$ $(n = 1, 2, 3, \ldots)$ is a family of complete orthogonal projections in X corresponding to the eigenvalues and eigenfunctions mentioned above.

- The linear operator $-D$ generates an analytic semigroup $\{e^{-Dt} : t \geq 0\}$ given by
$$e^{-Dt}x = \sum_{n=1}^{\infty} e^{-\lambda_n t} E_n x \qquad (t \geq 0, \ x \in X).$$

Recall that, for $r \in [0,1]$, the space of fractional power X^r is define by the norm

$$\|x\|_r = \|D^r x\| = \left\{ \sum_{n=1}^{\infty} \lambda_n^{2r} \|E_n x\|^2 \right\}^{1/2},$$

where D^r is the operator given by

$$D^r x = \sum_{n=1}^{\infty} \lambda_n^r E_n x$$

on the domain of definitio $\mathcal{D}(D^r) = X^r = \{x \in X : \|x\|_r < \infty\}$. Also, for $r \in [0,1]$ we consider the Hilbert space $Z_r = X^r \times X$, equipped with the natural norm

$$\left\| \begin{bmatrix} w \\ v \end{bmatrix} \right\|_{Z_r}^2 = \|w\|_r^2 + \|v\|^2.$$

Using the change of variables $w' = v$, the second order equation (26) can be written as a firs order system of ordinary differential equations in the Hilbert space $Z_{1/2} = \mathcal{D}(D^{1/2}) \times X = X^{1/2} \times X$

$$z'(t) = Az(t) + Bu(t) - f(t, z(t), u(t)) \qquad (t > 0, \;\; z \in Z_{1/2}), \tag{29}$$

where we use the notation

$$z = \begin{bmatrix} w \\ v \end{bmatrix}, \quad B = \begin{bmatrix} O \\ I \end{bmatrix}, \quad A = \begin{bmatrix} O & I \\ -\beta D & -\alpha I \end{bmatrix}, \tag{30}$$

and A is an unbounded linear operator with domain of definitio $\mathcal{D}(A) = \mathcal{D}(D) \times X$. Likewise, the function $f : [0, \tau] \times Z_{1/2} \times X \to Z_{1/2}$ in (29) is define by

$$f(t, z, u) = f(t, w, v, u) = \begin{bmatrix} 0 \\ g(t, u, w, v) \end{bmatrix}. \tag{31}$$

With this notation, equation (29) has precisely the form (3). Moreover, since the space $X^{1/2}$ is continuously imbedded into X, we conclude that

$$\|f(t, z_1, u_1) - f(t, z_2, u_2)\|_{1/2} \leq c \left\{ \|z_1 - z_2\|_{1/2} + \|u_1 - u_2\| \right\} \qquad (0 \leq t \leq \tau)$$

for some $c > 0$ and for all $z_1, z_2 \in Z_{1/2}$ and $u_1, u_2 \in X$, which is nothing else but the Lipschitz condition occurring in Lemma 1. For the remaining part of this section we will assume, without loss of generality, that $\alpha^2 < 4\beta\lambda_1$. The following proposition follows then from [8,9].

Proposition 3. *The matrix operator A given by (30) is the infinitesimal generator of a strongly continuous group $\{T(t) : -\infty < t < \infty\}$ in the space $Z_{1/2}$ which is given by*

$$T(t)z = \sum_{n=1}^{\infty} e^{A_n t} P_n z \qquad (t \in \mathbb{R}, \;\; z \in Z_{1/2}),$$

where $(P_n)_n$ *is the sequence of complete orthogonal projections given by* $P_n = diag\,(E_n, E_n)$ *on the Hilbert space* $Z_{1/2}$*, and*

$$A_n = B_n P_n, \quad B_n = \begin{bmatrix} 0 & 1 \\ -\beta\lambda_n & -\alpha \end{bmatrix} \quad (n = 1, 2, 3, \ldots).$$

This group decays exponentially to zero. More precisely, the estimate

$$\|T(t)\| \leq M(\alpha, \beta)e^{-ct/2} \qquad (t \geq 0)$$

holds, where

$$M(\alpha,\beta) = \sup_n \left\{ 4\sqrt{2}\left|\frac{\alpha \pm \sqrt{4\beta\lambda_n - \alpha^2}}{\sqrt{4\beta\lambda_n - \alpha^2}}\right|, (4+2\beta)\sqrt{2}\left|\frac{\sqrt{\lambda_n}}{\sqrt{4\beta\lambda_n - \alpha^2}}\right| \right\}.$$

Now we are in a position to give sufficien conditions for the exact controllability of both the linear system

$$\begin{cases} z'(t) = Az(t) + Bu(t) \quad \text{for } t > 0, \\ z(0) = z_0 \end{cases} \tag{32}$$

and the perturbed nonlinear system

$$\begin{cases} z'(t) = Az(t) + Bu(t) - f(t, z(t), u(t)) \quad \text{for } t > 0, \\ z(0) = z_0, \end{cases} \tag{33}$$

respectively, where A and B are given as in (30), and the nonlinearity f is define by (31). We summarize with the following

Theorem 6. *The linear system* (32) *is exactly controllable on* $[0, \tau]$*. Moreover, if the estimate*

$$c(M(\alpha,\beta)[1+c]e^{M(\alpha,\beta)c\tau}\tau + 1)M(\alpha,\beta)^2\|(LL^*)^{-1}\|\tau < 1 \tag{34}$$

is satisfied, then also the nonlinear system (33) *is exactly controllable on* $[0, \tau]$*.*

Proof. The firs statement is proved in the same way as Theorem 4.1 in [9]. The second statement follows from Theorem 5, observing that (34) implies (25), and the definitio (30) of B trivially implies that $\|B\| \leq 1$. □

6. Concluding Remarks

In this section we briefl indicate some more applications to the controllability of second order diffusion equations in Hilbert spaces like those considered in the previous sections. Thus, the results presented in this paper can be applied to the following class of second order diffusion system in Hilbert spaces

$$w'' + A_0 w = u(t) - g(t, w, u), \quad (t > 0, \ w \in W, \ u \in U), \tag{35}$$

where W and U are Hilbert spaces, and $A_0 : W \supset \mathcal{D}(A_0) \to W$ is an unbounded linear operator in W with the spectral decomposition

$$A_0 w = \sum_{j=1}^{\infty} \lambda_j E_j w, \qquad E_j w = \sum_{k=1}^{\gamma_j} \langle \phi_{k,j}, w \rangle \phi_{k,j}$$

Here the functions $\phi_{k,j}$ form a complete orthonormal set of eigenvectors of $-A_0$ corresponding to the eigenvalue sequence (28), but this time with multiplicity γ_j, and $-A_0$ generates the strongly continuous semigroup $\{T(t) : t \geq 0\}$ given by

$$T(t)w = \sum_{j=1}^{\infty} e^{-\lambda_j t} E_j w \qquad (t \geq 0, \quad w \in W).$$

The control function u is assumed to lie in the space $L^2(0, \tau; U)$, and the nonlinearity $f : [0, \tau] \times W \times U \to W$ satisfie some suitable hypotheses. Examples of this type include the following well-known systems of partial differential equations.

Example 2. Let Ω be a sufficientl smooth bounded domain in $\mathbb{R}^N$, $u \in L^2(0, \tau; L^2(\Omega))$, $\phi_0, \psi_0 \in L^2(\Omega)$, and f a suitable nonlinear function. Then the initial-boundary value problem for the wave equation

$$\begin{cases} \dfrac{\partial^2 w}{\partial t^2} - \Delta w = u(t,x) - g(t, u(t,x), w) & \text{for } t \geq 0 \text{ and } x \in \Omega, \\ w(t,x) = 0 & \text{for } t \geq 0 \text{ and } x \in \partial\Omega, \\ w(0,x) = \phi_0(x), \ \dfrac{\partial w}{\partial t}(0,x) = \psi_0(x) & \text{for } x \in \Omega \end{cases}$$

may be written as abstract evolution equation in the form (35).

Example 3. Let Ω, u, ϕ_0 ψ_0, and f as in Example 2. Then the initial-boundary value problem for the equation of a vibrating plate

$$\begin{cases} \dfrac{\partial^2 w}{\partial t^2} + \Delta^2 w = u(t,x) - g(t, u(t,x), w) & \text{for } t \geq 0 \text{ and } x \in \Omega, \\ w(t,x) = \Delta w(t,x) = 0 & \text{for } t \geq 0 \text{ and } x \in \partial\Omega, \\ w(0,x) = \phi_0(x), \ \dfrac{\partial w}{\partial t}(0,x) = \psi_0(x) & \text{for } x \in \Omega \end{cases}$$

may be written as abstract evolution equation in the form (35).

In a subsequent paper the authors will show how to apply the results of Section 4 to arrive at sufficien conditions for the exact controllability of such problems.

Acknowledgment

The authors express their gratitude to the referee for several valuable remarks which improved the paper.

References

[1] J. Appell, Intruducción a la teoría espectral de los operadores no lineales, *Notas de Matemáticas de ULA* (to appear).

[2] J. Appell, E.De Pascale, A. Vignoli, Nonlinear Spectral Theory, *DeGruyter Series in Nonlinear Analysis and Applications,* **10**, DeGruyter-Verlag, Berlin, 2004.

[3] R.F. Curtain, A.J. Pritchard, Infinit Dimensional Linear Systems, *Lecture Notes in Control and Information Sciences,* **8**, Springer-Verlag, Berlin, 1978.

[4] R.F. Curtain, H.J. Zwart, An Introduction to Infinit Dimensional Linear Systems Theory, *Text in Applied Mathematics,* **21**, Springer-Verlag, New York, 1995.

[5] W. Feng, A new spectral theory for nonlinear operators and its applications, *Abstr. Appl. Anal.,* **2** (1997), 163-183.

[6] M. Furi, M. Martelli, A. Vignoli, Contributions to the spectral theory for nonlinear operators in Banach spaces, *Annali Mat. Pura Appl.,* **118** (1978), 229-294.

[7] R.I. Kachurovskij, Regular points, spectrum and eigenfunctions of nonlinear operators [in Russian], Dokl. Akad. Nauk SSSR, **188** (1969), 274-277; Engl. transl.: *Soviet Math. Dokl.*, **10** (1969), 101-114.

[8] H. Leiva, A lemma on C_0-semigroups and applications to PDEs systems, *Quaestiones Mathematicae,* **26** (2003), 247-265.

[9] H. Leiva, Exact controllability of the suspension bridge model proposed by Lazer and McKenna, *J. Math. Anal. Appl.,* **309** (2005), 404-419.

[10] I.J. Maddox, A.W. Wickstead, The spectrum of uniformly Lipschitz mappings, *Proc. Royal Irish Acad.,* **89-A** (1989), 101-114.

[11] J.W. Neuberger, Existence of a spectrum for nonlinear transformations, *Pacific J. Math.*, **31** (1969), 157-159.

[12] M. Väth, The Furi-Martelli-Vignoli spectrum vs. the phantom, *Nonlin. Anal. TMA,* **47** (2001), 2237-2248.

In: Evolution Equations: Theories, Solutions and Functions ISBN 978-1-61324-090-8
Editor: Gaston M. N'Guerekata

Chapter 8

MATHEMATICAL MODELING ON THE DYNAMICS OF CD4$^+$T CELL INFECTION BY HIV THAT INCLUDES A TIME DELAY

***P. K. Roy*[1] *and Biplab Chattopadhyay*[2]**
[1]Department of Mathematics
Barasat Government College, 10 K. N. C. Road, Barasat,
Kolkata 700124, West Bengal, India
[2]3Department of Physics
Barasat Government College, 10 K. N. C. Road, Barasat,
Kolkata 700124, West Bengal, India

Abstract

In this paper we have considered a modifie version of the basic model of Herz et al., *(Proc. Natl. Acad. Sci. USA, 93, 7247-7251, 1996)* emulating the HIV dynamics in human blood. The modifie model includes a time delay in the death term of the infected CD4$^+$T-lymphocytic (immune) cells. as literature shows, there exists a positive discrete time lag between "the infection" and the death" of infected CD4$^+$T-cells and hence the inclusion of the delay. Delay differential model equations are solved theoretically as well as numerically. Our results show that delay induced instability are nonexistent and asymptotically stable solution, characteristic of a non delayed system, exists. The (asymptotic) concentrations of healthy CD4$^+$T cells and virus populations may significantl be altered with some crucial model parameters and controlling of these parameters may be useful for better management of the HIV inflicte disease AIDS.

Keywords and phrases: Immunodeficien y Virus, CD4$^+$T Cells, Cell Lysis, Time Delay, Time Series Solutions, Asymptotic Stability, Numerical Solutions.

1. Introduction

Human immunodeficien y virus (HIV) was firs confirme in the last decade of 20th century, primarily in the countries of North America, Western Europe and parts of the sub Saharan Africa. Within a span of merely two decades HIV infection has become a global pandemic having major epidemic foci in almost all the countries in the world. Manifestation of HIV in humans is recognized as acquired immune deficien y syndrome (AIDS) and in the present global scenario AIDS is the biggest challenge to the human race.

The healthiness of human immune system is identifie with the lymphocyte cells count in the human blood. Clinical research on HIV detection showed that $CD4^+T$ lymphocytes are predominant in constituting human immunity in the human blood. The attack of HIV on these lymphocytes leads to their depletion in blood and eventually an impairment or systematic degradation of the human immune system takes place. It is further clinically proved that an individual, having $CD4^+T$ cell count $< 200/\mu\, liter$, is considered to be infected with AIDS. An AIDS infected individual dies due to progressive occurrence of opportunistic diseases. It thus appears that, understanding of the $CD4^+T$ cells' dynamics together with a HIV population, is central to better management of the disease AIDS. A mathematical model comprising time differential equations for susceptible & infected $CD4^+T$ cells together with HIV viruses could be a relevant starting point. A number of such mathematical models (Nowak and May, 1993; Herz et al. 1996; Le Corfec et al., 1999) have emerged in the last few years. These models are analyzed extensively to understand and explain various clinically observed phenomenon of the disease AIDS. Out of the two strains of HIV viruses, the HIV-1 causes infection to the host population of $CD4^+T$ cells. In the present paper we will focus on a model where infection is only due to HIV-1.

HIV-1 viruses infect many types of immune cells in the human body but they preferentially replicate within the $CD4^+T$ lymphocytes (Rosemberg and Janossy, 1999). In the present model, we consider $CD4^+T$ cells as the host where replication of HIV-1 proceeds in two phases namely, (i) entry of HIV and (ii) synthesis and multiplication of HIV. Once replication is complete, assembled virons come out of the infected host T-cell by budding through the cell membrane (Hasse, 1999; Callaway and Perelson, 2002) and at the same time write off the involved host T-cell. A time delay thus naturally comes in the model for realistic emulation of the discrete biological process of infection and death of a T-lymphocytic cell.

In the present communication we intend to theoretically analyze the dynamical behaviors of a time-delayed model of HIV infection on the $CD4^+T$-cells in human blood. We also contemplate to carry out a rigorous numerical study of the model system to understand the dependence of the model dynamics on the various model parameters.

2. Formulation of the Mathematical Model

Let $x(t)$ and $y(t)$ be the susceptible and infected portions of the host $CD4^+T$ cells and $v(t)$ be the concentration of virus particles at a time t.

To formulate the mathematical model we make the following assumptions

(A1): Susceptible $CD4^+T$ cells are produced at a constant rate s and are removed from the system through a mortality rate μ per cell.

(A2): The process of infection to the susceptible host follows the law of mass action (i.e., the number of new infection is proportional to $x(t)v(t)$) under mixing-homogeneity. Note that, at start there exists only one or two virons but millions of host cells implying inhomogeneity. Nevertheless, as time progress viruses grow significantl setting the mixing homogeneity and law of mass action holds.

(A3): We assume a per capita death rate of infected CD4$^+$T cells denoted by α ($> \mu$). Here we note that $\alpha = e + \mu$ where e is the death rate constant due to cell lysis.

(A4): We assume a constant rate of production of free virons per infected CD4$^+$T cell due to cell lysis denoted by parameter c. We also take γ as the per capita death rate of free virons.

The assumptions $(A1) - (A4)$ yields a set of coupled differential model equations as

$$\begin{aligned} \frac{dx}{dt} &= s - \mu x - \beta x v \\ \frac{dy}{dt} &= \beta' x v - \alpha y \\ \frac{dv}{dt} &= c y - \gamma v, \end{aligned} \tag{1}$$

Here β is the rate of contact between the CD4$^+$T cell and virons and β' denotes the rate at which viron-attached $CD4^+T$ cells actually become infected. Some researchers (Tuckwell and Le Corfec, 1999) considered an additional virus removal through a term $-\beta x v$ in the third of equations (1). We argue that this removal can be through $-\beta v$ which can be shoved in $-\gamma v$ term in the third of equations (1), meaning an eventual increase of numerical value of γ. A schematic fl w diagram of the model is given in Fig. 1.

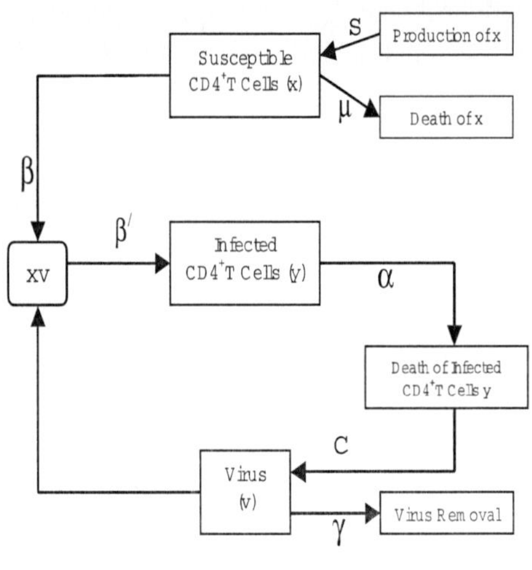

Figure 1. Schematic fl w diagram of the basic model as in equations (1).

The model (1) assumes an instantaneous death of infected host cells. But in actual case the infected CD4$^+$T cells rest in latency before death during which the viral-replication takes place inside them. Hence we consider a positive definit time delay in the death term of infected CD4$^+$T cells.

At an instant of time t the average effect of mentioned time delay in the second equation of (1) would mean the inclusion of a factor $F(t-u)$ with the term involving y, where $F(u)$ is a delay kernel (or memory) function. To get the average effect of delay at the time t, the whole term $y(u)F(t-u)du$ would have to be integrated from $-\infty$ *to* t. Under suitable conditions the Kernel takes the form of a delta function $F(u) = \delta(u-\tau)$, with τ (≥ 0) being a constant. Under such conditions, the model equation (1) are recast as the delay differential equations with a time delay factor τ (Culshaw et al., 2003)

$$\begin{aligned}\frac{dx}{dt} &= s - \mu x - \beta x v\\ \frac{dy}{dt} &= \beta' x v - \alpha y(t-\tau)\\ \frac{dv}{dt} &= cy - \gamma v.\end{aligned} \tag{2}$$

It is argued that, a definit latency for the infected CD4$^+$T cells during viral replication within them, would mean a nonzero finit time to complete the process of such replication. Thus one needs to include a delay factor, similar to that in the death of infected CD4$^+$T cells, to the lysis term in the third of equations (2). Thus the equations take the form

$$\begin{aligned}\frac{dx}{dt} &= s - \mu x - \beta x v\\ \frac{dy}{dt} &= \beta' x v - \alpha y(t-\tau)\\ \frac{dv}{dt} &= cy(t-\tau) - \gamma v.\end{aligned} \tag{3}$$

Theoretical analysis of the model system could bring out qualitative features of the solutions of model equations, for which, the inclusion of delay in the death involving term of the infected CD4$^+$T cells would suffice Thus, for the entire theoretical analysis, we remain focussed to model equations (2).

3. Theoretical Analysis: Local Stability and Delay Induced Instability

Right hand side of equation (2) is a smooth function of the variables (x, y, v) and the model parameters, as long as these quantities are non-negative. So, local existence and uniqueness properties hold in the positive octant. The model equation (2) has the following equilibria on all the co-ordinate planes, $E_0(0,0,0)$, $E_1(\frac{s}{\mu},0,0)$, and $E^*(x^*, y^*, v^*)$ where $x^* = \frac{\alpha\gamma}{c\beta}$, $y^* = \frac{s}{\alpha} - \frac{\mu\gamma}{c\beta}$, $v^* = \frac{sc}{\gamma\alpha} - \frac{\mu}{\beta}$. Note that E^* exists if $c > \frac{\mu\gamma\alpha}{s\beta}$.

Considering perturbation corrections to the variables $X(t) = x(t) - x'$, $Y(t) = y(t) - y'$, $V(t) = v(t) - v'$ the model equations are linearized. The linearized form of equations (2) at E^* is given by

$$\begin{aligned}\frac{dX}{dt} &= -\mu X - \beta v^* X - \beta x^* V\\ \frac{dY}{dt} &= \beta v^* X + \beta x^* V - \alpha Y(t-\tau)\\ \frac{dV}{dt} &= cY - \gamma V\end{aligned} \tag{4}$$

The characteristic equation of system (4) is given by

$$\lambda^3 + (A_1 + \alpha e^{-\lambda\tau})\lambda^2 + (A_2 + A_3 e^{-\lambda\tau})\lambda + A_4 e^{-\lambda\tau} - A_5 = 0 \tag{5}$$

where

$$A_1 = \mu + \gamma + \beta v^*$$
$$A_2 = \mu\gamma + \beta\gamma v^* - c\beta x^*$$
$$A_3 = \mu\alpha + \beta\alpha v^* + \alpha\gamma,$$
$$A_4 = (\mu + \beta v^*)\alpha\gamma$$
$$A_5 = c\mu\beta x^*$$

Now to determine the nature of the stability, we require the sign of the real parts of the roots of (5). Thus let us consider

$$\Phi(\lambda,\tau) = \lambda^3 + (A_1 + \alpha e^{-\lambda\tau})\lambda^2 + (A_2 + A_3 e^{-\lambda\tau})\lambda + A_4 e^{-\lambda\tau} - A_5 \quad (6)$$

For $\tau = 0$,

$$\Phi(\lambda, 0) = \lambda^3 + (A_1 + \alpha)\lambda^2 + (A_2 + A_3)\lambda + A_4 - A_5 = 0 \quad (7)$$

By Routh-Hurwitz criterion it is observed that the corresponding non-delayed system is locally asymptotically stable around the positive interior equilibrium if

$$c < \frac{\alpha\gamma(\mu + \beta v^*)}{\mu\beta x^*} \text{ and } \beta c v^* + 2\alpha\gamma < (\alpha + \gamma)(\mu + \beta v^*) \quad (8)$$

This condition biologically implies that the value of cell lysis constant (c) signifying viral replication should remain at moderate level, which is the actual case as reflecte by the reported values mentioned in Table.1.

To obtain the conditions for nonexistence of delay induced instability, we use the following original theorem of Gopalsamy (1992).

Theorem 1. *A set of necessary and sufficient conditions for the equilibrium E^* to be asymptotically stable for $\tau \geq 0$ is the following:*

(i) *The real parts of all the roots of $\phi(\lambda, 0) = 0$ are negative.*

(ii) *For all real m and any $\tau \geq 0$, the following holds*

$$\phi(im, \tau) \neq 0 \text{ where } i = \sqrt{-1}$$

Proof. Here $\phi(\lambda, 0) = 0$ has roots whose real parts are negative provided (8) holds.

Now for $m = 0$,

$$\phi(im,\tau) = \phi(0,\tau) = (\mu + \beta v^*)\alpha\gamma - c\mu\beta x^* \neq 0. \quad (9)$$

Now for $m \neq 0$,

$$\phi(im,\tau) = (im)^3 + (A_1 + \alpha e^{-im\tau})(im)^2 + (A_2 + A_3 e^{-im\tau})im + A_4 e^{-im\tau} - A_5 = 0. \quad (10)$$

Let $\phi(im, \tau) = 0$. Separating real and imaginary parts we get,

$$A_1 m^2 + A_5 = A_3 m \cos\left(\frac{\pi}{2} - m\tau\right) + A_4 \cos m\tau - \alpha m^2 \cos m\tau. \tag{11}$$

$$-m^3 + A_2 m = A_4 \sin m\tau - A_3 m \sin\left(\frac{\pi}{2} - m\tau\right) - \alpha m^2 \sin m\tau. \tag{12}$$

Squaring and adding the above two equations, we get

$$(A_1 m^2 + A_5)^2 + (-m^3 + A_2 m)^2 = A_3^2 m^2 + A_4^2 + \alpha^2 m^4 - 2A_4 \alpha m^2. \tag{13}$$

Let the right hand side of (13) be denoted by $f(m)$

Now for arbitrary real m, we get from (13)

$$f(m) \leq \alpha^2 m^4 + (A_3^2 - 2A_4\alpha)m^2 + A_4^2. \tag{14}$$

Therefore a sufficien condition for the non existence of a real number m satisfying $\phi(im, \tau) = 0$ can now be obtained from (13) and (14) as

$$m^6 + (A_1^2 - 2A_2 - \alpha^2)m^4 + (2A_1 A_5 + A_2^2 - A_3^2 + 2A_4\alpha)m^2 + A_5^2 - A_4^2 > 0. \tag{15}$$

The inequality we can write in the form of

$$m^6 + Pm^4 + Qm^2 + R > 0. \tag{16}$$

where

$$\begin{aligned} P &= A_1^2 - 2A_2 - \alpha^2 \\ Q &= 2A_1 A_5 + A_2^2 - A_3^2 + 2A_4\alpha \\ R &= A_5^2 - A_4^2 \end{aligned} \tag{17}$$

The sufficien condition can be obtained as if

$$\gamma > \alpha; \quad c < \frac{\alpha\gamma(\mu + \beta v^*)}{\mu + \beta x^*} \tag{18}$$

Therefore conditions (i) and (ii) of the above theorems are satisfie if (13) holds.

The firs part of the above condition (18) biologically means that the virus removal rate should always be greater than the death rate of infected CD4$^+$T cells, which is actually enforced in our case as in Table.1. The second part of (18) is same as the non-delayed system as in equation (8). □

4. Length of Delay and Preservation of Stability of Solutions

We fin that by continuity and for sufficientl small $\tau > 0$, all eigenvalues of the characteristic equation of the linearized form of the system have negative real parts, provided one can guarantee that no eigenvalue with positive real part bifurcates from infinit (which could happen since it is a retarded system). For stability analysis we use the Nyquist criterion (Freedman *et al.*, 1986). To do this, we consider the linearized form of the system and the space of real valued continuous functions define on $[\tau, \infty)$ satisfying the initial conditions.

Let $\bar{X}(L), \bar{Y}(L)$ and $\bar{V}(L)$ be the Laplace transform of $X(t)$, $Y(t)$ and $V(t)$ respectively.

The inverse Laplace transform of $\bar{V}(L)$ will have terms which exponentially increase with time, if $\bar{V}(L)$ has poles with positive real parts. Since E^* needs to be locally asymptotically stable, it is necessary and sufficien that all poles of $\bar{V}(L)$ have negative real parts. We shall employ the Nyquist criterion which states that if L is the arc length of a curve encircling the right half- plane, the curve $\bar{V}(L)$ will encircle the origin a number of times equal to the difference between the number of poles and the number of zeroes of $\bar{V}(L)$ in the right half-plane.

Thus conditions for local asymptotic stability of E^* as given by Freedman *et al.* (1986) are

$$Im\, F(in_0) > 0 \tag{19}$$

$$Re\, F(in_0) = 0 \tag{20}$$

where n_0 is the smallest positive root of equation (20).

To get an estimation on the length of delay, we utilize the following conditions,

$$-n^3 + \chi_2 n > \chi_3 \sin n\tau - n^2\alpha \sin n\tau - \alpha n \cos n\tau. \tag{21}$$

$$\chi_1 n^2 - \chi_4 = \alpha n \sin n\tau + \chi_3 \cos n\tau - \alpha n^2 \cos n\tau, \tag{22}$$

$$\begin{aligned} \text{where,} \quad & \chi_1 = \mu + \beta v^* + \gamma, \\ & \chi_2 = \mu + \beta v^* \gamma, \\ & \chi_3 = (\mu + \beta v^* + \gamma)(\mu + \beta v^*)\alpha, \\ & \chi_4 = c[\beta x^* \beta v^* - (\alpha + \mu + \beta v^*)\beta x^*]. \end{aligned} \tag{23}$$

Therefore, E^* will be stable if the inequality (21) holds at $n = n_0$, where n_0 is the firs positive root of equation (22). We shall now estimate an upper bound n_+ of n_0, independent of τ and then to estimate τ so that (21) holds for all values of n, $0 \leq n \leq n_+$, and hence in particular at $n = n_0$.

Since the right hand side of (22) is always less than or equal to $(\alpha^2 n^2 + \alpha^2 n^4 + \chi_3^2)^{\frac{1}{2}}$, the unique positive solution of $\chi_1 n^2 - \chi_4 = (\alpha^2 n^4 + \alpha^2 n^2 + \chi_3^2)^{\frac{1}{2}}$, denoted by n_+ is always greater than or equal to n. By straight forward calculation, one can determine that

$$n_+ = \left[\frac{(2\chi_1\chi_4 + \alpha^2) + \{\alpha^4 + 4\alpha^2\chi_4(\chi_1 + \chi_4) + 4\chi_3^2(\chi_1^2 - \alpha^2)\}^{1/2}}{2(\chi_1^2 - \alpha^2)}\right]^{1/2}. \tag{24}$$

We see that n_+ is independent of τ. Now we need an estimation on τ so that (21) holds for all $0 \leq n \leq n_+$. Now by suitable rearrangement and simplificatio of (21) and (22) we fin as

$$\tau_+ = \frac{-\rho + (\rho^2 + 4\xi\eta)^{\frac{1}{2}}}{2\xi}, \tag{25}$$

$$\begin{aligned} \text{where} \quad & \xi = (\chi_3 n - \alpha n^3 - \chi_1 \alpha n)\frac{n^2}{2}, \\ & \rho = \chi_1\chi_3 + \alpha n^2 - \chi_1 \alpha n^2 \end{aligned} \tag{26}$$

$$\text{and} \quad n(\chi_1\chi_2 - \chi_4 - \chi_3 + \alpha n^2 + \chi_1\alpha) \equiv \eta$$

It should be noted that τ_+ is considered as an unique positive root of η. Hence, for $\tau < \tau_+$, the Nyquist criterion holds and τ_+ is thus the estimate for the length of the delay τ for which stability is preserved.

5. Numerical Solutions of the Model Equations

The model equations (3) are solved numerically for a better understanding of the delayed system. Initially we set the model parameters to their standard reported values as in Table.1, but later we vary α and γ to fin out their effects on the dynamics of the time-delayed model.

Table 1. Values of parameters used for models dynamics calculations

Para-meter	Definitio	Reported Range	Default Value (day $^{-1}$)
s	Constant production rate of CD4$^+$T cells	0-10 $/mm^3$	2 cells per mm^3
μ	Death rate of Susceptible CD4$^+$T	0.0014-0.03	0.01
β	Rate of contact between x and v	0.00027-0.007 mm^3/vir	0.0006 mm^3/vir
β'	Rate of transformation of host into infected CD4+ T cells	0.4 β	0.00024 mm^3/vir
α	Total death rate of infected CD4+ T cells	0.3-0.5	0.3
c	Virus production rate through lysis	42-88	52
γ	Removal rate of virus	2-3	2
τ	Time delay in infected CD4+ T cells' death	1-5	varied in the range

Solutions of equations (1) show that inclusion of delay only in the infected CD4 $^+$T cells' death leads to the extinction of infected CD4$^+$T cells (y) and virus (v) at a small time t for all τ whatsoever. This is nonphysical and this result, in turn, justifie our argument to consider delay factor in the lysis term also yielding equations (3).

Fig. 2 consists population densities of susceptible and infected CD4 $^+$T cells and virus for situations of without and with the time delay. We observe that introduction of delay inflict strong oscillations in all the populations uptill a large value of t. However, asymptotically the solutions are stable. We also observed that time delay pushes the infected CD4$^+$T cell populations to a very low value within a very short time.

In Fig. 3 different populations (solutions) are plotted for $\tau = 3.0\&5.0$. Note that, incre-

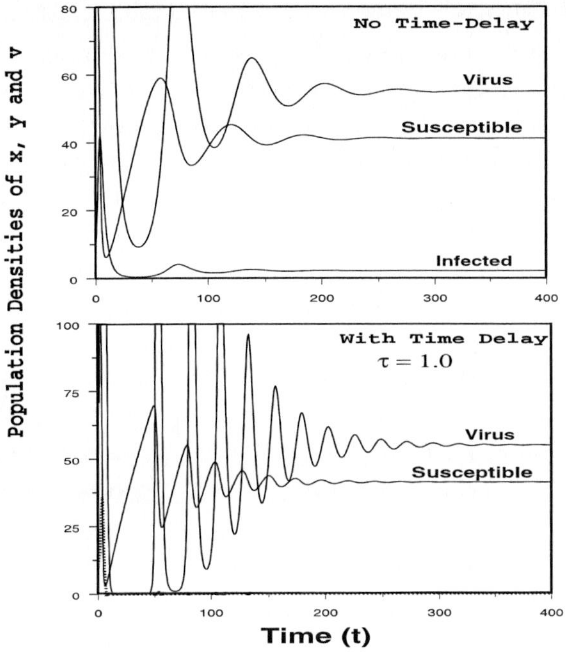

Figure 2. Solutions of the model system with initial values $[200, 10, 1]$ for the cases of without and with time delay. Value of τ is as noted in the plot.

ment of delay factor τ does not inflic any significan quantitative change in the solutions. This implies that the model system's response to delay is only qualitative and not signifi cantly quantitative.

To see the impact of time delay on the death rate of infected cells (α), we obtain stable solution for x, y, v as a function α keeping other parameters as in Table.1 and with delay factor $\tau = 2$. These stable solutions are plotted in Fig. 4 for cases of delay and no-delay. Introduction of delay leads to the extinction infected population very fast for small α, but the virus population keeps up with the susceptible CD4$^+$T cells. For large α (> 1) some oscillations in the stable solutions for virus and the susceptible CD4$^+$T cells are observed. In both the cases virus population goes to extinction for $\alpha > 1.5$ and only the susceptible population survives attaining the fi ed value $x = \frac{s}{\mu} = 200$ in the present case.

In Fig. 5 we plot stable solutions as a function of γ, the virus removal rate keeping other parameters as in Table.1 and $\tau = 2$. As in the previous case, we fin infected CD4$^+$T cells y) is pushed to extinction very fast for small γ, virus population persists till a large γ but moves to zero at $\gamma \geq 9.5$. We fin moderate oscillations in susceptible CD4$^+$T cells and the virus stable solutions for $\gamma > 7$. Around $\gamma = 9.5$, when the virus population extincts and the susceptible CD4$^+$T cells attain a globally stable value $x = \frac{s}{\mu} = 200$ (with $s = 2$ and $\mu = 0.01$). Attainment of this global stable value for susceptible CD4$^+$T cells population and can be understood easily.

In some parametric region, asymptotic solutions of model variables y and v become zero and we have $\dot{y} = \dot{v} = 0$ as well as $y = v = 0$. The situation also makes susceptible CD4$^+$T cells population x stable implying $\dot{x} = 0$. Thus, from the basic model equations

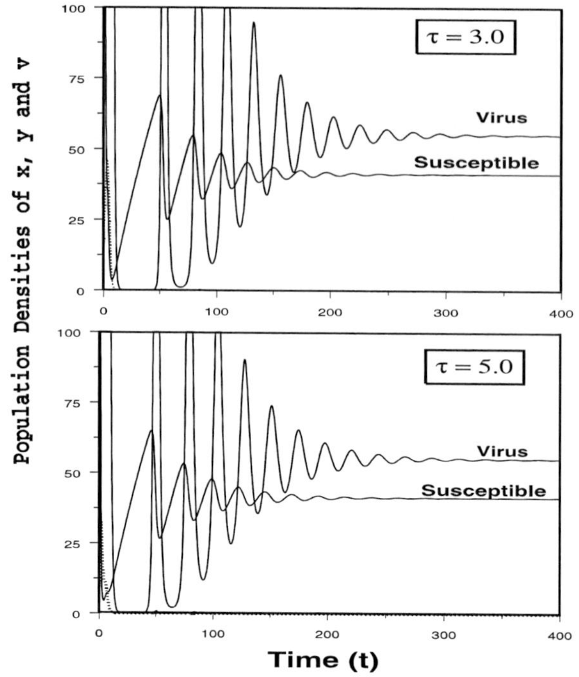

Figure 3. Time series Solutions of the delayed model with initial values $[200, 10, 1]$ and τ are as noted.

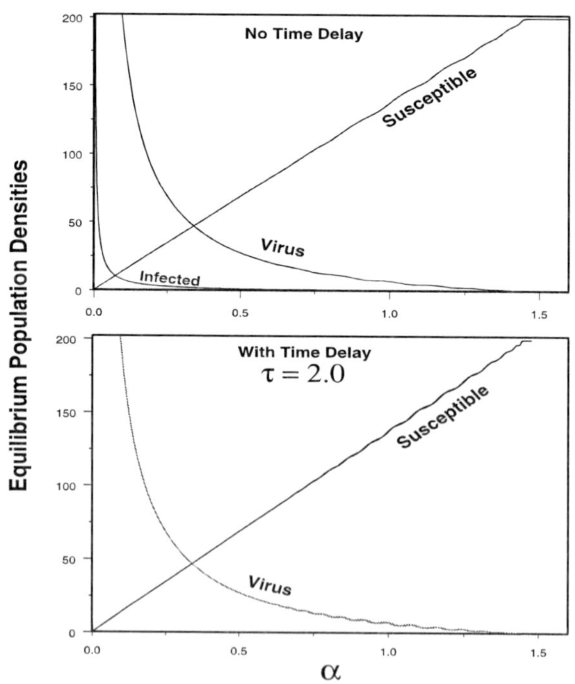

Figure 4. Stable Solutions of model variables are plotted with α, the death rate of infected $CD4^{+}T$ cells.

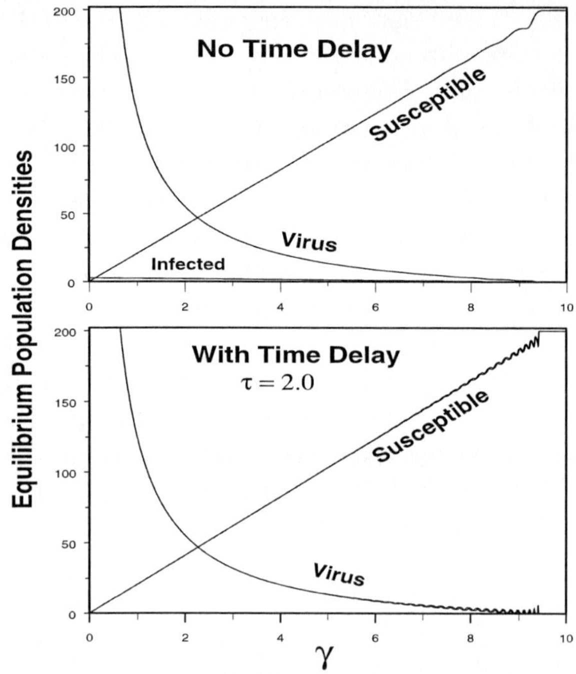

Figure 5. Stable Solutions of model variables as a function of virus death rate constant γ for the cases of with and without delay.

we have, $s - \mu x = \dot{x} = 0$, and it follows that $x = \frac{s}{\mu}$.

6. Discussion and Conclusions

We have considered a basic mathematical model depicting the interrelation between human immunodeficien y virus (HIV) and the CD4$^+$T lymphocytes in human blood. The virons cause infection to the healthy CD4$^+$T lymphocytic cells, replicate itself within the cell and come out in multiplied order as free virons from the infected cell marking its death. Here, we focus on the basic model including a finit definit time delay in the death term of infected CD4$^+$T cells.

Our analytical studies focusses on qualitative aspects of the model dynamics. Calculations show that the existence and uniqueness of the solutions of dynamical variables x, y and v holds good locally in the positive octant, provided all the model parameters are non negative including the delay factor. A rigorous local stability analysis enables us to obtain sufficien conditions for the nonexistence of delay induced instability. These conditions signify existence of asymptotic stability of interior equilibrium. An estimation of the length of delay, that preserve the stability of the solutions, has been made and is observed to assume a positive definit value. We fin that there exists definit parametric regions which preserve stability of the system as a whole.

Our numerical calculations show that partial inclusion of delay in CD4 $^+$T cells' death term (equations (2)) yields nonphysical solutions. Whereas inclusion of delay (in the virus replication term too) makes the solutions meaningful and biologically consistent. Inclusion of time delay is observed to inflic violent and persistent oscillations in the model solutions, which however become asymptotically stable reinforcing the conditions obtained analytically. Change in the delay factor τ produce no significan qualitative or quantitative changes in the asymptotic solutions. We also studied the delayed model by varying death rates of infected $CD4^{+}T$ lymphocytes (α) and the virus (γ). We fin that for both the cases time delay suppresses infected $CD4^{+}T$ population very fast, but a stable viral population sustains uptill significan values of α and γ and then decreases monotonically with further increase of these parameters. At the respective large values of α and γ, stable viral populations completely die down. As soon as virons are completely removed, susceptible $CD4\ ^{+}T$ cell population achieves global stability. This procedure biologically implies a selective removal of infected host cells and virus from the system. The resulting feature is indicative of a disease free $CD4^{+}T$ lymphocytes and hence implies eradication of AIDS.

Acknowledgments

Its a pleasure to acknowledge Joydev Chattopadhyay and Shovanlal Roy for useful discussions. The authors thank the referee for valuable comments and suggestions.

References

[1] D.S. Callaway, A.S. Perelson, HIV-1 infection and low virul loads, *Bull. Math. Biol.* **64** (2002), 29-64.

[2] R.V. Culshaw, S. Ruan, G. Webb, A mathematical model of cell to cell spread of HIV-1 that includes a time delay, *J. Math. Biol.,* **46** (2003), 425-444.

[3] H.I. Freedman, L.H. Erbe, V.S.H. Rao, Three species food chain models with mutual interference and time delays, *Math. Biosci.,* **80**: 57-80, 1986.

[4] K. Gopalsamy, *Stability and Oscillations in Delay Differential Equations of Populayion dynamics*, Kluwer Academic, 1992.

[5] A.T. Hasse, Population biology of HIV-1 infection: viral and CD4+ T cell demographics and dynamics in lymphatic tissues, *Annual Review of Immunology.* **17** (1999), 625-656.

[6] A.V.M. Herz, S. Bonhoeffer, R.M. Anderson, R.M. May, M.A. Nowak, Viral dynamics in vivo; limitations on stemates of intracellular delay and virus decay., *Proc. Natl. Acad. Sci. USA.,* **93** (1996), 7247-7251.

[7] E.Le Corfec, F.Le Pont, H.C. Tuckwell, C. Rouzious, D. Costagliola, Direct HIV testing in a blood donations; variation of the yield with dection threshold and pool size. *Transfusion.* **39** (1999), 1141-1144.

[8] E.Le Corfec, H.C. Tuckwell, Variability in early HIV-1 population dynamics. *AIDS.,* **12** (1998), 960-962.

[9] M.A. Nowak, R.M. Bangham, Population dynamics of immune response to persistent viruses, *Sciece,* **272** (1996), 74-79.

[10] M.A. Nowak, R.M. May, Mathematical biology of HIV infections antigenic variation and diversity threshold, *Math. Biosc.* **106** (1991), 1-21.

[11] M.A. Nowak, R.M. May, AIDS pathogenis: Mathematical models of HIV and SIV infections. *AIDS.* **7** (1993), S3-S18.

[12] M.A. Nowak, R.M. May, Virus dynamics, *Cambridge University Press, Cambridge, UK.*, 2000.

[13] Y. Rosenberg, G. Janossy, The importance of lymphocyte traffickin in regulating blood lymphocyte levels during HIV and SIV infections, *Seminars in Immunol.,* **11** (1991), 139-154.

[14] M.A. Stafford, L. Corey, Y. Cao, et al., Modelling plasma virus concentration during primary HIV infection, *J. Theor. Biol.,* **23** (2000), 285-301.

[15] W.Y. Tan, H. Wu, Stochastic modelling of the dynamics of CD4 T cell infection by HIV and some Monte Carlo studies. *Math. Biosci.,* **147** (1998), 173-205.

In: Evolution Equations: Theories, Solutions and Functions ISBN 978-1-61324-090-8
Editor: Gaston M. N'Guerekata

Chapter 9

ON THE GLOBAL CONTROLLABLE FOR SET DIFFERENTIAL EQUATIONS

Nguyen Dinh Phu*
Faculty of Mathematics and Computer Science,
VNU Ho Chi Minh City,Vietnam

Abstract

Recently, the fiel of differential equations has been studying in a very abstract method. Instead of considering the behaviour of one solution of a differential equation, one studies its sheaf-solution (see [10]- [12]). Instead of studying a differential equation, one studies differential inclusion (see [8, 9]). Especially, one studies fuzzy differential equation (FDE) (a differential equation whose variables and derivative are fuzzy sets) see [1]- [6]. In [13]- [15] the authors have studied the fuzzy control differential equation (FCDE). In [7] the authors have investigated the set differential equation (SDE). In [16]- [18] the authors have studied the set control differential equation (SCDE).

In this paper, we present the problem of global controllable (GC) for set control differential equation (SCDE). This paper is a continuation of some works in this direction (see [16]- [18]).

2000 Mathematics Subject Classification: 34G10, 47D06.

Keywords and phrases: Differential equations, control theory, set differential equations.

1. Introduction

Before we proceed to investigate problem of Global Controllable for set control differential equations, let's note the following facts:

In [7], Prof.V. Lakshmikantham and the other authors considered set differential equations (SDE) and had some important results on existence, comparison and stability criteria for SDE

$$D_H X(t) = F(t, X(t)) \tag{1.1}$$

*E-mail address: ndphu_dhtn@yahoo.com.vn

where $X(t_0) = X_0$, $X(t) \in K_c(\mathbb{R}^n)$, $t \in [t_0, T] = I \subset \mathbb{R}_+$ and $F : I \times K_c(\mathbb{R}^n) \to K_c(\mathbb{R}^n)$.

In [16]- [18], we have considered a set control differential equation (SCDE) as following

$$D_H X(t) = F(t, X(t), U(t)) \tag{1.2}$$

where $X(t_0) = X_0, X(t) \in K_c(\mathbb{R}^n), U(t) \in K_c(\mathbb{R}^p), t \in [t_0, T] = I \subset \mathbb{R}_+$ and $F : I \times K_c(\mathbb{R}^n) \times K_c(\mathbb{R}^p) \to K_c(\mathbb{R}^n)$, and had some important results on existence, comparison, boundedness and stability criteria for SCDE.

This paper is organized as follows: in Section 2, we recall some basic concepts and notations which are useful in next sections. In Section 3, we present the global controllable (GC) for SCDE, and some examples for the global controllable (GC) and uncontrollable for SCDE. In Section 4, we give conclusion of this paper.

2. Preliminaries

We recall some notations and concepts presented in detail in recent series works of V. Lakshmikantham et al (see [7]).

Let $K_c(\mathbb{R}^n)$ denote the collection of all nonempty compact convex subsets of $\mathbb{R}^n$. Given $A, B \in K_c(\mathbb{R}^n)$, the Hausdorff distance between A and B is define by

$$D[A, B] = \max\{\sup_{a \in A} \inf_{b \in B} \|a - b\|_{\mathbb{R}^n}, \sup_{b \in B} \inf_{a \in A} \|b - a\|_{\mathbb{R}^n}\} \tag{2.1}$$

where $\|.\|_{\mathbb{R}^n}$ denotes the Euclidean norm in $\mathbb{R}^n$.

We defin the magnitude of a nonempty subset of A:

$$D[A, \theta^n] = \|A\| = \sup\{\|a\|_{\mathbb{R}^n}, a \in A\} \tag{2.2}$$

where θ^n is the zero element of $\mathbb{R}^n$ which is regarded as a one point set. $\|A\| = D[A, \theta^n]$ - norm in $K_c(\mathbb{R}^n)$ is finit when the supremum in (2.2) is attained with $A \in K_c(\mathbb{R}^n)$.
The Hausdorff metric (2.1) satisfie some below properties.

$$\begin{aligned}
&D[A + C, B + C] = D[A, B] \quad \text{and} \quad D[A, B] = D[B, A],\\
&D[\lambda A, \lambda B] = \lambda D[A, B],\\
&D[A, B] \le D[A, C] + D[C, B],\\
&D[A + A', B + B'] \le D[A, B] + D[A', B']
\end{aligned}$$

for all $A, B, C \in K_c(\mathbb{R}^n)$ and $\lambda \in \mathbb{R}_+$.

If $\alpha, \beta \in \mathbb{R}$ and $A, B \in K_c(\mathbb{R}^n)$, then

$$\alpha(A + B) = \alpha A + \alpha B, \quad \alpha(\beta A) = (\alpha\beta)A, \quad 1.A = A \tag{2.3}$$

It is known that $(K_c(\mathbb{R}^n), D)$ is a complete metric space and if the space $K_c(\mathbb{R}^n)$ is equipped with the natural algebraic operations of addition and nonnegative scalar multiplication, then $K_c(\mathbb{R}^n)$ becomes a semilinear metric space which can be embedded as a complete cone into a corresponding Banach space.

Let $A, B \in K_c(\mathbb{R}^n)$. The set $C \in K_c(\mathbb{R}^n)$ satisfying $A = B + C$ is called the Hausdorff difference (the geometric difference) of the sets A and B and is denoted by the symbol $A - B$. Given an interval I in $\mathbb{R}_+$. We say that the set mapping $F : I \rightarrow K_c(\mathbb{R}^n)$ has a Hukuhara derivative $D_H F(t_0)$ at a point $t_0 \in I$, if

$$\lim_{h\rightarrow 0^+} \frac{F(t_0 + h) - F(t_0)}{h} \quad \text{and} \quad \lim_{h\rightarrow 0^+} \frac{F(t_0) - F(t_0 - h)}{h} \tag{2.4}$$

exist in the topology of $K_c(\mathbb{R}^n)$ and are equal to $D_H F(t_0)$.

By embeding $K_c(\mathbb{R}^n)$ as a complete cone in a corresponding Banach space and taking into account result on the differentiation of Bochner integral, we fin that if

$$F(t) = X_0 + \int_{t_0}^{t} \Phi(s)ds, \quad X_0 \in K_c(\mathbb{R}^n), \tag{2.5}$$

where $\Phi : I \rightarrow K_c(\mathbb{R}^n)$ is integrable in the sense of Bochner, then $D_H F(t)$ exists and the equality $D_H F(t) = \Phi(t)$ a.e on I holds.

The Hukuhara integral of F is given by

$$\int_I F(s)ds \Big[\int_I f(s)ds : \text{f is a continuous selector of F} \Big]$$

for any compact set $I \subset \mathbb{R}_+$.

Some properties of the Hukuhara integral are in [7].

If $F : I \rightarrow K_c(\mathbb{R}^n)$ is integrable, one has

$$\int_{t_0}^{t_2} F(s)ds = \int_{t_0}^{t_1} F(s)ds + \int_{t_1}^{t_2} F(s)ds, \quad t_0 \geq t_1 \geq t_2 \tag{2.6}$$

and

$$\int_{t_0}^{t} \lambda F(s)ds = \lambda \int_{t_0}^{t} F(s)ds, \quad \lambda \in \mathbb{R}. \tag{2.7}$$

If $F, G : I \rightarrow K_c(\mathbb{R}^n)$ are integrable, then $D[F(.), G(.)] : I \rightarrow \mathbb{R}$ is integrable and

$$D\Big[\int_{t_0}^{t} F(s)ds, \int_{t_0}^{t} G(s)ds \Big] \leq \int_{t_0}^{t} D[F(s), G(s)]ds. \tag{2.8}$$

3. Main Results

We consider SCDE:

$$D_H X(t) = F(t, X(t), U(t)), \quad X(t_0) = X_0 \in K_c(\mathbb{R}^n) \tag{3.1}$$

where $F \in C[I \times K_c(\mathbb{R}^n) \times K_c(\mathbb{R}^p), K_c(\mathbb{R}^n)]$, $t \in I \subset \mathbb{R}_+$, $U(t) \in K_c(\mathbb{R}^p)$, whose set solution is of the form $X(t) = X(t_0, X_0, t, U(t)) \in K_c(\mathbb{R}^n)$.

Definition 3.1. A state pair of set solutions $(X_0, X_1) \in K_c(\mathbb{R}^n)$ will be a controllable if after time t_1 we shall fin a set control $U(t) \in K_c(\mathbb{R}^p)$ such that

$$X(t_1) = X(t_0, X_0, t_1, U(t_1)) = X_1 \in K_c(\mathbb{R}^n). \tag{3.2}$$

Definition 3.2. The SCDE (3.1) is said to be

(GC) Global controllable if every state pair of set solutions $(X_0, X_1) \in K_c(\mathbb{R}^n)$ will be a controllable;

(GA) Global achievable if for every $X_1 \in K_c(\mathbb{R}^n)$ we have state pair of set solutions $(\theta^n, X_1) \in K_c(\mathbb{R}^n)$ that will be a controllable;

(GAZ) Global achievable to $\theta^n \in K_c(\mathbb{R}^n)$ if for every $X_1 \in K_c(\mathbb{R}^n)$ we have state pair of set solutions $(X_1, \theta^n) \in K_c(\mathbb{R}^n)$ that will be a controllable.

Suppose that, the set global achievable (GA) from $X_0 \in K_c(\mathbb{R}^n)$ after time t:

$$i. \quad R_t(X_0) = \Big\{X(t) \in K_c(\mathbb{R}^n) \,|\exists U(t) \in K_c(\mathbb{R}^p) : X(t_0, X_0, t, U(t)) = X(t)\Big\}; \tag{3.3}$$

$$ii. \quad R(X_0) = \bigcup_{t>o} R_t(X_0). \tag{3.4}$$

According to definitio 3.2, we have:

Theorem 3.3. *i. The SCDE (3.1) is (GC) if for every $X_0 \in K_c(\mathbb{R}^n)$ we have $R(X_0) = K_c(\mathbb{R}^n)$;*

ii. The SCDE (3.1) is (GA) if we have $R(\theta^n) = K_c(\mathbb{R}^n)$;

iii. The SCDE (3.1) is (GAZ) if for every $X_0 \in K_c(\mathbb{R}^n)$ we have $\theta^n \in R(X_0)$.

In the following we consider many kinds of SCDE :

A. Stationary linear set control differential equation (SLSCDE)

We consider SLSCDE:

$$D_H X(t) = AX(t) + BU(t), \quad X(t_0) = X_0 \in K_c(\mathbb{R}^n) \tag{3.5}$$

where A, B are operators: $A : K_c(\mathbb{R}^n) \to K_c(\mathbb{R}^n)$, $B : K_c(\mathbb{R}^p) \to K_c(\mathbb{R}^n)$, $t \in I \subset \mathbb{R}_+$, $U(t) \in K_c(\mathbb{R}^p)$ whose set solution is of the form $X(t) = X(t_0, X_0, t, U(t)) \in K_c(\mathbb{R}^n)$. Let the Cauchy operator for $D_H X(t) = AX(t)$ be: $W_0(t, s) = W_0(t, \tau)W_0(\tau, s)$ then set solution of SLSCDE (3.5) is rewrited in the form:

$$X(t) = X(t_0, X_0, t, U(t)) = W_0(t, t_0)X_0 + \int_{t_0}^{t} W_0(t, s)BU(s)ds. \tag{3.6}$$

If we put $L_t = \int_{t_0}^{t} W_0(t, s)W_0(t, t_0)ds$, then $L_{t_1} = \int_{t_0}^{t_1} W_0(t_1, s)W_0(t_1, t_0)ds$.

Theorem 3.4. *The SLSCDE (3.5) is (GC) if and only if exist B^{-1} and $L_{t_1}^{-1}$.*

Proof. a/ Sufficien condition: By $L_t = \int_{t_0}^{t} W_0(t, s)W_0(t, t_0)ds$, for every $t_1 > t_0$ we fin the control $U(t) \in K_c(\mathbb{R}^p)$ in type:

$$U(s) = -B^{-1}W_0(t_1, s)L_{t_1}^{-1}[W_0(t_1, t_0)X_0 - X_1]. \tag{3.7}$$

Replacing this control (3.7) into (3.6) at moment $t_1 > t_0$, we have

$$\begin{aligned} X(t_1) &= X(t_0, X_0, t_1, U(t_1)) = W_0(t_1, t_0)X_0 + \int_{t_0}^{t_1} W_0(t, s)BU(s)ds \\ &= W_0(t_1, t_0)X_0 - \int_{t_0}^{t_1} W_0(t, s)B\{B^{-1}W_0(t_1, s)L_{t_1}^{-1}[W_0(t_1, t_0)X_0 - X_1]\}ds \\ &= W_0(t_1, t_0)X_0 - \int_{t_0}^{t_1} W_0(t, s)BB^{-1}W_0(t_1, s)L_{t_1}^{-1}[W_0(t_1, t_0)X_0 - X_1]ds \\ &= W_0(t_1, t_0)X_0 - L_{t_1}L_{t_1}^{-1}[W_0(t_1, t_0)X_0 - X_1] \\ &= W_0(t_1, t_0)X_0 - [W_0(t_1, t_0)X_0 - X_1] = X_1. \end{aligned}$$

That means the state pair $(X_0, X_1) \in K_c(\mathbb{R}^n)$ will be controllable or SLSCDE (3.5) is (GC).

b/ Necessary condition: Let SLSCDE (3.5) is (GC) we can prove that exist B^{-1} and $L_{t_1}^{-1}$. The necessity of Theorem 3.4 is obvious. If given operator B has not B^{-1} that means B is decline then SLSCDE (3.5) isn't controllable. If $L_{t_1}^{-1}$ isn't existed then we have not the feedback control $U(t)$, so that SLSCDE (3.5) isn't (GC). □

B. Stationary linear set control differential equation with perturbation (SLSCDEP).

We consider SLSCDEP:

$$D_H X(t) = AX(t) + BU(t) + R(t, X(t), U(t)), \quad X(t_0) = X_0 \in K_c(\mathbb{R}^n) \tag{3.8}$$

where A, B are operators: $A : K_c(\mathbb{R}^n) \to K_c(\mathbb{R}^n)$, $B : K_c(\mathbb{R}^p) \to K_c(\mathbb{R}^n)$, $t \in I \subset \mathbb{R}_+$, $U(t) \in K_c(\mathbb{R}^p)$ whose set solution is form $X(t) = X(t_0, X_0, t, U(t)) \in K_c(\mathbb{R}^n)$. Let the Cauchy operator for $D_H X(t) = AX(t)$ is: $W_0(t, s) = W_0(t, \tau)W_0(\tau, s)$ then set solution is rewritten in the form:

$$X(t) = X(t_0, X_0, t, U(t)) = W_0(t, t_0)X_0 + \int_{t_0}^{t} W_0(t, s)BU(s)ds + \int_{t_0}^{t} R(t, X(s), U(s))ds. \tag{3.9}$$

If we put $L_t = \int_{t_0}^{t} W_0(t, s)W_0(t, t_0)ds$, then $L_{t_1} = \int_{t_0}^{t_1} W_0(t_1, s)W_0(t_1, t_0)ds$

$$X(t) = X(t_0, X_0, t, U(t)) = W_0(t, t_0)X_0 + \int_{t_0}^{t} W_0(t, s)BU(s)ds + \int_{t_0}^{t} R(t, X(s), U(s))ds.$$

Let

$$\lim_{\|X\| \to 0} \frac{\|R(t, X, U)\|}{\|X\|} = 0. \tag{3.10}$$

Theorem 3.5. *The SLSCDEP* (3.8) *with* (3.10) *is (GC) if and only if exist* B^{-1} *and* $L_{t_1}^{-1}$.

C. Unstationary linear set control differential equation (USLSCDE).

We consider USLSCDE:

$$D_H X(t) = A(t)X(t) + B(t)U(t), \quad X(t_0) = X_0 \in K_c(\mathbb{R}^n) \tag{3.11}$$

where A, B are operators, which defend on time t, $t \in I \subset \mathbb{R}_+$, $U(t) \in K_c(\mathbb{R}^p)$ whose set solution is form $X(t) = X(t_0, X_0, t, U(t)) \in K_c(\mathbb{R}^n)$. Let the Cauchy operator for $D_H X(t) = AX(t)$ is: $W_1(t,s) = W_1(t,\tau)W_1(\tau,s)$ then set solution of SLSCDE (3.11) is rewritted in the form:

$$X(t) = X(t_0, X_0, t, U(t)) = W_1(t, t_0)X_0 + \int_{t_0}^{t} W_1(t,s)B(s)U(s)ds \qquad (3.12)$$

If we put $\Lambda_t = \int_{t_0}^{t} W_1(t,s)W_1(t,t_0)ds$, then $\Lambda_{t_1} = \int_{t_0}^{t_1} W_1(t,s)W_1(t_1,t_0)ds$.

Theorem 3.6. *The SLSCDE* (3.11) *is (GC) if and only if exist* B^{-1} *and* $\Lambda_{t_1}^{-1}$.

Proof. a/ Sufficien condition: By $\Lambda_{t_1} = \int_{t_0}^{t_1} W_1(t,s)W_1(t_1,t_0)ds$, for every $t_1 > t_0$ we fin the control $U(t) \in K_c(\mathbb{R}^p)$ in type:

$$U(s) = -B^{-1}(s)W_1(t,s)\Lambda_{t_1}^{-1}[W_1(t,t_0)X_0 - X_1]. \qquad (3.13)$$

Replacing this control (3.13) in (3.12) at moment $t_1 > t_0$, we have

$$\begin{aligned}
X(t_1) &= X(t_0, X_0, t_1, U(t_1)) = W_1(t_1,t_0)X_0 + \int_{t_0}^{t_1} W_1(t_1,s)B(s)U(s)ds \\
&= W_1(t_1,t_0)X_0 - \int_{t_0}^{t_1} W_1(t_1,s)B(s)\{B^{-1}(s)W_1(t_1,s)\Lambda_{t_1}^{-1}[W_1(t_1,t_0)X_0 - X_1]\}ds \\
&= W_1(t_1,t_0)X_0 - \int_{t_0}^{t_1} W_1(t_1,s)B(s)B^{-1}(s)W_1(t_1,s)\Lambda_{t_1}^{-1}[W_1(t_1,t_0)X_0 - X_1]ds \\
&= W_1(t_1,t_0)X_0 - \Lambda_{t_1}\Lambda_{t_1}^{-1}[W_1(t_1,t_0)X_0 - X_1] \\
&= W_1(t_1,t_0)X_0 - [W_1(t_1,t_0)X_0 - X_1] = X_1.
\end{aligned}$$

That means the state pair $(X_0, X_1) \in K_c(\mathbb{R}^n)$ will be controllable or USLSCDE (3.11) is (GC).

b/ Necessary condition: Let USLSCDE (3.11) is (GC) we can poof that exist B^{-1} and $\Lambda_{t_1}^{-1}$. The necessity of Theorem 3.6 is obvious. If given operator $B(t)$ has not $B^{-1}(t)$ that means $B(t)$ is decline then USLSCDE (3.11) isn't controllable. If $\Lambda_{t_1}^{-1}$ isn't existed then we have not the feedback control, so that USLSCDE (3.11) isn't (GC). □

D. Unstationary linear set control differential equation with perturbation (USLSCDEP).

We consider USLSCDEP:

$$D_H X(t) = A(t)X(t) + B(t)U(t) + R(t, X(t), U(t)), \quad X(t_0) = X_0 \in K_c(\mathbb{R}^n) \qquad (3.14)$$

where $A(t), B(t)$ are operators, which defend on time t, $t \in I \subset \mathbb{R}_+, U(t) \in K_c(\mathbb{R}^p)$ and $R(t, X(t), U(t))$ is perturbation set.

The set solution of (3.14) is form $X(t) = X(t_0, X_0, t, U(t)) \in K_c(\mathbb{R}^n)$. Let the Cauchy operator for USLSCDE $D_H X(t) = A(t)X(t)$ is: $W_1(t,s) = W_1(t,\tau)W_1(\tau,s)$, then set solution of USLSCDEP (3.14) is rewritten in the form:

$$X(t) = X(t_0, X_0, t, U(t)) = W_1(t,t_0)X_0 + \int_{t_0}^{t} W_1(t,s)B(s)U(s)ds + \int_{t_0}^{t} R(t, X(s), U(s))ds \qquad (3.15)$$

Let $R(t, X(t), U(t))$ - a perturbation set that satisfie (3.10)

Theorem 3.7. *The USLSCDEP* (3.14) *with (3.10) is GC iff exist* $B^{-1}(s)$ *and* $\Lambda_{t_1}^{-1}$.

E. Nonlinear set control differential equation (NLSCDE).

We consider NLSCDE (3.1), that is:

$$D_H X(t) = F(t, X(t), U(t)), \quad X(t_0) = X_0 \in K_c(\mathbb{R}^n) \tag{3.16}$$

where $F \in C[I \times K_c(\mathbb{R}^n) \times K_c(\mathbb{R}^p), K_c(\mathbb{R}^n)], t \in I \subset \mathbb{R}_+, U(t) \in K_c(\mathbb{R}^p)$.

The set solution of NLSCDE (3.16) is writted in the form:

$$X(t) = X(t_0, X_0, t, U(t)) = X_0 + \int_{t_0}^{t} F(s, X(s), U(s))ds.$$

And the set solution NLSCDE (3.1) at t_1 is:

$$X(t_1) = X(t_0, X_0, t_1, U(t_1)) = X_0 + \int_{t_0}^{t_1} F(s, X(s), U(s))ds.$$

Lemma 3.8. *Suppose that, the Lyapunov-like function* $V(t, X(t), U(t))$ *(see* [18]*) satisfies the followings:*

$i.\quad V \in C[\mathbb{R}_+ \times K_c(\mathbb{R}^n) \times K_c(\mathbb{R}^p), K_c(\mathbb{R}^n)];$

$ii.\quad |V(t, X_1, U_1) - V(t, X_2, U_2)| \le L\big(D[X_1, X_2] + D[U_1, U_2]\big);$

$iii.\quad D^+V(t, X, U) \le 0;$

$iv.\quad b(\|X\|) \le V(t, X, U) \le a(t, \|X\|);$

for $t, L \in \mathbb{R}_+$, $X, X_1, X_2 \in K_c(\mathbb{R}^n)$, $U, U_1, U_2 \in K_c(\mathbb{R}^p)$, *where* $b(.), a(t, .) \in \kappa = \{\sigma \in C[\mathbb{R}_+, \mathbb{R}_+] : \sigma(\omega)$ *is increasing in* ω *and* $\sigma(\omega) \to \infty$ *as* $\omega \to \infty\}$ *then for every* $\alpha > 0$, $t_0 \in \mathbb{R}_+ \exists \beta = \beta(t, \alpha) > 0$ *such that*

$$\|X_0\| < \alpha \text{ implies } \|X(t)\| < \beta, \ \forall t > t_0. \tag{3.17}$$

Proof. Let $\alpha > 0, t_0 \in \mathbb{R}_+$ be given.

We can choose $\beta = \beta(t, \alpha) > 0$ such that

$$a(t_0, \alpha) < b(\beta) \tag{3.18}$$

then (3.17) holds. If this is not true, there would exists $X(t) = X(t_0, X_0, t, U(t)) = X_0 + \int_{t_0}^{t} F(s, X(s), U(s))ds$, which set solution of NLSCDE (3.16) and $\exists t_1 > t_0$ such that $\|X(t_1)\| = \beta$. Because the Lyapunov-like function $V(t, X(t), U(t))$ is decreasing then

$$V(t, X(t), U(t)) \le V(t_1, X(t_1), U(t_1)), \ \forall t > t_1$$

or $\|X(t)\| \le \beta$, as results, condition iv/ of Lemma 3.8 and (3.18) yield

$$\begin{aligned} b(\beta) &\le b(\|X(t_1)\|) \le V(t_1, X(t_1), U(t_1)) \le V(t_0, X(t_0), U(t_0)) \\ &\le a(t_0, \|X_0\|) < a(t_0, \alpha) < b(\beta). \end{aligned}$$

This contradiction proves (3.17), completing the proof. □

Theorem 3.9. *Suppose that for NLSCDE* (3.16) *satisfy the followings:*

i. *The Lyapunov-like function* $V(t, X(t), U(t))$ *satisfies the conditions of Lemma* 3.1*;*

ii. $F \in C[I \times K_c(\mathbb{R}^n) \times K_c(\mathbb{R}^p), K_c(\mathbb{R}^n)]$, $\|F(t, X(t), U(t))\| \leq \bar{L}(\|X\| + \|U\|), \bar{L} > 0$*;*

iii. *For every pair of set states* $(X_0, X_1) \in K_c(\mathbb{R}^n), \forall t_1 > t_0, \exists U(t) \in K_c(\mathbb{R}^p)$ *such that*

$$\int_{t_0}^{t_1} \|U(s)\| ds = \frac{\|X_1\| \exp[-\bar{L}(t_1 - t_0)] - \|X_0\|}{\bar{L}} \tag{3.19}$$

then NLSCDE (3.16) *is (GC).*

Proof. Before we choose the controllable set $U(t) \in K_c(\mathbb{R}^p)$ satisfie (3.19).

Because the set solution of (3.16) is written in the form

$$X(t) = X(t_0, X_0, t, U(t)) = X_0 + \int_{t_0}^{t} F(s, X(s), U(s)) ds.$$

then $X(t_1) = X(t_0, X_0, t_1, U(t_1)) = X_0 + \int_{t_0}^{t_1} F(s, X(s), U(s)) ds$ implies that

$$\begin{aligned} \|X(t_1)\| &\leq \|X_0\| + \int_{t_0}^{t_1} \|F(s, X(s), U(s))\| ds \\ &\leq \|X_0\| + \int_{t_0}^{t_1} \bar{L}(\|X(s)\| + \|U(s)\|) ds \\ &\leq \|X_0\| + \bar{L} \int_{t_0}^{t_1} \|X(s)\| ds + \bar{L} \int_{t_0}^{t_1} \|U(s)\| ds \end{aligned} \tag{3.20}$$

Replacing $\int_{t_0}^{t_1} \|U(s)\| ds = \frac{\|X_1\| \exp[-\bar{L}(t_1 - t_0)] - \|X_0\|}{\bar{L}}$ from (3.19) into (3.20) we received $\|X(t_1)\| \leq \|X_1\|$.

The proof is complete, if we prove that $\|X(t_1)\| = \|X_1\|$. If this is not true, that if $\|X(t_1)\| < \|X_1\|$. But in this case we have by Lemma 3.8, $\|X(t_1)\| = \beta$ implies:

$$b(\beta) \leq b(\|X(t_1)\|) \leq V(t_1, X(t_1), U(t_1)) \leq V(t_1, X_1, U(t_1)) \leq a(t_1, \|X_1\|) < a(t_1, \beta) < b(\beta)$$

This contradiction proves $\|X(t_1)\| = \|X_1\|$. That means the state pair $(X_0, X_1) \in K_c(\mathbb{R}^n)$ will be controllable, the NLSCDE (3.16) is (GC). □

Theorem 3.10. *Suppose that for NLSCDE* (3.16) *satisfy the followings:*

i. *The Lyapunov-like function* $V(t, X(t), U(t))$ *satisfies the conditions of Lemma* 3.1*;*

ii. $F \in C[I \times K_c(\mathbb{R}^n) \times K_c(\mathbb{R}^p), K_c(\mathbb{R}^n)]$, $\|F(t, X(t), U(t))\| \leq \bar{L}(\|X\| + \|U\|), \bar{L} > 0$*;*

iii. *For every pair of set states* $(X_0, X_1) \in K_c(\mathbb{R}^n), \forall t_1 > t_0, \exists U(t) \in K_c(\mathbb{R}^p)$ *such that* $U(t) = \lambda(t) X_1$*, where the control operator* $\lambda(t)$ *satisfies:*

$$\int_{t_0}^{t_1} \|\lambda(s)\| ds = \frac{\|X_1\| \exp[-\bar{L}(t_1 - t_0)] - \|X_0\|}{\bar{L}\|X_1\|} \tag{3.21}$$

then NLSCDE (3.16) *is (GC).*

Proof. Before we choose the controllable set: $U(t) = \lambda(t)X_1$, whose $\lambda(t)$ satisfie (3.21).

Because the set solution of (3.16) is

$$X(t) = X(t_0, X_0, t, U(t)) = X_0 + \int_{t_0}^{t} F(s, X(s), U(s))ds.$$

then

$$X(t_1) = X(t_0, X_0, t_1, U(t_1)) = X_0 + \int_{t_0}^{t_1} F(s, X(s), U(s))ds$$

implies that

$$\begin{aligned}\|X(t_1)\| &\le \|X_0\| + \int_{t_0}^{t_1} \|F(s, X(s), U(s))\|ds \\ &\le \|X_0\| + \int_{t_0}^{t_1} \bar{L}(\|X(s)\| + \|U(s)\|)ds \\ &\le \|X_0\| + \bar{L}\int_{t_0}^{t_1} \|X(s)\|ds + \bar{L}\int_{t_0}^{t_1} \|U(s)\|ds \end{aligned} \tag{3.22}$$

Replacing

$$\int_{t_0}^{t_1} \|\lambda(s)\|ds = \frac{\|X_1\| \exp[-\bar{L}(t_1 - t_0)] - \|X_0\|}{\bar{L}\|X_1\|}$$

from (3.21) into (3.22) we received $\|X(t_1)\| \le \|X_1\|$.The proof is complete, if we prove that $\|X(t_1)\| = \|X_1\|$.

If this is not true, that if $\|X(t_1)\| < \|X_1\|$. But in this case we have by Lemma 3.8, $\|X(t_1)\| = \beta$ implies:

$$b(\beta) \le b(\|X(t_1)\|) \le V(t_1, X(t_1), U(t_1)) \le V(t_1, X_1, U(t_1)) \le a(t_1, X_1) < a(t_1, \beta) < b(\beta)$$

This contradiction proves $\|X(t_1)\| = \|X_1\|$. That means the state pair $(X_0, X_1) \in K_c(\mathbb{R}^n)$ will be controllable, the NLSCDE (3.16) is (GC). □

Corollary 3.11. *In the case, when $p = n$ we must choose the control operator $\lambda(t)$ as real function such that for every pair of set states $(X_0, X_1) \in K_c(\mathbb{R}^n), \forall t_1 > t_0, \exists U(t) \in K_c(\mathbb{R}^p)$ with $U(t) = \lambda(t)X_1$, where this function $\lambda(t)$ satisfies:*

$$\int_{t_0}^{t_1} |\lambda(s)|ds = \frac{\|X_1\| \exp[-L(t_1 - t_0)] - \|X_0\|}{L\|X_1\|} \tag{3.23}$$

then NLSCDE (3.16) *is (GC).*

Corollary 3.12. *i. If approximation of $F(t, X(t), U(t)) \approx AX + BU + R(t, X, U)$ then problem (GC) of NLSCDE* (3.16) *will be a problem (GC) of USLSCDEP* (3.8)*;*

ii. If approximation of $F(t, X(t), U(t)) \approx A(t)X + B(t)U + R(t, X, U)$ then problem (GC) of NLSCDE (3.16) will be a problem (GC) of USLSCDEP (3.14).

We present some examples for two cases:

Example 3.1. Global controllable case: The movement of the oil trace on the sea will be controllable. Because for every pair of set states $(X_0, X_1) \in K_c(\mathbb{R}^n)$, $\forall t_1 > t_0$, $\exists U(t) \in K_c(\mathbb{R}^p)$, such that $U(t) = \lambda(t)X_1$, where the control operator $\lambda(t)$ satisfies

$$\int_{t_0}^{t_1} \|\lambda(s)\|ds = \frac{\|X_1\| \exp[-L(t_1 - t_0)] - \|X_0\|}{\|LX_1\|}$$

and NLSCDE (3.16) will be (GC).

Example 3.2. Uncontrollable case: We consider the problem of transfer from one place to another of chemical concentrations in environment. This motion of the chemical cloud on the air will uncontrollable, because it moves somewhere else and we can't fin any feedback $U(t) = h(X(t))$.

4. Conclusion

In this paper we give a concept of Problem of Global controllable for SCDE and study it with many kinds of feedback controls (3.7), (3.13), (3.19) and (3.21). Some more results on partial differential operators in Haussdorff metric space $K_c(\mathbb{R}^n)$ and there applications will be presented in next works.

Acknowledgments

The author wish to thank the referee for his (her) careful reading and valuable remarks which improve the presentation of the paper.

References

[1] O. Kaleva, Fuzzy differential equations, *J. Fuzzy Sets and Systems* **24** (1987), 301–317.

[2] O. Kaleva, Fuzzy differential equations, *J. Fuzzy Sets and Systems* **35** (1990), 389–396.

[3] O. Kaleva, Fuzzy differential equations, *J. Fuzzy Sets and Systems* **24** 1987), 301–317.

[4] V. Lakshmikantham Set differential equations versus fuzzy differential equations, *J. Applied Mathematics and Computation* **164** (2005) , 287–294.

[5] V. Lakshmikantham and S. Leela, Fuzzy differential systems and the new concept of stability, *J. Nonlinear Dynamics and Systems Theory* **1(2)** (2001), 111–119.

[6] V. Lakshmikantham and R. Mohapatra, *Theory of fuzzy differential equations and inclusions*, Taylor Francis, London, 2003.

[7] V. Lakshmikantham, T. Gnana Bhaskar and J. Vasundhara Devi, *Theory of set differential equations in metric spaces*, Cambridge Scientifi Publisher, UK, 2006.

[8] N. D. Phu, *Genaral views in theory of systems*, VNU – Publishing House, HCM City, 2003.

[9] N. D. Phu and N.T. Huong, *Multivalued Differential Equations*, VNU – Publishing House, HCM City, 2005.

[10] N. D. Phu and T. T. Tung, The comparison of sheaf-solutions in fuzzy control problems, *J. Science and Technology Development* **9 (2)** (2006), 5–10.

[11] N. D. Phu and T. T. Tung, Some properties of sheaf-solutions of sheaf fuzzy control problems, *E. Journal of Differential Equations* **2006**, N. 108, 1–8.

[12] N. D. Phu and T. T. Tung, Some results on sheaf solutions of sheaf set control problems, *J. Nonlinear Analysis TMA* **9** (2007), 1309–1315.

[13] N. D. Phu and T. T. Tung, Sheaf optimal control problems in fuzzy type, *J. Science and Technology Development* **8 (12)** (2005), 5–11.

[14] N. D. Phu and T. T. Tung, Existence of solutions of fuzzy control differential equations, *J. Science and Technology Development* **10 (5)** (2007), 5–12.

[15] N. D. Phu and T. T. Tung, Some new results on the fuzzy control differential equations, *J. Science and Technology Development* **11 (1)** (2008), 5–20.

[16] N. D. Phu and T. T. Tung, Existence of solutions of set control differential equations, *J. Science and Technology Development* **10 (6)** (2007), 5–14.

[17] N. D. Phu, L. T. Quang and T. T. Tung, Stability criteria for set control differential equations, *J. Nonlinear Analysis TMA* **69** (2008), 3715–3721.

[18] N. D. Phu, L. T. Quang and T. T. Tung, Criteria for boundedness of solutions for set control differential equations, *J. Evolution Equations* (accepted).

In: Evolution Equations: Theories, Solutions and Functions
Editor: Gaston M. N'Guerekata
ISBN 978-1-61324-090-8

Chapter 10

Some Results about Uniform Exponential Stability of Linear Skew-Evolution Semiflows

Pham Viet Hai*
Faculty of Mathematics, Mechanics and Informatics,
College of Science, Viet Nam National University, Ha Noi,
334, Nguyen Trai Road, Thanh Xuan Dist., Ha Noi, Viet Nam

Abstract

In this paper we study the concept "linear skew-evolution semifl ws" and give several characterizations for the property of the uniform exponential stability of linear skew-evolution semifl ws on Banach and Hilbert spaces. By extending some of previous techniques, we obtain some generalization of notable results due to R. Datko, S. Rolewicz, Zabczyk and J.M.A.M van Neerven.

2000 Mathematics Subject Classification: 93D20, 34E05, 39A12.

Keywords and phrases: Exponential stability, linear skew-evolution semifl ws, function spaces.

1. Introduction

In the last three decades a great number of papers about the qualitative behaviour was published. We will summarize some characterizations of uniform exponential stability as it will be presented below.

The firs important result due to Datko, Pazy and Rolewicz and is given by:

Theorem 1.1. *C_0-semigroup $(T(t))_{t\geq 0}$ is uniformly exponentially stable if and only if there exists $p \in [1, \infty)$ such that:*

$$\int_0^\infty \|T(\tau)x\|^p \, d\tau < \infty$$

*E-mail address: phamviethai86@gmail.com

for all $x \in X$.

Datko proved this result for the case $p = 2$ in Hilbert space ([6]) and then Pazy extended for the case $p \in [1, \infty)$ in Banach space ([17]). Ichikawa proved for the case nonlinear semigroups ([18]).

The second important result is given by

Theorem 1.2. *C_0-semigroup $(T(t))_{t\geq 0}$ is uniformly exponentially stable if and only if there exists a strictly positive, non-decreasing function φ such that:*

$$\int_0^\infty \varphi(\|T(\tau)x\|)\, d\tau \ < \ \infty$$

for all $x \in X$.

Rolewicz proved the result in [13]. And Litman gave another proof of this theorem in [19]. Then R. Datko and S. Rolewicz proved for evolution family $\mathcal{U} = \{U(t, s)\}_{t \geq s \geq 0}$, with uniform exponential growth.

A new idea has been presented by Neerven in [9] in which he used Banach function spaces to study the asymptotic behaviour of C_0-semigroups.

Theorem 1.3. *Let E be a Banach function space over $\mathbb{R}_+$ with*

$$\lim_{t\to\infty} \varphi_E(t) \ = \ \infty$$

where $\varphi_E(t) \ := \ \left\|\mathcal{X}_{[0,\, t)}\right\|_E$ and C_0-semigroup $(T(t))_{t\geq 0}$ on a Banach space X. If for all $x \in X$, the maps $t \ \to \ \|T(t)x\|$ belongs to E, then $(T(t))_{t\geq 0}$ is uniformly exponentially stable.

This interesting idea can be generalized to study the uniform exponential stability of linear skew-product semifl ws. In [12], M. Megan, A.L. Sasu, B. Sasu proved necessary and sufficien conditions for the uniform exponential stability of evolution equation on Banach spaces, by emphasizing techniques from the domain of linear skew-product semifl ws and of some function spaces which is stated as follows.

Theorem 1.4. *The linear skew product semiflow $\pi_0 = (\Phi_0, \sigma_0)$ is uniformly exponentially stable if and only if there exist $B \in \mathcal{B}(\mathbb{N})$ and a sequence (t_n) of positive real numbers satisfying the following properties*

1. $$\sup_{n\in \mathbb{N}} \mid t_{n+1} - t_n \mid < \infty$$

2. *The function $\varphi_{x,\,\theta}(.) : \mathbb{N} \ \to \ \mathbb{R}_+$ defined by*

   $$\varphi_{x,\,\theta}(n) \ = \ \|\Phi_0(\theta, t_n)x\|$$

   belongs to B.

3. *There exists* $K : X \to (0, \infty)$ *such that:*

$$\|\varphi_{x,\,\theta}\|_B \leq K(x)$$

for all $(x, \theta) \in \mathcal{E}$.

Theorem 1.5. *The linear skew-product semiflow* $\pi_0 = (\Phi_0, \sigma_0)$ *is uniformly exponentially stable if and only if there exist a non-decreasing function* $N : \mathbb{R}_+ \to \mathbb{R}_+$, *a sequence* $(t_n) \in \mathbb{R}_+$ *and a constant* K *with the following properties:*

1. $N(0) = 0$ *and* $N(t) > 0$ *for all* $t > 0$

2.

$$\sup_{n \in \mathbb{N}} \mid t_{n+1} - t_n \mid < \infty$$

3. *for each* $x \in X$ *there exists* $\alpha(x) > 0$ *such that:*

$$\sum_{n=0}^{\infty} N(\alpha(x) \, \|\Phi_0(\theta, t_n)x\|) \leq K, \quad \theta \in \ominus$$

In next section, we study the concept of linear skew-evolution semifl ws which can be considered generalizations for evolution operators and skew-product semifl ws. And then we give some conditions for the uniform exponential stability of linear skew-evolution semifl ws on Banach and Hilbert spaces.

2. Preliminaries

2.1. Linear Skew Evolution Semiflow

Let X a Banach space, $(\ominus, d)$ a metric space. We denote by $\mathcal{L}(X)$ be the Banach algebra of all bounded linear operators acting on X; $T = \{(t, s) \in \mathbb{R}_+^2 \mid t \geq s\}$ and $\Delta = \{(m, n) \in \mathbb{N}^2 \mid m \geq n\}$

Definition 2.1. A continous mapping $\sigma : T \times \ominus \to \ominus$ is called a evolution semifl w on $\ominus$ if:

1. $\sigma(t, t, \theta) = \theta$

2. $\sigma(t, s, \sigma(s, r, \theta)) = \sigma(t, r, \theta)$, for all $t \geq s \geq r \geq 0$; $\theta \in \ominus$

Definition 2.2. A pair $\pi = (\Phi, \sigma)$ is called a linear skew-evolution semifl w on $\mathcal{E} = X \times \ominus$ if σ is a evolution semifl w on $\ominus$ and $\Phi : T \times \ominus \to \mathcal{L}(X)$ satisfie the following conditions:

1. $\Phi(t, t, \theta) = I$ the identity operator on X, for all $(t, \theta) \in \mathbb{R}_+ \times \ominus$.

2. $\Phi(t, r, \theta) = \Phi(t, s, \sigma(s, r, \theta))\Phi(s, r, \theta)$ for all $t \geq s \geq r \geq 0$; $\theta \in \ominus$.

3. There are two constants M, ω such that:

$$\|\Phi(t,s,\theta)x\| \leq Me^{\omega(t-s)}\|x\|$$

for all $((t,s),\theta,x) \in T \times \Theta \times X$.

Example 2.3. Let Θ be a metric space, σ a evolution semifl w on Θ. $\mathcal{W} = \{W(t)\}_{t\geq 0}$ is a C_0-semigroup on X and $A : D(A) \subset X \to X$ its infinitesima generator. It is well known that the Cauchy problem

$$\begin{cases} u'(t) &= Au(t), \quad t \geq 0 \\ u(0) &= x \end{cases} \tag{2.1}$$

is well-posed and the mild solution of (2.1) is given by $u(t) = W(t)x$. Then $\pi = (\Phi, \sigma)$, where $\Phi(t,s,\theta) = W(t-s)$ for all $((t,s),\theta) \in T \times \Theta$ is a linear skew-evolution semifl w on $\mathcal{E} = X \times \Theta$ which is called the linear skew-evolution semifl w generated by the C_0-semigroup $\mathcal{W}$ and the evolution semifl w σ.

Following examples will emphasize that the skew-evolution semifl ws are generalizations of the evolution operators and of the skew-product semifl ws.

Example 2.4. If $\mathcal{U} = \{U(t,s)\}_{t\geq s\geq 0}$ is the evolution family and σ is a evolution semifl w then we can defin

$$\Phi(t,s,\theta) := U(t,s)$$

for all $\theta \in \Theta$. Thus $\pi = (\Phi, \sigma)$ is a linear skew-evolution semifl w.

Example 2.5. It is well known that if X is a Banach space, Θ is a compact metric space, σ_0 is a fl w on Θ and $A_0 : \Theta \to \mathcal{L}(X)$ is a continous mapping, then $\Phi_0(\theta,t)$, the solution operator of the linear differential equation

$$\begin{cases} u'(t) &= A_0(\sigma_0(\theta,t))u(t), \quad t \geq 0 \\ u(0) &= x \end{cases} \tag{2.2}$$

is a cocycle and the pair $\pi_0 = (\Phi_0, \sigma_0)$ is a linear skew-product semifl w. Often, (2.2) aries from the linearization of nonlinear equations. Formore details about linear skew-product semifl w we refer to [2], [23], [11], [12],.... We can defin a linear skew-evolution semifl w:

$$\sigma(t,s,\theta) := \sigma_0(\theta, t-s)$$
$$\Phi(t,s,\theta) := \Phi_0(\theta, t-s)$$

Example 2.6. Let $\{U(t,s)\}_{t\geq s\geq 0}$ be an evolution family, with uniform exponential growth. If there exists a bounded strongly continous family of idempotent operators $\{P(\theta)\}_{\theta\in\Theta}$, with the property that

$$P(\theta)U(t,s) = U(t,s)P(\theta), \quad t \geq s \geq 0, \quad \theta \in \Theta$$

then the pair $\pi = (\Phi, \sigma)$ define by

$$\sigma(t,s,\theta) = \theta, \quad \Phi(t,s,\theta) = P(\theta)U(t,s)$$

is a linear skew-evolution semifl w.

Example 2.7. Let Θ be a compact metric space, σ a evolution semifl w on Θ, X a Banach space and $A : \Theta \to \mathcal{L}(X)$ a continous map. If $\Phi(t, t_0, \theta)x$ is the solution of the Cauchy problem

$$u'(t) = A(\sigma(t, t_0, \theta))u(t), \quad t \geq t_0 \tag{2.3}$$

then the pair $\pi = (\Phi, \sigma)$ is a linear skew-evolution semifl w. Equation (2.3) is the starting point of our paper.

Definition 2.8. The linear skew-evolution semifl w $\pi = (\Phi, \sigma)$ is said to be uniformly exponentially stable if there are two constants $K > 0$ and $\nu > 0$ such that:

$$\|\Phi(t, s, \theta)x\| \leq Ke^{-\nu(t-s)} \|x\|$$

for all $((t, s), \theta, x) \in T \times \Theta \times X$.

2.2. Function Spaces, Sequence Spaces

This section shall present some definitions notations and results about normed function and sequence spaces. Formore details about function and sequence spaces we refer to [9] (page 219), [15], [16], [20]. Let (Ω, Σ, μ) be a positive σ-finit measure space. $M(\mu)$ is the linear space of μ-measure functions $f : \Omega \to \mathbb{C}$, identifying the functions which are equal μ-a.e.

Definition 2.9. A Banach function norm is a function $N : M(\mu) \to [0; \infty]$ with the following properties:

1. $N(f) = 0$ if and only if $f = 0$ μ-a.e.
2. If $| f | \leq | g |$ μ-a.e. then $N(f) \leq N(g)$
3. $N(af) = | a | N(f)$ for all $a \in \mathbb{C}$ and $f \in M(\mu)$ with $N(f) < \infty$
4. $N(f + g) \leq N(f) + N(g)$ for all $f, g \in M(\mu)$

Let $B = B_N$ be the set

$$B := \{f \in M(\mu) : \|f\|_B := N(f) < \infty\}$$

It is easy to see that $(B; \|.\|_B)$ is a normed linear space. If B is complete then B is called Banach function space over Ω.

If $| f | < | g |$ μ-a.e. and $g \in B$ then also $f \in B$ and:

$$\|f\|_B \leq \|g\|_B$$

Let $(\Omega, \Sigma, \mu) = (\mathbb{N}, \mathcal{P}(\mathbb{N}), \mu_c)$ where μ_c is the countable measure and B be a Banach function space over $\mathbb{N}$ (or Banach sequence space). We defin the fundamental function of B by

$$F_B : \mathbb{N}^* \to \mathbb{R}_+ \cup \{\infty\}$$

$$F_B(n) = \begin{cases} \left\|\mathcal{X}_{\{0\,;1\,;\ldots;\,n-1\}}\right\|_B \; ; \; \mathcal{X}_{\{0\,;1\,;\ldots;\,n-1\}} \in B \\ \infty \; ; \; \mathcal{X}_{\{0\,;\,1;\ldots;\,n-1\}} \notin B \end{cases}$$

$\mathcal{B}(\mathbb{N})$ is the set of all Banach sequence spaces B with following properties

$$\lim_{n \to \infty} F_B(n) = \infty$$

and

$$\inf_{n \in \mathbb{N}} \|\mathcal{X}_{\{n\}}\|_B > 0$$

Example 2.10. If $p \in [1, \infty)$ then $B = l^p$ with:

$$\|s\|_p := \left(\sum_{n=0}^{\infty} | s(n) |^p\right)^{\frac{1}{p}}$$

is a Banach sequence space which belongs to $\mathcal{B}(\mathbb{N})$.

Example 2.11 (Orlicz sequence spaces). Let $g : \mathbb{R}_+ \to \mathbb{R}_+ \cup \{\infty\}$ be a non-decreasing, left continous which is not identically 0 or ∞ on $(0; \infty)$. We defin the Young function associated to g:

$$Y_g(t) := \int_0^t g(\tau)\, d\tau$$

For each $s : \mathbb{N} \to \mathbb{C}$, we put:

$$M_g(s) := \sum_{n=0}^{\infty} Y_g(| s(n) |)$$

The set O_g of all sequences with the property that there exists $k > 0$ such that $M_g(ks) < \infty$. It is obvious that O_g is a linear space. With respect to the norm:

$$\|s\|_g = \inf\{k > 0 : M_g\left(\frac{1}{k}s\right) \leq 1\}$$

it is a Banach sequence space called Orlicz sequence space.

3. Main Results

Lemma 3.1. *If there are two constants* $p \in \mathbb{N}^*$ *and* $c \in (0; 1)$ *(doesn't belong to* (x, θ, m)*) such that:*

$$\|\Phi(p + m, m, \theta)x\| \leq c\,\|x\|$$

for all $(x, \theta, m) \in X \times \Theta \times \mathbb{N}$*, then the linear skew-evolution semiflow* $\pi = (\Phi, \sigma)$ *is uniformly exponentially stable.*

Proof. For each $(t, s) \in \mathbb{R}_+ \times \mathbb{R}_+$:
If $t \in [0; 1]$:

$$\begin{aligned} \|\Phi(t + s, s, \theta)x\| &\leq Me^{\omega t}\,\|x\| \\ &\leq Me^{\omega}\,\|x\| \end{aligned}$$

If $t \geq 1$:

$$t + s \;\geq\; [t] + [s] \;\geq\; 1 + [s] \;>\; s$$

There exist $k \in \mathbb{N}$ and $r \in [0, p)$ such that:

$$[t] - 1 \;=\; kp + r$$

$$k \;>\; \frac{[t] - 1}{p} - 1 \;>\; \frac{t - 2}{p} - 1$$

It is clear that:

$$\begin{aligned}
\|\Phi(t + s, s, \theta)x\| &\leq M e^{\omega(t+s-[t]-[s])} \|\Phi(kp + r + 1 + [s], s, \theta)x\| \\
&\leq M e^{\omega(t+s-[t]-[s])} M e^{\omega r} \|\Phi(kp + 1 + [s], s, \theta)x\| \\
&\leq M^2 e^{\omega(t+s-[t]-[s])} e^{\omega p} c^k \|\Phi(1 + [s], s, \theta)x\| \\
&\leq M^3 e^{\omega(t+s-[t]-[s])} e^{\omega p} c^k e^{\omega(1+[s]-s)} \|x\| \\
&\leq M^3 e^{\omega p + 2\omega} c^{\frac{t-2}{p} - 1} \|x\|
\end{aligned}$$

where M, ω in Definitio 2.2. So Lemma is proved. □

Throughout the paper we shall denote:

$$U \;=\; \{x \in X : \; \|x\| \;=\; 1\}$$

Theorem 3.2. *The linear skew-evolution semiflow* $\pi = (\Phi, \sigma)$ *is uniformly exponentially stable if and only if there are* $B \in \mathcal{B}(\mathbb{N})$ *and a constant* $L > 0$ *such that:*

1. *The mapping:* $\varphi(x, \theta, m, .) : \mathbb{N} \rightarrow \mathbb{R}_+$

   $$\varphi(x, \theta, m, n) \;:=\; \|\Phi(m + n, m, \theta)x\|$$

   $\varphi(x, \theta, m, .) \in B$ *for all* $(x, \theta, m) \in U \times \ominus \times \mathbb{N}$

2. $\|\varphi(x, \theta, m, .)\|_B \leq L$ *for all* $(x, \theta, m) \in U \times \ominus \times \mathbb{N}$

Proof. Necessity: $\pi = (\Phi, \sigma)$ is uniformly exponentially stable. There are $K, \nu > 0$ such that:

$$\|\Phi(m + n, m, \theta)x\| \;\leq\; K e^{-\nu n}$$

for all $(m, n, x) \in \mathbb{N} \times \mathbb{N} \times U$.

$$\begin{aligned}
\sum_{n=0}^{\infty} \|\Phi(m + n, m, \theta)x\| &\leq K \sum_{n=0}^{\infty} c^{-\nu n} \\
&= \frac{K}{1 - e^{-\nu}}
\end{aligned}$$

So $B := l^1$ and $\varphi(x, \theta, m, .) \in B$.

Sufficien y: Let $x \in U$ and $B \in \mathcal{B}(\mathbb{N})$

$$\begin{cases} \lim_{n \to \infty} F_B(n) &= \infty \\ c := \inf_{n \in \mathbb{N}} \|\mathcal{X}_{\{n\}}\|_B &> 0 \end{cases}$$

We see:

$$\begin{aligned} \varphi(x, \theta, m, n)\mathcal{X}_{\{n\}} &\leq \varphi(x, \theta, m, .) \\ L &\geq \|\varphi(x, \theta, m, .)\|_B \\ &\geq \varphi(x, \theta, m, n) \|\mathcal{X}_{\{n\}}\|_B \\ &\geq \varphi(x, \theta, m, n)c \\ \varphi(x, \theta, m, n) &\leq \frac{L}{c} \end{aligned}$$

For $j \in \{0; ...; n\}$ and $\theta_1 := \sigma(m + j, m, \theta)$ we see:

$$\begin{aligned} \|\Phi(m+n, m, \theta)x\| &= \|\Phi(m+n, m+j, \theta_1)\Phi(m+j, m, \theta)x\| \\ &\leq \frac{L}{c} \|\Phi(m+j, m, \theta)x\| \\ \varphi(x, \theta, m, n)\mathcal{X}_{\{0;...;n\}} &\leq \frac{L}{c}\varphi(x, \theta, m, .) \\ \varphi(x, \theta, m, n)F_B(n+1) &\leq \frac{L}{c} \|\varphi(x, \theta, m, .)\|_B \\ &\leq \frac{L^2}{c} \end{aligned}$$

$$\lim_{n \to \infty} \varphi(x, \theta, m, n) = 0$$

There exists $n_0 \in \mathbb{N}$ such that:

$$\|\Phi(m+n_0, m, \theta)x\| \leq \frac{1}{2}$$

By Lemma 3.1, Theorem is proved. □

Corollary 3.3. *The linear skew-evolution semiflow $\pi = (\Phi, \sigma)$ is uniformly exponentially stable if and only if there are $p \in [1; \infty)$ and a constant $K > 0$ such that:*

$$\sum_{n=0}^{\infty} \|\Phi(m+n, m, \theta)x\|^p \leq K$$

for all $(m, \theta, x) \in \mathbb{N} \times \ominus \times U$.

Proof. The necessity is obvious.

The sufficien y: Theorem 3.2 for $B = l^p$. □

Let $\mathcal{N}$ be the set of all non-decreasing continuous functions:

$$N : [0;\infty) \rightarrow [0;\infty)$$

with properties:

$$\begin{cases} N(0) = 0 \\ N(t) > 0 \, ; \, t > 0 \end{cases}$$

Theorem 3.4. *The linear skew-evolution semiflow* $\pi = (\Phi, \sigma)$ *is uniformly exponentially stable if and only if there are* $N \in \mathcal{N}$ *and a constant L such that:*

$$\sum_{j=0}^{\infty} N(\|\Phi(m+j,m,\theta)x\|) \leq L$$

for all $(m, \theta, x) \in \mathbb{N} \times \ominus \times U$.

Proof. The necessity is obvious for $N(t) = t$

The sufficien y: $M; \omega$ in Definitio 2.2.

Step 1: We prove there is L_1 such that:

$$\|\Phi(m+n,m,\theta)\| \leq L_1$$

for all $((m,n),\theta) \in \mathbb{N}^2 \times \ominus$. Indeed:

Let $n_0 \in \mathbb{N}$ such that:

$$L \leq n_0 N(1)$$

We put $y = \dfrac{x}{Me^{\omega n_0}}$ where $x \in U$.

If $n \leq n_0$:

$$\|\Phi(m+n,m,\theta)\| \leq Me^{\omega n} \leq Me^{\omega n_0}$$

If $n \geq n_0$, we put $\theta_j = \sigma(m+j,m,\theta)$:

$$\begin{aligned}
& n_0 N(\|\Phi(m+n,m,\theta)y\|) \\
= & \sum_{j=n-n_0+1}^{n} N(\|\Phi(m+n,m,\theta)y\|) \\
= & \sum_{j=n-n_0+1}^{n} N(\|\Phi(m+n,m+j,\theta_j)\Phi(m+j,m,\theta)y\|) \\
\leq & \sum_{j=n-n_0+1}^{n} N(Me^{\omega(n-j)} \|\Phi(m+j,m,\theta)y\|) \\
\leq & \sum_{j=n-n_0+1}^{n} N(\|\Phi(m+j,m,\theta)x\|) \\
\leq & \; L
\end{aligned}$$

$$\begin{aligned} L &\geq n_0 N(\|\Phi(m+n,m,\theta)y\|) \\ &\geq \frac{L}{N(1)} N(\|\Phi(m+n,m,\theta)y\|) \\ 1 &\geq \|\Phi(m+n,m,\theta)y\| \\ Me^{\omega n_0} &\geq \|\Phi(m+n,m,\theta)x\| \\ Me^{\omega n_0} &\geq \|\Phi(m+n,m,\theta)\| \end{aligned}$$

Step 2: We prove there are B; L which satisfy Theorem 3.2. Indeed, we put:

$$\beta := \min\left\{1; \frac{1}{LL_1}\right\}$$

and

$$\begin{aligned} z &:= \beta x \; ; \; x \in U \\ \varphi(x,\theta,m,n) &:= \|\Phi(m+n,m,\theta)x\| \\ Y_N(t) &:= \int_0^t N(\tau)\, d\tau \leq tN(t) \end{aligned}$$

$$\begin{aligned} Y_N(\varphi(z,\theta,m,n)) &\leq \varphi(z,\theta,m,n)N(\varphi(z,\theta,m,n)) \\ &= \beta \|\Phi(m+n,m,\theta)x\| \, N(\varphi(z,\theta,m,n)) \\ &\leq \frac{1}{LL_1} \|\Phi(m+n,m,\theta)x\| \, N(\varphi(z,\theta,m,n)) \\ &\leq \frac{1}{L} N(\beta\varphi(x,\theta,m,n)) \\ &\leq \frac{1}{L} N(\|\Phi(m+n,m,\theta)x\|) \end{aligned}$$

Hence:

$$\begin{aligned} \sum_{n=0}^{\infty} Y_N(\varphi(z,\theta,m,n)) &\leq \frac{1}{L}\sum_{n=0}^{\infty} N(\|\Phi(m+n,m,\theta)x\|) \\ M_N(\varphi(z,\theta,m,.)) &\leq 1 \end{aligned}$$

$$\varphi(z,\theta,m,.) \in O_N$$

Or:

$$\varphi(x,\theta,m,.) \in O_N$$

So $B = O_N$. Then by Theorem 3.2, it results that π is uniformly exponentially stable. □

By extending some of techniques in [12], we obtain some generalizations for the case linear skew-evolution semifl ws of important results due to M. Megan, B. Sasu, A.L. Sasu (Theorem 3.1, 3.2 in [12]).

In next section, we give a characterization for the uniform exponential stability on Hilbert spaces. By using Corollary 3.3, we will establish the connections between the uniform exponential stability and some discrete equation.

4. Application for the Case Real Hilbert Spaces

First, we give some definitions

Definition 4.1. A map $H : \Delta \times \circleddash \to \mathcal{L}(X)$ is called positive if:

$$\langle H(m,n,\theta)x, x\rangle \geq 0$$

for all $(m,n) \in \Delta$ and $x \in X; \theta \in \circleddash$.

Definition 4.2. A map $H : \Delta \times \circleddash \to \mathcal{L}(X)$ is called uniformly positive if there exists the constant $a > 0$ such that:

$$\langle H(m,n,\theta)x, x\rangle \geq a\,\|x\|^2$$

for all $(m,n) \in \Delta$ and $x \in X; \theta \in \circleddash$.

Theorem 4.3. *The linear skew-evolution semiflow* $\pi = (\Phi, \sigma)$ *is uniformly exponentially stable if and only if there exists* $H : \Delta \times \circleddash \to \mathcal{L}(X)$ *uniformly positive such that the equation*

$$\begin{aligned} 0 &= \Phi^*(n+m,m,\theta)W(m+n,m,\theta)\Phi(n+m,m,\theta) \\ &- W(m,m,\theta) \\ &+ \sum_{k=0}^{n-1}\Phi^*(k+m,m,\theta)H(m,m,\theta)\Phi(k+m,m,\theta) \end{aligned}$$

(*where* $*$ *denotes the conjugate transpose*) *has a positive solution* $W : \Delta \times \circleddash \to \mathcal{L}(X)$ *with*

$$\sup_{((m,n),\theta)\in\,\Delta\times\circleddash} \|W(m,n,\theta)\| < \infty$$

Proof. Necessity. There are $M, \nu > 0$ such that:

$$\|\Phi(t,s,\theta)x\| \leq Me^{-\nu(t-s)}\,\|x\|$$

Let $H, W : \Delta \times \circleddash \to \mathcal{L}(X)$ given by:

$$H(m,n,\theta) = I$$

$$W(m,n,\theta) = \sum_{k=0}^{\infty}\Phi^*(k+m,m,\sigma(m,n,\theta))\Phi(k+m,m,\sigma(m,n,\theta))$$

It is clear that:

$$\langle H(m,n,\theta)x, x\rangle \geq \|x\|^2 \text{ for all } (m,n) \in \Delta, \theta \in \circleddash \text{ and all } x \in X$$

and

$$\begin{aligned} \|W(m,n,\theta)\| &\leq \sum_{k=0}^{\infty}\|\Phi(k+m,m,\sigma(m,n,\theta))\|^2 \\ &\leq \sum_{k=0}^{\infty}M^2e^{-2\nu k} = \frac{M^2}{1-e^{-2\nu}} \end{aligned}$$

On the other hand, W and H satisfy the equation; W is a positive solution with:

$$\sup_{((m,n),\theta)\in\,\Delta\times\ominus} \|W(m,n,\theta)\| < \infty$$

Sufficien y. H is uniformly positive such that the equation has a positive solution $W : \Delta\times\ominus\to\mathcal{L}(X)$ with:

$$K = \sup_{((m,n),\theta)\in\,\Delta\times\ominus} \|W(m,n,\theta)\|$$

a is the constant in Definitio 4.2, this means:

$$\langle H(m,n,\theta)x,x\rangle \geq a\,\|x\|^2$$

for all $(m,n)\in\Delta;\ x\in X;\ \theta\in\ominus$

$$\begin{aligned}
& \sum_{k=0}^{n-1}\|\Phi(k+m,m,\theta)x\|^2 \\
\leq\ & \sum_{k=0}^{n-1}\frac{1}{a}\langle H(m,m,\theta)\Phi(k+m,m,\theta)x,\Phi(k+m,m,\theta)x\rangle \\
=\ & \langle\sum_{k=0}^{n-1}\frac{1}{a}\Phi^*(k+m,m,\theta)H(m,m,\theta)\Phi(k+m,m,\theta)x,x\rangle \\
=\ & \frac{1}{a}\langle W(m,m,\theta)x,x\rangle \\
-\ & \frac{1}{a}\langle\Phi^*(n+m,m,\theta)W(m+n,m,\theta)\Phi(n+m,m,\theta)x,x\rangle \\
\leq\ & \frac{1}{a}\langle W(m,m,\theta)x,x\rangle \leq \frac{K}{a}\|x\|^2
\end{aligned}$$

Thus:

$$\sum_{k=0}^{\infty}\|\Phi(k+m,m,\theta)x\|^2 \leq \frac{K}{a}\|x\|^2 < \infty$$

By Corollary 3.3, Theorem is proved. However, we can give a simple proof as it will be presented below.

It is easy to see that:

$$\|\Phi(k+m,m,\theta)\|^2 \leq \frac{K}{a}$$

Next, we consider:

$$\begin{aligned}
& (n+1)\,\|\Phi(n+m,m,\theta)x\|^2 \\
=\ & \sum_{k=0}^{n}\|\Phi(n+m,m,\theta)x\|^2 \\
=\ & \sum_{k=0}^{n}\|\Phi(n+m,k+m,\sigma(k+m,m,\theta))\Phi(k+m,m,\theta)x\|^2
\end{aligned}$$

$$\leq \frac{K}{a}\sum_{k=0}^{n}\|\Phi(k+m,m,\theta)x\|^2$$
$$\leq \frac{K}{a}\sum_{k=0}^{\infty}\|\Phi(k+m,m,\theta)x\|^2$$
$$\leq \frac{K}{a}\frac{K}{a}\|x\|^2$$

and so:

$$\|\Phi(n+m,m,\theta)\| \leq \frac{K}{a\sqrt{n+1}}$$

$$\lim_{n\to\infty}\frac{K}{a\sqrt{n+1}} = 0$$

and hence, there exists $n_0 \in N$ such that:

$$\|\Phi(n_0+m,m,\theta)\| \leq \frac{1}{2}$$

for all $(\theta, m) \in \ominus \times \mathbb{N}$. By Lemma 3.1, π is uniformly exponentially stable □

Acknowledgment

The author is grateful to the referee for carefully reading the paper and valuable suggestions.

References

[1] N. V. Minh, F. Räbiger and R. Schnaubelt, Exponential stability, exponential expanssiveness and exponential dichtomy of evolution equations on the half line, *Integral Equations Operator Theory* **32** (1998), 332–353.

[2] N. V. Minh, Semigroups and nonautonomous linear differnetial equations on the half-line, *Dokl. Belarussian. Acad. Sci.* **37** (1993), No. 1, 19–22.

[3] B. Aulbach, N. V. Minh and P. P. Zabreiko, A generalization of the monodromy operator for non-periodic linear differential equations, *Diff. Eqns. Dyn. Syst.* 1993, 211–222.

[4] Y. Latushkin and A. M. Stepin, Linear skew-product fl ws and semigroups of weighted composition operators, *Lecture Notes in Math.,* **1486**, 98–111, Springer-Verlag, New-York, 1991.

[5] R. Datko, Uniform asymtotic stability of evolutionary processes in Banach spaces, *SIAM J. Math. Anal.* **3** (1972), 428–445.

[6] R. Datko, Extending a theorem of Lyapunov to Hilbert spaces, *J. math. Anal. Appl.* **32** (1970), 610–616.

[7] Z. Zabczyk, Remarks on the control of discrete-time distributed parameter systems, *SIAM J. Control Optim.* **12** (1971), 721–735.

[8] J.M.A.M Van Neerven, Exponential stability of operators and operator semigroups, *J. Funct. Anal.* **130** (1995), 293–309.

[9] J.M.A.M Van Neerven, The asymptotic behaviour of semigroups of linear operators, *Theory, Advances and Applications,* Vol. 88, Birkhäuser, Boston, 1996.

[10] M. Megan, B. Sasu and A. L. Sasu, On uniform exponential stability of evolution families, *Rivista di Matematica della Universita di Parma* **4** (2001), 27–43.

[11] M. Megan, A. L. Sasu and B. Sasu, Exponential stability and exponential instability for linear skew-product fl ws, *Mathematica Bohemica* **129** (2004), 225–243.

[12] M. Megan, A. L. Sasu and B. Sasu, On uniform exponential stability of linear skew-product semifl ws in Banach spaces, *Bulletin of The Belgian Mathematical Society – Simon Stevin* **9** (2002), 143–154.

[13] S. Rolewicz, On uniform N-equistability, *J. Math. Anal. Appl.* **115** (1986), 434–441.

[14] K. M. Przyuski and S. Rolewicz, On stability of linear time-varying infinite dimensional discrete-time systems, *Systems Control Lett.* **4** (1984), 307–315.

[15] C. Bennett and S. Sharpley, *Interpolation of operators*, Academic Press, Boston, 1988.

[16] S. G. Krein, Yu. I. Petunin and E. M. Semeonov, Interpolation of linear operators, *Transl. Math. Monogr.* **54** (Amer. Math. Soc., Providence), 1982.

[17] A. Pazy, *Semigroups of linear operators and applications to partial differnetial equations*, Springer, Berlin, 1983.

[18] A. Ichikawa, Equivalence of L^p and uniform exponential stability for a class of non-linear semigroups, *Nonlinear Anal. Theory Appl.* **8** (1984), 805–810.

[19] W. Litman, A generalization of the theorem Datko-Pazy, *Lecture Notes in Control and Information Science* **130** (1983), 318–323.

[20] P. Meyer-Nieberg, *Banach Lattices*, Springer Verlag, Berlin, Heidelberg, New York, 1991.

[21] A. L. Sasu and B. Sasu, Exponential stability for linear skew-product fl ws, *Bulletin des Sciences Mathematiques* **128** (2004), 727–738.

[22] B. Sasu, Discrete orbits and exponential stability of evolution families, *An. Univ. Timisoara Ser. Mat.-Inform.* **42** (2004), No. 1, 129–139.

[23] S. N. Chow and H. Leiva, Existence and roughness of the exponential dichotomy for linear skew product semifl w in Banach spaces, *J. Differential Equations,* **102** (1995), 429–477.

In: Evolution Equations: Theories, Solutions and Functions ISBN 978-1-61324-090-8
Editor: Gaston M. N'Guerekata

Chapter 11

GLOBAL WELL-POSEDNESS OF A CLASS OF NONLINEAR WAVE EQUATIONS AND HEAT EQUATIONS

***Wu Weining*[1], *Xu Runzhang*[2] *and Zhang Wenying*[2,*]**
[1] School of Computer Science and Technology, Harbin Institute of Technology, Harbin 150001, People's Republic of China
[2] College of Science, Harbin Engineering University, Harbin 150001, People's Republic of China

Abstract

In this paper we study the initial boundary value problem of wave equations and reaction-diffusion equations with two nonlinear source terms of different sign. By introducing a family of potential wells W_δ and corresponding outside sets V_δ of W_δ we firs obtain the invariant sets and vacuum isolating of solutions. Then we get the threshold result of global existence and nonexistence of solutions. Finally we prove the global existence of solutions for the problem with critical initial conditions.

Keywords and phrases: Semilinear wave equations, reaction-diffusion equations, potential wells, global existence, nonexistence, vacuum isolating.

1. Introduction

As is well known the most important models of semilinear wave equations are

$$u_{tt} - \Delta u = -|u|^{p-1}u$$

and

$$u_{tt} - \Delta u = |u|^{p-1}u.$$

In each physical system described by one of above equations, there is only one source, where in firs equation the sign of source term and u are different and in second equation

*E-mail address: xurunzh@yahoo.com.cn

the sign of source term and u are the same. There are a lot of results [1-14] on the global existence and nonexistence of solutions for above equations and their generalizations. Recently in [10] by introducing a family of potential wells, the initial boundary value problem for above second equation was studied. And some new results on global existence of solutions and the invariant sets were obtained. In particular,the vacuum isolating behaviors of solutions were discovered in [10].

If in a physical system there are two nonlinear source of different sign, then the following wave equations with two nonlinear source terms of different sign should be considered

$$u_{tt} - \Delta u = f(u) \equiv a|u|^{p-1}u - b|u|^{q-1}u, \quad x \in \Omega, \ t > 0, \tag{1.1}$$

where $p \neq q$,$p > 1$,$q > 1$, a and b are two positive constants. To our knowledge, equation (1.1) can not be included by any known results. For example, in [14] Tsutsumi studied the global existence and finit time blow up of solutions for the following semilinear hyperbolic equations

$$u_{tt} + Au = f(u).$$

He required $f(u)$ to be homogeneous of degree $p > 2$, i.e.

$$f(\lambda u) = \lambda^p f(u), \quad \forall \lambda > 0.$$

Clearly $f(u)$ in equation (1.1) does not satisfy this assumption. In [4] Glassey studied the finit time blow up of solutions for semilinear hyperbolic equations

$$u_{tt} - \Delta u = f(u),$$

where $f(u)$ must satisfy the condition: there exists a convex function $g(u)$ such that $f(u) \geq g(u)$ for $-\infty < u < \infty$. Hence this condition is not applicable for $f(u)$ in equation (1.1). In [11] Payne and Sattinger also studied the finit time blow up of solutions for above semilinear hyperbolic equations by using potential well method. Where $f(u)$ must satisfy the condition: a) $f(u)$ is monotone and convex for $u > 0$, concave for $u < 0$; or b) $f(u)$ is convex for $-\infty < u < \infty$. Clearly $f(u)$ in equation (1.1) does not satisfy any one of above two conditions, even if for $u > 0$.

In this paper we study the initial boundary value problem of equation (1.1) with initial boundary conditions

$$u(x,0) = u_0(x), \ u_t(x,0) = u_1(x), \quad x \in \Omega, \tag{1.2}$$

$$u(x,t) = 0, \quad x \in \partial\Omega, \ t \geq 0 \tag{1.3}$$

and the initial boundary value problem of reaction-diffusion equation with two nonlinear source terms of different sign

$$u_t - \Delta u = f(u) \equiv a|u|^{p-1}u - b|u|^{q-1}u, \quad x \in \Omega, \ t > 0 \tag{1.4}$$

with

$$u(x,0) = u_0(x), \ x \in \Omega \tag{1.5}$$

and (1.3), where $\Omega \subset \mathbf{R}^n$ is a bounded domain, $a > 0$, $b > 0$, p and q satisfy

(H) $1 < q < p < \infty$ if $n = 1, 2$; $1 < q < p \leq \dfrac{n+2}{n-2}$ if $n \geq 3$.

First we introduce a family of potential wells W_δ and corresponding outside sets V_δ of W_δ and give a series of the properties of W_δ and V_δ. Then by using W_δ and V_δ we prove the invariance of some sets under the fl w of (1.1)–(1.3) and vacuum isolating of solutions. After that we obtain a threshold result of global existence and nonexistence of solutions. Finally we discuss the global existence of solutions for the problem with critical initial conditions $I(u_0) \geq 0$, $E(0) = d$ (or $J(u_0) = d$) where d is the depth of potential well $W = W_1$ (see Definitio 2.1).

In this paper we also denote $\|\cdot\|_{L^p(\Omega)}$ by $\|\cdot\|_p$, $\|\cdot\| = \|\cdot\|_2$ and $(u, v) = \int\limits_\Omega uv\mathrm{d}x$.

2. Introduction of the Family of Potential Wells

In this paper we always assume that p and q satisfy (H). For problem (1.1)–(1.3) and problem (1.4), (1.5), (1.3) we defin

$$J(u) = \frac{1}{2}\|\nabla u\|^2 - \frac{a}{p+1}\|u\|_{p+1}^{p+1} + \frac{b}{q+1}\|u\|_{q+1}^{q+1},$$
$$I(u) = \|\nabla u\|^2 - a\|u\|_{p+1}^{p+1} + b\|u\|_{q+1}^{q+1}$$

and

$$I_\delta(u) = \delta\|\nabla u\|^2 - a\|u\|_{p+1}^{p+1} + b\|u\|_{q+1}^{q+1}, \quad \delta \geq 0.$$

Lemma 2.1. *Let* $\|\nabla u\| \neq 0$. *Then*

(i) $\lim\limits_{\lambda\to 0} J(\lambda u) = 0$, $\lim\limits_{\lambda\to+\infty} J(\lambda u) = -\infty$.

(ii) *In the interval* $0 < \lambda < \infty$ *there exists a unique* $\bar{\lambda} = \bar{\lambda}(u)$ *such that*

$$\frac{\mathrm{d}}{\mathrm{d}\lambda}J(\lambda u)\bigg|_{\lambda=\bar{\lambda}} = 0.$$

(iii) $J(\lambda u)$ *is increasing on* $0 \leq \lambda \leq \bar{\lambda}$, *decreasing on* $\bar{\lambda} \leq \lambda < \infty$ *and takes the maximum at* $\lambda = \bar{\lambda}$.

(iv) $I(\lambda u) > 0$ *for* $0 < \lambda < \bar{\lambda}$, $I(\lambda u) < 0$ *for* $\bar{\lambda} < \lambda < \infty$ *and* $I(\bar{\lambda}u) = 0$.

Proof. First $\|\nabla u\| \neq 0$ gives $\|u\|_{p+1} \neq 0$, $\|u\|_{q+1} \neq 0$.

(i) The conclusion follows from

$$J(\lambda u) = \frac{\lambda^2}{2}\|\nabla u\|^2 - \frac{a\lambda^{p+1}}{p+1}\|u\|_{p+1}^{p+1} + \frac{b\lambda^{q+1}}{q+1}\|u\|_{q+1}^{q+1}.$$

(ii) First note that

$$\frac{\mathrm{d}}{\mathrm{d}\lambda}J(\lambda u) = \lambda\|\nabla u\|^2 - a\lambda^p\|u\|_{p+1}^{p+1} + b\lambda^q\|u\|_{q+1}^{q+1} = 0, \quad \lambda > 0$$

is equivalent to

$$a\lambda^{p-1}\|u\|_{p+1}^{p+1} - b\lambda^{q-1}\|u\|_{q+1}^{q+1} = \|\nabla u\|^2. \tag{2.1}$$

Let

$$h(\lambda) = a\lambda^{p-1}\|u\|_{p+1}^{p+1} - b\lambda^{q-1}\|u\|_{q+1}^{q+1} = \lambda^{q-1}\left(a\lambda^{p-q}\|u\|_{p+1}^{p+1} - b\|u\|_{q+1}^{q+1}\right),$$

$$\lambda_0 = \left(\frac{b\|u\|_{q+1}^{q+1}}{a\|u\|_{p+1}^{p+1}}\right)^{\frac{1}{p-q}}.$$

Then $h(\lambda) < 0$ for $0 < \lambda < \lambda_0$, $h(\lambda) > 0$ for $\lambda_0 < \lambda < \infty$ and $h(\lambda)$ is increasing for $\lambda > \lambda_0$. Hence for any $\|\nabla u\| > 0$ there exists a unique $\bar{\lambda} > \lambda_0$ such that (2.1) holds.

(iii) Note that

$$\frac{\mathrm{d}}{\mathrm{d}\lambda}J(\lambda u) = \lambda\left(\|\nabla u\|^2 - h(\lambda)\right).$$

By the proof of (ii) it follows that if $0 < \lambda \le \lambda_0$, then $h(\lambda) \le 0$; if $\lambda_0 < \lambda < \bar{\lambda}$, then $0 < h(\lambda) < \|\nabla u\|^2$; if $\bar{\lambda} < \lambda < \infty$, then $h(\lambda) > \|\nabla u\|^2$. Hence we have $\frac{\mathrm{d}}{\mathrm{d}\lambda}J(\lambda u) > 0$ for $0 < \lambda < \bar{\lambda}$, $\frac{\mathrm{d}}{\mathrm{d}\lambda}J(\lambda u) < 0$ for $\bar{\lambda} < \lambda < \infty$. From this, the conclusion of (ii) follows.

(iv) The conclusion follows from the proof of (iii) and

$$I(\lambda u) = \lambda^2\|\nabla u\|^2 - a\lambda^{p+1}\|u\|_{p+1}^{p+1} + b\lambda^{q+1}\|u\|_{q+1}^{q+1} = \lambda\frac{\mathrm{d}}{\mathrm{d}\lambda}J(\lambda u).$$ □

Lemma 2.2. *Let $\delta > 0$. If $0 < \|\nabla u\| \le r(\delta)$, then $I_\delta(u) > 0$. In particular, if $0 < \|\nabla u\| \le r(1)$, then $I(u) > 0$, where*

$$r(\delta) = \left(\frac{\delta}{aC_p^{p+1}}\right)^{\frac{1}{p-1}}, \quad C_p = \sup_{u\in H_0^1(\Omega)} \frac{\|u\|_{p+1}}{\|\nabla u\|}.$$

Proof. From $0 < \|\nabla u\| < r(\delta)$ we have $\|u\|_{q+1} > 0$ and by

$$\begin{aligned} a\|u\|_{p+1}^{p+1} - b\|u\|_{q+1}^{q+1} &< a\|u\|_{p+1}^{p+1} \le aC_p^{p+1}\|\nabla u\|^{p+1} \\ &= aC_p^{p+1}\|\nabla u\|^{p-1}\|\nabla u\|^2 \le \delta\|\nabla u\|^2 \end{aligned}$$

we get $I_\delta(u) > 0$. □

Lemma 2.3. *Let $\delta > 0$. If $I_\delta(u) < 0$, then $\|\nabla u\| > r(\delta)$. In particular, if $I(u) < 0$, then $\|\nabla u\| > r(1)$.*

Proof. $I_\delta(u) < 0$ gives

$$\delta\|\nabla u\|^2 < a\|u\|_{p+1}^{p+1} - b\|u\|_{q+1}^{q+1} < a\|u\|_{p+1}^{p+1} \le aC_p^{p+1}\|\nabla u\|^{p-1}\|\nabla u\|^2. \tag{2.2}$$

Eq. (2.2) implies $\|\nabla u\| > r(\delta)$. □

Lemma 2.4. *Let $\delta > 0$. If $I_\delta(u) = 0$, then $\|\nabla u\| > r(\delta)$ or $\|\nabla u\| = 0$. In particular, if $I(u) = 0$, then $\|\nabla u\| > r(1)$ or $\|\nabla u\| = 0$.*

Proof. If $\|\nabla u\| = 0$, then $I_\delta(u) = 0$. If $I_\delta(u) = 0$ and $\|\nabla u\| \neq 0$, then by

$$\delta\|\nabla u\|^2 = a\|u\|_{p+1}^{p+1} - b\|u\|_{q+1}^{q+1} < a\|u\|_{p+1}^{p+1} \leq aC_p^{p+1}\|\nabla u\|^{p-1}\|\nabla u\|^2$$

we get $\|\nabla u\| > r(\delta)$. □

Definition 2.1. For $\delta \geq 0$ we defin

$$d(\delta) = \inf J(u)$$

subject to the conditions $u \in H_0^1(\Omega),\ I_\delta(u) = 0,\ \ \|\nabla u\| \neq 0$ and $d = d(1)$.

Lemma 2.5. *As a function of* δ $d(\delta)$ *possesses the following properties*

(i) $d(\delta) > a(\delta)r^2(\delta)$ *for* $0 < \delta < \dfrac{p+1}{2}$, $a(\delta) = \dfrac{1}{2} - \dfrac{\delta}{p+1}$*;*

(ii) $\lim\limits_{\delta\to 0} d(\delta) = d(0) > 0$ *and there exists a unique* $\delta_0 > \dfrac{p+1}{2}$ *such that* $d(\delta_0) = 0$ *and* $d(\delta) > 0$ *for* $0 < \delta < \delta_0$.

(iii) $d(\delta)$ *is increasing on* $0 \leq \delta \leq 1$, *decreasing on* $1 \leq \delta \leq \delta_0$ *and takes the maximum* $d = d(1)$ *at* $\delta = 1$.

(iv) $d(\delta)$ *is continuous for* $0 < \delta < \delta_0$.

Proof. (i) If $I_\delta(u) = 0$ and $\|\nabla u\| \neq 0$, then by Lemma 2.4 we have $\|\nabla u\| > r(\delta)$, and

$$\begin{aligned} J(u) &= \frac{1}{2}\|\nabla u\|^2 - \frac{a}{p+1}\|u\|_{p+1}^{p+1} + \frac{b}{q+1}\|u\|_{q+1}^{q+1} \\ &> \frac{1}{2}\|\nabla u\|^2 - \frac{1}{p+1}\left(a\|u\|_{p+1}^{p+1} - b\|u\|_{q+1}^{q+1}\right) \\ &= \left(\frac{1}{2} - \frac{\delta}{p+1}\right)\|\nabla u\|^2 + \frac{1}{p+1}I_\delta(u) = a(\delta)\|\nabla u\|^2 > a(\delta)r^2(\delta). \end{aligned}$$

(ii) First $I_0(u) = 0$ implies $a\|u\|_{p+1}^{p+1} = b\|u\|_{q+1}^{q+1}$ and

$$\begin{aligned} J(u) &= \frac{1}{2}\|\nabla u\|^2 - \frac{a}{p+1}\|u\|_{p+1}^{p+1} + \frac{b}{q+1}\|u\|_{q+1}^{q+1} \\ &= \frac{1}{2}\|\nabla u\|^2 + a\frac{p-q}{(p+1)(q+1)}\|u\|_{p+1}^{p+1} \\ &= \frac{1}{2}\|\nabla u\|^2 + b\frac{p-q}{(p+1)(q+1)}\|u\|_{q+1}^{q+1}. \end{aligned}$$

Hence we have $d(0) \geq 0$. If $d(0) = 0$, then there exist $\{u_m\} \in H_0^1(\Omega)$ such that $\|\nabla u_m\| \neq 0$, $m = 1, 2, \ldots$ and $\lim\limits_{m\to\infty}\|\nabla u_m\| = \lim\limits_{m\to\infty}\|u_m\|_{p+1} = \lim\limits_{m\to\infty}\|u_m\|_{q+1} = 0$ and $a\|u_m\|_{p+1}^{p+1} = b\|u_m\|_{q+1}^{q+1}$. This is impossible from $\|u_m\|_{q+1} = O\left(\|u_m\|_{p+1}\right)$. For example, for any $u \in H_0^1(\Omega)$, $\|\nabla u\| \neq 0$ and $\lambda_m \to 0$, $m \to \infty$ we have $\lim\limits_{m\to\infty}\|\nabla(\lambda_m u)\| = \lim\limits_{m\to\infty}\|\lambda_m u\|_{p+1} = \lim\limits_{m\to\infty}\|\lambda_m u\|_{q+1} = 0$, but $a\|\lambda_m u\|_{p+1}^{p+1} = b\|\lambda_m u\|_{q+1}^{q+1}$ can not hold for sufficientl large m.

Next for any $u \in H_0^1(\Omega)$, $\|\nabla u\| \neq 0$ and $\delta > 0$, we defin $\lambda = \lambda(\delta)$ such that

$$\delta\|\nabla(\lambda u)\|^2 = a\|\lambda u\|_{p+1}^{p+1} - b\|\lambda u\|_{q+1}^{q+1}. \tag{2.3}$$

Then $I_\delta(\lambda u) = 0$ and

$$\delta\|\nabla u\|^2 = a\lambda^{p-1}\|u\|_{p+1}^{p+1} - b\lambda^{q-1}\|u\|_{q+1}^{q+1}. \tag{2.4}$$

From the proof of (ii) of Lemma 2.1 it follows that for any $\delta > 0$ there exists a unique $\lambda = \lambda(\delta)$ such that (2.3) and (2.4) hold. Again by the proof of Lemma 2.1 we have $\lim\limits_{\delta\to+\infty} \lambda(\delta) = +\infty$. Hence by (i) of Lemma 2.1 we get $\lim\limits_{\delta\to+\infty} J(\lambda u) = \lim\limits_{\lambda\to+\infty} J(\lambda u) = -\infty$. From this and (i) of this Lemma it follows that there exists a unique $\delta_0 > \dfrac{p+1}{2}$ such that $d(\delta_0) = 0$ and $d(\delta) > 0$ for $0 \leq \delta < \delta_0$.

(iii) We prove that $d(\delta') < d(\delta'')$ for any $0 < \delta' < \delta'' < 1$ or $1 < \delta'' < \delta' < \delta_0$. Clearly it is enough to prove that for any $0 < \delta' < \delta'' < 1$ or $1 < \delta'' < \delta' < b$ and any $u \in H_0^1(\Omega)$, $I_{\delta''}(u) = 0$ and $\|\nabla u\| \neq 0$ there exists a $v \in H_0^1(\Omega)$, $I_{\delta'}(v) = 0$ and $\|\nabla v\| \neq 0$ and a constant $\varepsilon(\delta', \delta'') > 0$ such that $J(v) < J(u) - \varepsilon(\delta', \delta'')$. In fact, for above u we also defin $\lambda(\delta)$ by (2.3), then $I_\delta(\lambda(\delta)u) = 0$, $\lambda(\delta'') = 1$ and (2.4) holds. Let $g(\lambda) = J(\lambda u)$. Then

$$\begin{aligned}\frac{\mathrm{d}}{\mathrm{d}\lambda} g(\lambda) &= \frac{1}{\lambda}\Big(\|\nabla(\lambda u)\|^2 - a\|\lambda u\|_{p+1}^{p+1} + b\|\lambda u\|_{q+1}^{q+1}\\ &= \frac{1}{\lambda}\Big((1-\delta)\|\nabla(\lambda u)\|^2 + I_\delta(\lambda u)\Big) = (1-\delta)\lambda\|\nabla u\|^2.\end{aligned}$$

Take $v = \lambda(\delta')u$, then $I_{\delta'}(v) = 0$ and $\|\nabla v\| \neq 0$.

If $0 < \delta' < \delta'' < 1$, then

$$J(u) - J(v) = g(1) - g(\lambda(\delta')) > (1-\delta'')r^2(\delta'')\lambda(\delta')(1-\lambda(\delta')) \equiv \varepsilon(\delta', \delta'').$$

If $1 < \delta'' < \delta' < b$, then

$$J(u) - J(v) = g(1) - g(\lambda(\delta')) > (\delta''-1)r^2(\delta'')\lambda(\delta'')(\lambda(\delta')-1) \equiv \varepsilon(\delta', \delta'').$$

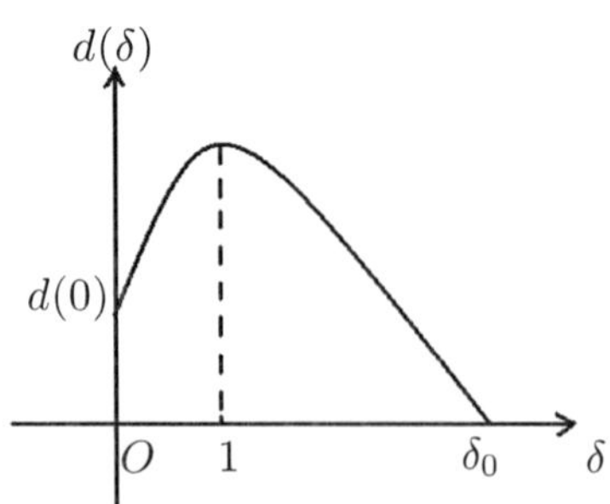

Figure 1.

(iv) Clearly it is enough to prove that for any $\delta_1 \in (0,1)$ and $\delta_2 \in (1,\delta_0)$,$d(\delta)$ is uniformly continuous on $[\delta_1, 1]$ and $[1, \delta_2]$ respectively.

First we prove that for any $\delta_1 \in (0,1)$, $d(\delta)$ is uniformly continuous on $[\delta_1, 1]$. For any given $0 < \varepsilon < d$ and $\delta' \in [\delta_1, 1)$ take a $u_{\delta'}(x) \in H_0^1(\Omega)$ such that $I_{\delta'}(u_{\delta'}) = 0$, $\|\nabla u_{\delta'}\| \neq 0$ and

$$0 \leq J(u_{\delta'}) - d(\delta') < \frac{\varepsilon}{2}. \tag{2.5}$$

From the proof of Lemma 2.1 it follows that for each $\delta \in [\delta', 1]$ there exists a unique $\lambda = \lambda_{\delta'}(\delta)$ satisfying

$$\delta\|\nabla(\lambda u_{\delta'})\|^2 = a\|\lambda u_{\delta'}\|_{p+1}^{p+1} - b\|\lambda u_{\delta'}\|_{q+1}^{q+1}, \tag{2.6}$$

which implies that

$$I_\delta\left(\lambda_{\delta'}(\delta)u_{\delta'}\right) = 0, \ \ \lambda_{\delta'}(\delta') = 1$$

and

$$\delta = \frac{a\lambda^{p-1}\|u_{\delta'}\|_{p+1}^{p+1} - b\lambda^{q-1}\|u_{\delta'}\|_{q+1}^{q+1}}{\|\nabla u_{\delta'}\|^2} \stackrel{\text{def}}{\equiv} \varphi_{\delta'}(\lambda), \ \ 1 \leq \lambda \leq \lambda_{\delta'}(1). \tag{2.7}$$

Again by the proof of Lemma 2.1 we know that $\varphi_{\delta'}(\lambda)$ is increasing and continuous on $[1, \lambda_{\delta'}(1)]$, hence its inverse function $\lambda = \lambda_{\delta'}(\delta)$ is increasing and continuous on $[\delta', 1]$. Let $g_{\delta'}(\lambda) = J(\lambda u_{\delta'}) = J(\lambda_{\delta'}(\delta)u_{\delta'})$ for $\delta' \leq \delta \leq 1$. Then as in (iii) of this lemma we can get

$$\frac{\mathrm{d}}{\mathrm{d}\lambda} g_{\delta'}(\lambda) = (1-\delta)\lambda\|\nabla u_{\delta'}\|^2 \geq 0, \ \ \delta' \leq \delta \leq 1.$$

Hence for $\delta' \leq \delta \leq 1$ and $\lambda = \lambda_{\delta'}(\delta)$ we get

$$\begin{aligned} 0 \leq J(\lambda u_{\delta'}) - J(u_{\delta'}) &= g_{\delta'}(\lambda) - g_{\delta'}(1) = (1-\tilde{\delta})\tilde{\lambda}\|\nabla u_{\delta'}\|^2(\lambda - 1) \\ &\leq (1-\delta_1)\|\nabla u_{\delta'}\|^2 \lambda(\lambda-1), \ \ \delta' \leq \tilde{\delta} \leq \delta, \ 1 \leq \tilde{\lambda} \leq \lambda. \end{aligned}$$

From

$$\begin{aligned} \left(\frac{1}{2} - \frac{1}{p+1}\right)\|\nabla u_{\delta'}\|^2 &\leq \left(\frac{1}{2} - \frac{\delta'}{p+1}\right)\|\nabla u_{\delta'}\|^2 + \frac{1}{p+1} I_{\delta'}(u_{\delta'}) \\ &\leq J(u_{\delta'}) < d(\delta') + \frac{\varepsilon}{2} < 2d \end{aligned}$$

we get

$$\|\nabla u_{\delta'}\|^2 \leq 4\frac{p+1}{p-1} d, \ \ \delta_1 \leq \delta' \leq 1.$$

Hence for $\delta_1 \leq \delta' \leq \delta \leq 1$ and $\lambda = \lambda_{\delta'}(\delta)$ we have

$$0 \leq J\left(\lambda_{\delta'}(\delta)u_{\delta'}\right) - J(u_{\delta'}) \leq 4(1-\delta_1)\frac{p+1}{p-1} d\lambda(\lambda - 1).$$

Take $\lambda_1 > 1$ such that

$$4(1-\delta_1)\frac{p+1}{p-1} d\lambda_1(\lambda_1 - 1) = \frac{\varepsilon}{2}.$$

Then from

$$\begin{aligned}\varphi'_{\delta'}(\lambda) &= \frac{1}{\|\nabla u_{\delta'}\|^2}\left((p-1)\lambda^{p-2}a\|u_{\delta'}\|_{p+1}^{p+1} - (q-1)\lambda^{q-2}b\|u_{\delta'}\|_{q+1}^{q+1}\right)\\ &\geq \frac{p-1}{\|\nabla(\lambda u_{\delta'})\|^2}\left(a\|\lambda u_{\delta'}\|_{p+1}^{p+1} - b\|\lambda u_{\delta'}\|_{q+1}^{q+1}\right) = (p-1)\delta' \geq (p-1)\delta_1,\end{aligned}$$

we get

$$0 \leq \varphi_{\delta'}(\lambda_1) - \varphi_{\delta'}(1) \geq (p-1)\delta_1(\lambda_1 - 1) \stackrel{\text{def}}{=} \Delta_1 \text{ for all } \delta' \in [\delta_1, 1].$$

Hence when $\delta'' \in [\delta', 1]$ and $0 \leq \delta'' - \delta' < \Delta_1$ we have $0 \leq \lambda_{\delta'}(\delta'') - 1 < \lambda_1 - 1$ and

$$0 \leq J\left(\lambda_{\delta'}(\delta'')u_{\delta'}\right) - J(u_{\delta'}) < \frac{\varepsilon}{2}.$$

Therefore we get

$$\begin{aligned}0 &\leq d(\delta'') - d(\delta') \leq J\left(\lambda_{\delta'}(\delta'')u_{\delta'}\right) - d(\delta')\\ &= J\left(\lambda_{\delta'}(\delta'')u_{\delta'}\right) - J(u_{\delta'}) + J(u_{\delta'}) - d(\delta') < \frac{\varepsilon}{2} + \frac{\varepsilon}{2} = \varepsilon.\end{aligned}$$

Next we prove that for any $\delta_2 \in (1, \delta_0)$, $d(\delta)$ is uniformly continuous on $[1, \delta_2]$. For any given $\varepsilon > 0$ and $\delta' \in [1, \delta_2]$ take a $u_{\delta'}(x) \in H_0^1(\Omega)$ such that $I_{\delta'}(u_{\delta'}) = 0$, $\|\nabla u_{\delta'}\| \neq 0$ and satisfie (2.5). Again by the proof of Lemma 2.1 it follows that for each $\delta \in [1, \delta']$ there exists a unique $\lambda = \lambda_{\delta'}(\delta)$ satisfying (2.6), $I_\delta\left(\lambda_{\delta'}(\delta)u_{\delta'}\right) = 0$ and $\lambda_{\delta'}(\delta') = 1$. Again denote the function in (2.7) by $\varphi_{\delta'}(\lambda)$, where $\lambda_{\delta'}(1) \leq \lambda \leq 1$, then $\varphi_{\delta'}(\lambda)$ on $[\lambda_{\delta'}(1), 1]$ and $\lambda_{\delta'}(\delta)$ on $[1, \delta']$ are increasing and continuous respectively. Again for $\delta \in [1, \delta']$ and $\lambda = \lambda_{\delta'}(\delta)$ defin $g_{\delta'}(\lambda) = J(\lambda u_{\delta'})$,then we have

$$\frac{\mathrm{d}}{\mathrm{d}\lambda}g_{\delta'}(\lambda) = (1-\delta)\lambda\|\nabla u_{\delta'}\|^2 \leq 0, \quad 1 \leq \delta \leq \delta'$$

and

$$\begin{aligned}0 &\leq J(\lambda u_{\delta'}) - J(u_{\delta'}) = g_{\delta'}(\lambda) - g_{\delta'}(1) = \left(\tilde{\tilde{\delta}} - 1\right)\tilde{\tilde{\lambda}}\|\nabla u_{\delta'}\|^2(1-\lambda)\\ &\leq (\delta_2 - 1)\|\nabla u_{\delta'}\|^2(1-\lambda), \quad \delta \leq \tilde{\tilde{\delta}} \leq \delta', \ \lambda \leq \tilde{\tilde{\lambda}} \leq 1.\end{aligned}$$

Now we prove that $\|\nabla u_{\delta'}\|$ are bounded for $1 \leq \delta' \leq \delta_2$. First in (i) of Lemma 2.1 let $v = \lambda u$ we get

$$\lim_{\|\nabla v\| \to +\infty} J(v) = -\infty.$$

From this and

$$J(u_{\delta'}) \geq d(\delta') \geq d(\delta_2) > 0 \text{ for all } \delta' \in [1, \delta_2]$$

we can obtain $\|\nabla u_{\delta'}\|$ are bounded for $1 \leq \delta' \leq \delta_2$. Let $\|\nabla u_{\delta'}\|^2 \leq M$. then we have

$$0 \leq J(\lambda u_{\delta'}) - J(u_{\delta'}) \leq (\delta_2 - 1)M(1-\lambda), \quad \lambda_{\delta'}(1) \leq \lambda \leq 1.$$

Take λ_2 such that $0 < \lambda_2 < 1$ and

$$(\delta_2 - 1)M(1 - \lambda_2) = \frac{\varepsilon}{2}.$$

As above we can get

$$\varphi'_{\delta'}(\lambda) \geq (p-1)\delta' \geq p-1$$

and

$$0 \leq \varphi_{\delta'}(1) - \varphi_{\delta'}(\lambda_2) \geq (p-1)(1-\lambda_2) \stackrel{\text{def}}{=} \Delta_2 \text{ for all } \delta' \in [1, \delta_2].$$

Hence when $\delta'' \in [1, \delta']$ and $0 \leq \delta' - \delta'' < \Delta_2$ we have $0 \leq 1 - \lambda_{\delta'}(\delta'') < 1 - \lambda_2$ and

$$0 \leq J\left(\lambda_{\delta'}(\delta'')u_{\delta'}\right) - J(u_{\delta'}) < \frac{\varepsilon}{2}.$$

Finally we obtain

$$\begin{aligned} 0 &\leq d(\delta'') - d(\delta') \leq J\left(\lambda_{\delta'}(\delta'')u_{\delta'}\right) - d(\delta') \\ &= J\left(\lambda_{\delta'}(\delta'')u_{\delta'}\right) - J(u_{\delta'}) + J(u_{\delta'}) - d(\delta') < \frac{\varepsilon}{2} + \frac{\varepsilon}{2} = \varepsilon. \end{aligned}$$ □

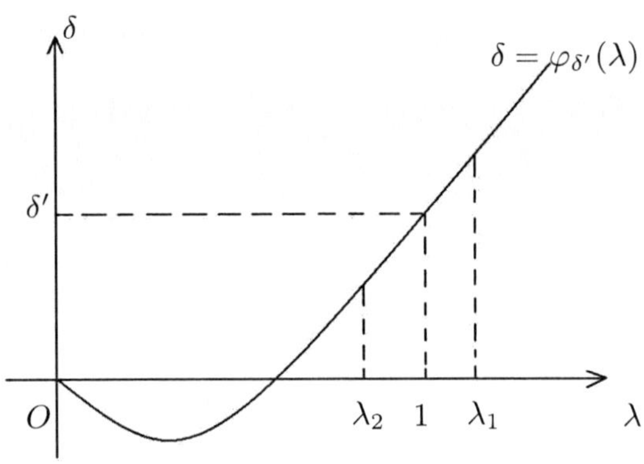

Figure 2.

Lemma 2.6. *Let* $0 < \delta < \frac{p+1}{2}$. *Assume that* $J(u) \leq d(\delta)$ *and* $I_\delta(u) > 0$. *Then*

$$0 < \|\nabla u\|^2 < \frac{d(\delta)}{a(\delta)}.$$

In particular, if $J(u) \leq d$ *and* $I(u) > 0$, *then*

$$0 < \|\nabla u\|^2 < \frac{2(p+1)}{p-1}d.$$

Proof. This lemma follows from

$$a(\delta)\|\nabla u\|^2 + \frac{1}{p+1}I_\delta(u) < J(u) \leq d(\delta). \qquad \square \tag{2.8}$$

Again by (2.8) we can obtain the following Lemma 2.7 and Lemma 2.8.

Lemma 2.7. *Let* $0 < \delta < \dfrac{p+1}{2}$. *Assume that* $J(u) \leq d(\delta)$ *and* $\|\nabla u\|^2 \geq \dfrac{d(\delta)}{a(\delta)}$. *Then* $I_\delta(u) < 0$. *In particular, if* $J(u) \leq d$ *and* $\|\nabla u\|^2 \geq \dfrac{2(p+1)}{p-1}d$, *then* $I(u) < 0$.

Lemma 2.8. *Let* $0 < \delta < \dfrac{p+1}{2}$. *Assume that* $J(u) \leq d(\delta)$ *and* $I_\delta(u) = 0$. *Then* $\|\nabla u\|^2 < \dfrac{d(\delta)}{a(\delta)}$. *In particular, if* $J(u) \leq d$ *and* $I(u) = 0$, *then* $\|\nabla u\|^2 < \dfrac{2(p+1)}{p-1}d$.

Remark 2.1. The results of Lemma 2.2–Lemma 2.4 and Lemma 2.6–Lemma 2.8 show that the space $H_0^1(\Omega)$ is divided into two parts $I_\delta(u) > 0$ and $I_\delta(u) < 0$ by the surface $I_\delta(u) = 0$ ($\|\nabla u\| \neq 0$). The inside part of $I_\delta(u) = 0$ is $I_\delta(u) > 0$ and the outside part of $I_\delta(u) = 0$ is $I_\delta(u) < 0$. Sphere $\|\nabla u\| = r(\delta)$ lies inside of $I_\delta(u) > 0$ and sphere $\|\nabla u\|^2 = \dfrac{d(\delta)}{a(\delta)}$ lies inside of $I_\delta(u) < 0$ (see Fig. 3).

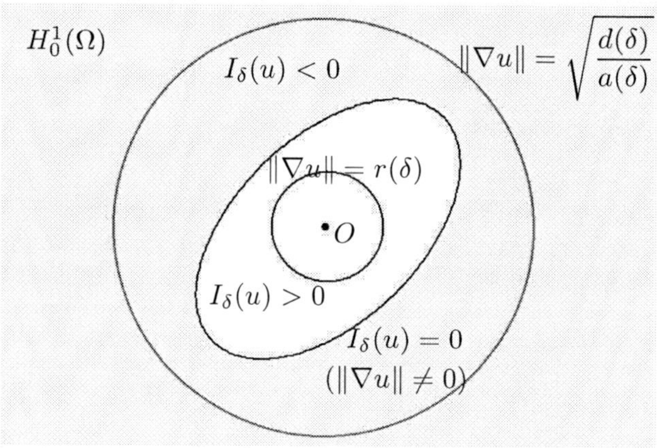

Figure 3.

Now we can defin a family of potential wells as follows

$$W = \{u \in H_0^1(\Omega) \mid I(u) > 0,\ J(u) < d\} \cup \{0\};$$
$$W_\delta = \{u \in H_0^1(\Omega) \mid I_\delta(u) > 0,\ J(u) < d(\delta)\} \cup \{0\},\quad 0 < \delta < \delta_0;$$
$$\overline{W}_\delta = W_\delta \cup \partial W_\delta = \{u \in H_0^1(\Omega) \mid I_\delta(u) \geq 0,\ J(u) \leq d(\delta)\},\quad 0 < \delta < \delta_0.$$

In addition we defin

$$V = \{u \in H_0^1(\Omega) \mid I(u) < 0,\ J(u) < d\};$$
$$V_\delta = \{u \in H_0^1(\Omega) \mid I_\delta(u) < 0,\ J(u) < d(\delta)\}\quad 0 < \delta < \delta_0;$$
$$B_\delta = \{u \in H_0^1(\Omega) \mid \|\nabla u\| < r(\delta)\}\quad 0 < \delta < \delta_0;$$
$$\overline{B}_\delta = B_\delta \cup \partial B_\delta = \{u \in H_0^1(\Omega) \mid \|\nabla u\| \leq r(\delta)\}\quad 0 < \delta < \delta_0;$$
$$B_\delta^c = \{u \in H_0^1(\Omega) \mid \|\nabla u\| > r(\delta)\}\quad 0 < \delta < \delta_0.$$

Lemma 2.9. *Let* $0 < \delta < \dfrac{p+1}{2}$. *Then*

$$B_{r_1(\delta)} \subset W_\delta \subset B_{r_2(\delta)}, \quad V_\delta \subset B_\delta^c$$

where

$$B_{r_1(\delta)} = \{u \in H_0^1(\Omega) \mid \|\nabla u\|^2 < \min\{r^2(\delta),\ r_0^2(\delta)\},$$
$$B_{r_2(\delta)} = \left\{u \in H_0^1(\Omega) \,\middle|\, \|\nabla u\|^2 < \frac{d(\delta)}{a(\delta)}\right\},$$

and $r_0(\delta)$ *is the unique real root of equation*

$$\frac{1}{2}r^2 + \frac{b}{q+1}C_q^{q+1}r^{q+1} = d(\delta), \quad C_q = \sup_{u\in H_0^1(\Omega)} \frac{\|u\|_{q+1}}{\|\nabla u\|}.$$

Proof. First $\|\nabla u\| < r(\delta)$ gives $\|\nabla u\| = 0$ or $I_\delta(u) > 0$. On the other hand,

$$J(u) \le \frac{1}{2}\|\nabla u\|^2 + \frac{b}{q+1}\|u\|_{q+1}^{q+1} \le \frac{1}{2}\|\nabla u\|^2 + \frac{b}{q+1}C_q^{q+1}\|\nabla u\|^{q+1}$$

and $\|\nabla u\|^2 < r_0^2(\delta)$ yield $J(u) < d(\delta)$. Hence we have $B_{r_1(\delta)} \subset W_\delta$. The remainder of this lemma follows from Lemma 2.3 and Lemma 2.6. □

From the definitio of W_δ, V_δ and Lemma 2.5 we can obtain the following

Lemma 2.10.

(i) *If* $0 < \delta' < \delta'' \le 1$, *then* $W_{\delta'} \subset W_{\delta''}$.
(ii) *If* $1 \le \delta'' < \delta' < \delta_0$, *then* $V_{\delta'} \subset V_{\delta''}$.

Lemma 2.11. *Let* $d(0) \le J(u) < d$ *for some* $u \in H_0^1(\Omega)$, $\delta_1 < \delta_2$ *be the two roots of equation* $d(\delta) = J(u)$. *Then the sign of* $I_\delta(u)$ *is unchangeable for* $\delta_1 < \delta < \delta_2$.

Proof. First $J(u) > 0$ implies $\|\nabla u\| \ne 0$. If the sign of $I_\delta(u)$ is changed for $\delta_1 < \delta < \delta_2$, then there exists a $\bar\delta \in (\delta_1, \delta_2)$ such that $I_{\bar\delta}(u) = 0$. Thus by the definitio of $d(\delta)$ we have $J(u) \ge d(\bar\delta)$ which contradicts

$$J(u) = d(\delta_1) = d(\delta_2) < d(\bar\delta).$$ □

Lemma 2.12. *Let* $0 < J(u) < d(0)$ *for some* $u \in H_0^1(\Omega)$, δ_2 *be the unique real root of equation* $d(\delta) = J(u)$. *Then the sign of* $I_\delta(u)$ *is unchangeable for* $0 < \delta < \delta_2$.

Proof. First $J(u) > 0$ implies $\|\nabla u\| \ne 0$. If the sign of $I_\delta(u)$ is changed for $0 < \delta < \delta_2$, then there exists a $\bar\delta \in (0, \delta_2)$ such that $I_{\bar\delta}(u) = 0$. Hence by Lemma 2.5 we have

$$J(u) \ge d(\bar\delta) > d(0) > d(\delta_2) \text{ if } 0 < \bar\delta \le 1$$

or

$$J(u) \ge d(\bar\delta) > d(\delta_2) \text{ if } 1 \le \bar\delta < \delta_2$$

which contradicts $J(u) = d(\delta_2)$. □

3. Invariant Sets and Vacuum Isolating of Solutions

In this section we discuss the invariance of some sets under the fl w of (1.1)–(1.3) and vacuum isolating behavior of solutions for problem (1.1)–(1.3). First for problem (1.1)–(1.3) we defin the energy

$$E(t) = \frac{1}{2}\|u_t\|^2 + \frac{1}{2}\|\nabla u\|^2 - \frac{a}{p+1}\|u\|_{p+1}^{p+1} + \frac{b}{q+1}\|u\|_{q+1}^{q+1} \equiv \frac{1}{2}\|u_t\|^2 + J(u).$$

Theorem 3.1. *Let $u_0(x) \in H_0^1(\Omega)$, $u_1(x) \in L^2(\Omega)$. Assume that $d(0) \le e < d$, $\delta_1 < \delta_2$ are the two roots of equation $d(\delta) = e$. Then*

(i) *All solutions of problem (1.1)–(1.3) with $E(0) = e$ belong to W_δ for $\delta_1 < \delta < \delta_2$, provided $I(u_0) > 0$ or $\|\nabla u_0\| = 0$.*

(ii) *All solutions of problem (1.1)–(1.3) with $E(0) = e$ belong to V_δ for $\delta_1 < \delta < \delta_2$, provided $I(u_0) < 0$.*

Proof. (i) Let $u(t)$ be any solution of problem (1.1)–(1.3) with $E(0) = e$ and $I(u_0) > 0$ or $\|\nabla u_0\| = 0$, T be the existence time of $u(t)$. If $\|\nabla u_0\| = 0$, then $u_0(x) \in W_\delta$ for $0 < \delta < \delta_0$. If $I(u_0) > 0$, then by the definitio of $d(\delta)$ and

$$\frac{1}{2}\|u_1\|^2 + J(u_0) = E(0) = d(\delta_1) = d(\delta_2) < d(\delta), \quad \delta_1 < \delta < \delta_2, \tag{3.1}$$

it follows that $I_\delta(u_0) > 0$ and $J(u_0) < d(\delta)$ *i.e.* $u_0(x) \in W_\delta$ for $\delta_1 < \delta < \delta_2$. Next we prove $u(t) \in W_\delta$ for $\delta_1 < \delta < \delta_2$ and $0 < t < T$. If it is false, then must have a $t_0 \in (0, T)$ such that $u(t_0) \in \partial W_\delta$ for some $\delta \in (\delta_1, \delta_2)$, *i.e.* $I_\delta(u(t_0)) = 0$, $\|\nabla u(t_0)\| \neq 0$ or $J(u(t_0)) = d(\delta)$. From the energy equality

$$\frac{1}{2}\|u_t\|^2 + J(u) = E(0) < d(\delta), \quad \delta_1 < \delta < \delta_2,\ 0 < t < T, \tag{3.2}$$

we see that $J(u(t_0)) = d(\delta)$ is impossible. On the other hand, if $I_\delta(u(t_0)) = 0$ and $\|\nabla u(t_0)\| \neq 0$, then by the definitio of $d(\delta)$ we have $J(u(t_0)) \ge d(\delta)$ which contradicts (3.2).

(ii) Let $u(t)$ be any solution of problem (1.1)–(1.3) with $E(0) = e$ and $I(u_0) < 0$, T be the existence time of $u(t)$. First by (3.1) we can obtain $I_\delta(u_0) < 0$ and $J(u_0) < d(\delta)$, *i.e.* $u_0(x) \in V_\delta$ for $\delta_1 < \delta < \delta_2$. Next we prove $u(t) \in V_\delta$ for $\delta_1 < \delta < \delta_2$ and $0 < t < T$. If it is false, let $t_0 \in (0, T)$ be the firs time such that $u(t) \in V_\delta$ for $0 \le t < t_0$ and $u(t_0) \in \partial V_\delta$, *i.e.* $I_\delta(u(t_0)) = 0$ or $J(u(t_0)) = d(\delta)$ for some $\delta \in (\delta_1, \delta_2)$. Again (3.2) shows that $J(u(t_0)) = d(\delta)$ is impossible. If $I_\delta(u(t_0)) = 0$, then $I_\delta(u(t)) < 0$ for $0 < t < t_0$ and Lemma 2.3 yield $\|\nabla u(t)\| > r(\delta)$ and $\|\nabla u(t_0)\| \ge r(\delta)$. Hence by the definitio of $d(\delta)$ we get $J(u(t_0)) \ge d(\delta)$ which contradicts (3.2). □

From Theorem 3.1 and Lemma 2.5 we can obtain the following

Corollary 3.2. *If in Theorem 3.1 the assumption $E(0) = e$ is replaced by $d(0) \le E(0) \le e$, then the conclusion of Theorem 3.1 also holds.*

Theorem 3.3. *Let* $u_0(x) \in H_0^1(\Omega)$, $u_1(x) \in L^2(\Omega)$. *Assume that* $0 < e < d(0)$, δ_2 *is the unique real root of equation,* $d(\delta) = e$. *Then*

(i) *All solutions of problem* (1.1)–(1.3) *with* $E(0) = e$ *belong to* W_δ *for* $0 < \delta < \delta_2$, *provided* $I(u_0) > 0$ *or* $\|\nabla u_0\| = 0$.

(ii) *All solutions of problem* (1.1)–(1.3) *with* $E(0) = e$ *belong to* V_δ *for* $0 < \delta < \delta_2$, *provided* $I(u_0) < 0$.

Proof. The proof of this theorem is similar to the proof of Theorem 3.1 where δ_1 is replaced by 0 and (3.1) and (3.2) are replaced by

$$\frac{1}{2}\|u_1\|^2 + J(u_0) = E(0) < d(\delta), \quad 0 < \delta < \delta_2 \tag{3.1'}$$

and

$$\frac{1}{2}\|u_t\|^2 + J(u) = E(0) < d(\delta), \quad 0 < \delta < \delta_2, \ 0 < t < T. \quad \Box \tag{3.2'}$$

Corollary 3.4. *If in Theorem* 3.3 *the assumption* $E(0) = e$ *is replaced by* $0 < E(0) \leq e$, *then the conclusion of Theorem* 3.3 *also holds.*

Corollary 3.5. *If in Corollary* 3.2 *the assumption* $d(0) \leq E(0) < e$ *is replaced by* $0 < E(0) \leq e$, *then the conclusion of Corollary* 3.2 *also holds.*

Theorem 3.6. *Let* $u_i(x)$ $(i = 0, 1)$, e *and* δ_i $(i = 1, 2)$ *be the same as those in Theorem* 3.1. *Then for any* $\delta \in (\delta_1, \delta_2)$ *both sets* W_δ *and* V_δ *are invariant, thereby both sets*

$$W_{\delta_1\delta_2} = \bigcup_{\delta_1<\delta<\delta_2} W_\delta, \quad \text{and} \quad V_{\delta_1\delta_2} = \bigcup_{\delta_1<\delta<\delta_2} V_\delta$$

are invariant respectively under the flow of (1.1)–(1.3), *provided* $0 < E(0) \leq e$.

Theorem 3.7. *Let* $u_i(x)$ $(i = 0, 1)$, e *and* δ_2 *be the same as those in Theorem* 3.3. *Then for any* $\delta \in (0, \delta_2)$ *both sets* W_δ *and* V_δ *are invariant, thereby both sets*

$$W_{0,\delta_2} = \bigcup_{0<\delta<\delta_2} W_\delta, \quad \text{and} \quad V_{0,\delta_2} = \bigcup_{0<\delta<\delta_2} V_\delta$$

are invariant respectively under the flow of (1.1)–(1.3), *provided* $0 < E(0) \leq e$.

From the proof of Theorem 3.1 and Theorem 3.3 we see that if $0 < E(0) < d$, then $I(u_0) = 0$ and $\|\nabla u_0\| \neq 0$ is impossible. So the results of Corollary 3.2 and Corollary 3.4 show that for the set of all solutions of problem (1.1)–(1.3) with $0 < E(0) \leq e$ there is a vacuum region

$$U_e = \{u \in H_0^1(\Omega) \mid \|\nabla u\| \neq 0, \ I_\delta(u) = 0, \ \delta_1 < \delta < \delta_2\} \text{ if } d(0) \leq e < d$$

or

$$U_e = \{u \in H_0^1(\Omega) \mid \|\nabla u\| \neq 0, \ I_\delta(u) = 0, \ 0 < \delta < \delta_2\} \text{ if } 0 < e < d(0)$$

such that there is no any solution of problem (1.1)–(1.3)(see Fig. 4).

The vacuum region U_e become bigger and bigger with decreasing of e. As the limit case we obtain

$$U_0 = \{u \in H_0^1(\Omega) \mid \|\nabla u\| \neq 0, \ I_\delta(u) = 0, \ 0 < \delta < \delta_0\}.$$

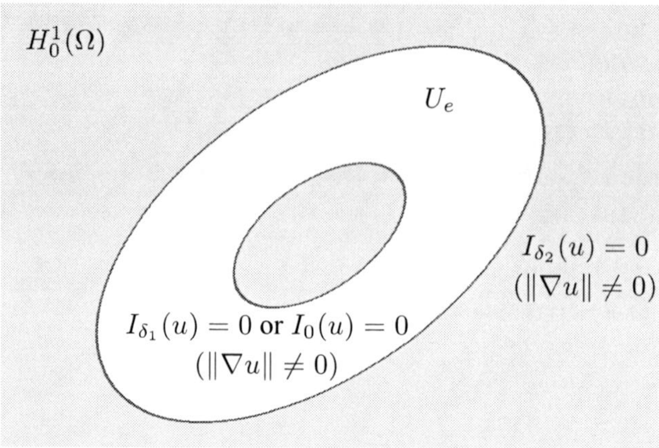

Figure 4.

In fact, we have the following

Theorem 3.8. *All nontrivial solutions $u(t)$ of problem (1.1)–(1.3) with $E(0) = 0$ belong to*

$$B_{r_0}^c = \left\{ u \in H_0^1(\Omega) \mid \|\nabla u\| > r_0 = \left(\frac{p+1}{2aC_*^{p+1}} \right)^{\frac{1}{p-1}} \right\}$$

in the existence interval of $u(t)$.

Proof. Let $u(t)$ be any solution of problem (1.1)–(1.3) with $E(0) = 0$, T be the existence time of $u(t)$. From the energy equality

$$\frac{1}{2}\|u_t\|^2 + J(u) \equiv E(0) = 0$$

we get $J(u) \leq 0$ for $0 \leq t < T$. Hence by

$$\begin{aligned} \frac{1}{2}\|\nabla u\|^2 &\leq \frac{a}{p+1}\|u\|_{p+1}^{p+1} - \frac{b}{q+1}\|u\|_{q+1}^{q+1} \leq \frac{a}{p+1}\|u\|_{p+1}^{p+1} \\ &\leq \frac{a}{p+1}C_*^{p+1}\|\nabla u\|^{p-1}\|\nabla u\|^2, \quad 0 \leq t < T \end{aligned}$$

we see that one must have either $\|\nabla u\| = 0$ or $\|\nabla u\| \geq r_0$. If $\|\nabla u_0\| = 0$, then $\|\nabla u\| \equiv 0$ for $0 \leq t < T$. Otherwise there exists a $t_0 \in (0, T)$ such that $0 < \|\nabla u(t_0)\| < r_0$. By a similar argument we can prove that if $\|\nabla u_0\| \geq r_0$, then $\|\nabla u\| \geq r_0$ for $0 < t < T$. □

Theorem 3.9. *Let $u_0(x) \in H_0^1(\Omega)$, $u_1(x) \in L^2(\Omega)$. Assume that $E(0) < 0$ or $E(0) = 0$, $\|\nabla u_0\| \neq 0$. Then all solutions of problem (1.1)–(1.3) belong to V_δ for $0 < \delta < \frac{p+1}{2}$.*

Proof. Let $u(t)$ be any solution of problem (1.1)–(1.3) with $E(0) < 0$ or $E(0) = 0$, $\nabla u_0 \neq 0$, T be the existence time of $u(t)$. The energy equality gives

$$\frac{1}{2}\|u_t\|^2 + a(\delta)\|\nabla u\|^2 + \frac{1}{p+1}I_\delta(u) \leq \frac{1}{2}\|u_t\|^2 + J(u) = E(0) \ \ 0 < \delta < \frac{p+1}{2} \tag{3.3}$$

From (3.3) it follows that if $E(0) < 0$, then $I_\delta(u) < 0$, $J(u) < 0 < d(\delta)$ for $0 < \delta < \frac{p+1}{2}$; if $E(0) = 0$, $\|\nabla u_0\| \neq 0$, then by Theorem 3.8 we have $\|\nabla u\| > r_0$ for $0 \leq t < T$. Again by (3.3) we get $I_\delta(u) < 0$, $J(u) < 0 < d(\delta)$ for $0 < \delta < \frac{p+1}{2}$. Hence for above two cases we always have $u(t) \in V_\delta$ for $0 < \delta < \frac{p+1}{2}$, $0 \leq t < T$. □

Corollary 3.10. *Under the conditions of Theorem* 3.9 *all solutions of problem* (1.1)–(1.3) *belong to* $\bar{B}^c_{\frac{p+1}{2}}$.

Proof. Let $u(t)$ be any solution of problem (1.1)–(1.3) with $E(0) < 0$ or $E(0) = 0$, $\|\nabla u_0\| \neq 0$, T be the existence time of $u(t)$. Then Theorem 3.9 gives

$$u(t) \in V_\delta \ \text{ for } \ 0 < \delta < \frac{p+1}{2}, 0 \leq t < T.$$

From this and Lemma 2.3 we get $\|\nabla u\| > r(\delta)$ for $0 < \delta < \frac{p+1}{2}$, $0 \leq t < T$. Letting $\delta \to \frac{p+1}{2}$ we obtain $\|\nabla u\| \geq r\left(\frac{p+1}{2}\right)$ for $0 \leq t < T$. □

4. Global Existence and Blow up of Solutions for Problem (1.1)–(1.3)

In this section we shall prove the global existence and finit time blow up of solutions for problem (1.1)–(1.3) by using the family of potential wells introduced above.

Theorem 4.1. *Let* $u_0(x) \in H^1_0(\Omega)$, $u_1(x) \in L^2(\Omega)$. *Assume that* $0 < E(0) < d$ *and* $I(u_0) > 0$ *or* $\|\nabla u_0\| = 0$. *Then problem* (1.1)–(1.3) *admits a global weak solution* $u(t) \in L^\infty(0, \infty; H^1_0(\Omega))$ *with* $u_t(t) \in L^\infty(0, \infty; L^2(\Omega))$ *and* $u(t) \in W$ *for* $0 \leq t < \infty$.

Proof. Construct the approximate solutions $u_m(x, t)$ of problem (1.1)–(1.3) as we did in [10]. Then by the same arguments used in the proof of Theorem 3.2 in [10] we can obtain

$$\frac{1}{2}\|u_{mt}\|^2 + J(u_m) = E_m(0) < d, \quad 0 \leq t < \infty \tag{4.1}$$

and $u_m(t) \in W$ for sufficientl large m and $0 \leq t < \infty$. From (4.1) and

$$\begin{aligned} J(u_m) &\geq \frac{1}{2}\|\nabla u_m\|^2 - \frac{1}{p+1}\left(a\|u_m\|^{p+1}_{p+1} - b\|u_m\|^{q+1}_{q+1}\right) \\ &= \left(\frac{1}{2} - \frac{1}{p+1}\right)\|\nabla u_m\|^2 + \frac{1}{p+1}I(u_m) \geq \frac{p-1}{2(p+1)}\|\nabla u_m\|^2 \end{aligned}$$

it follows that

$$\frac{1}{2}\|u_{mt}\|^2 + \frac{p-1}{2(p+1)}\|\nabla u_m\|^2 < d, \quad 0 \leq t < \infty \tag{4.2}$$

for sufficientl large m. Eq. (4.2) gives

$$\|\nabla u_m\|^2 < \frac{2(p+1)}{p-1}d, \quad 0 \leq t < \infty; \tag{4.3}$$

$$\|u_m\|^2_{r_k+1} \le C_r^2 \|\nabla u_m\|^2 < C_r^2 \frac{2(p+1)}{p-1} d, \quad r = p, q, \ 0 \le t < \infty; \tag{4.4}$$

$$\|u_{mt}\|^2 < 2d, \quad 0 \le t < \infty. \tag{4.5}$$

From (4.3)–(4.5) and compactness method it follows that problem (1.10–(1.3) admits a global weak solution $u(t) \in L^\infty\left(0, \infty; H_0^1(\Omega)\right)$ with $u_t(t) \in L^\infty\left(0, \infty; L^2(\Omega)\right)$. Finally by Theorem 3.1 and Theorem 3.3 we have $u(t) \in W$ for $0 \le t < \infty$. □

Corollary 4.2. *Under the conditions of Theorem* 4.1 *we further have* $u(t) \in W_\delta$ *for* $\delta_1 < \delta < \delta_2$, $0 \le t < \infty$, *provided* $d(0) \le E(0) < d$, *where* $\delta_1 < \delta_2$ *are the two roots of equation* $d(\delta) = E(0)$.

Corollary 4.3. *Under the conditions of Theorem* 4.1 *we further have* $u(t) \in W_\delta$ *for* $0 < \delta < \delta_2$, $0 \le t < \infty$, *provided* $0 < E(0) < d(0)$, *where* δ_2 *is the unique root of equation* $d(\delta) = E(0)$.

Corollary 4.4. *If in Corollary* 4.2 *and Corollary* 4.3 *the assumption* $I(u_0) > 0$ *is replaced by* $I_{\delta_2}(u_0) > 0$, *then the conclusions of Corollary* 4.2 *and Corollary* 4.3 *also hold.*

Theorem 4.5. *Let* $u_0(x) \in H_0^1(\Omega)$, $u_1(x) \in L^2(\Omega)$. *Assume that* $0 < E(0) < d$, $\|\nabla u\| \le r(\delta_2)$, *where* δ_2 *is the larger real root (if* $d(0) \le E(0) < d$*) or the unique real root (if* $0 < E(0) < d(0)$*) of equation* $d(\delta) = E(0)$. *Then problem* (1.1)–(1.3) *admits a global weak solution* $u(t) \in L^\infty(0, \infty; H_0^1(\Omega))$ *with* $u_t(t) \in L^\infty(0, \infty; L^2(\Omega))$ *and* $\|\nabla u(t)\|^2 < \frac{d(\delta)}{a(\delta)}$ *for* $\delta_1 < \delta < \min\left\{\delta_2, \frac{p+1}{2}\right\}$ *(if* $d(0) \le E(0) < d$*) or* $\|\nabla u(t)\|^2 < \frac{d(\delta)}{a(\delta)}$ *for* $0 < \delta < \min\left\{\delta_2, \frac{p+1}{2}\right\}$ *(if* $0 < E(0) < d(0)$*) and* $\|u_t(t)\|^2 \le 2E(0)$, $0 \le t < \infty$.

Proof. First from $\|\nabla u_0\| \le r(\delta_2)$ and Lemma 2.2 we have $I_{\delta_2}(u_0) > 0$ or $\|\nabla u_0\| = 0$. Hence from Corollary 4.4 it follows that problem (1.1)–(1.3) admits a global weak solution $u(t) \in L^\infty(0, \infty; H_0^1(\Omega))$ with $u_t(t) \in L^\infty(0, \infty; L^2(\Omega))$ and $u(t) \in W_\delta$ for $\delta_1 < \delta < \delta_2$ (if $d(0) \le E(0) < d$) or $0 < \delta < \delta_2$ (if $0 < E(0) < d(0)$). Finally Theorem 4.5 follows from Lemma 2.9 and (3.2). □

Corollary 4.6. *Under the conditions of Theorem* 4.5 *we further have* $\|\nabla u(t)\|^2 \le \frac{d(\delta_1)}{a(\delta_1)}$ *for* $d(0) \le E(0) < d$ *or* $\|\nabla u(t)\|^2 \le \frac{d(0)}{a(0)} = 2d(0)$ *for* $0 < E(0) < d(0)$.

Theorem 4.7. *Let* $u_0(x) \in H_0^1(\Omega)$, $u_1(x) \in L^2(\Omega)$. *Assume that* $E(0) < d$ *and* $I(u_0) < 0$. *Then the solution of problem* (1.1)–(1.3) *must blow up in finite time, i.e. there exists a* $T > 0$ *such that*

$$\lim_{t \to T} \|u(t)\| = +\infty. \tag{4.6}$$

Proof. Let $u(t)$ be any solution of problem (1.1)–(1.3) with $E(0) < d$ and $I(u_0) < 0$, $F(t) = \|u(t)\|^2$. Then $\dot{F}(t) = 2(u_t, u)$,

$$\ddot{F}(t) = 2\|u_t\|^2 + 2(u_{tt}, u) = 2\|u_t\|^2 - 2I(u). \tag{4.7}$$

Hence

$$\begin{aligned}\ddot{F}(t) &\geq 2\|u_t\|^2 - 2\|\nabla u\|^2 + 2(p+1)\left(\frac{a}{p+1}\|u\|_{p+1}^{p+1} - \frac{b}{q+1}\|u\|_{q+1}^{q+1}\right)\\ &= (p+3)\|u_t\|^2 + (p-1)\|\nabla u\|^2 - 2(p+1)E(0)\\ &\geq (p+3)\|u_t\|^2 + (p-1)\lambda_1 F(t) - 2(p+1)E(0),\end{aligned} \tag{4.8}$$

where $\lambda_1 > 0$ is the firs eigenvalue of problem $\Delta\varphi + \lambda\varphi = 0$, $\varphi|_{\partial\Omega} = 0$.

(i) If $E(0) \leq 0$, then (4.8) gives

$$\ddot{F}(t) \geq (p+3)\|u_t\|^2. \tag{4.9}$$

(ii) If $0 < E(0) < d$, then from Theorem 3.1 and Theorem 3.3 it follows that $u(t) \in V_\delta$ for $1 < \delta < \delta_2$ and $t > 0$, where δ_2 is the same as those in Theorem 3.1 or Theorem 3.3. Hence $I_\delta(u) < 0$ and (by Lemma 2.3) $\|\nabla u\| > r(\delta)$ for $1 < \delta < \delta_2$ and $t > 0$. So we get $I_{\delta_2}(u) \leq 0$ and $\|\nabla u\| \geq r(\delta_2)$ for $t > 0$. And by (4.7) we obtain

$$\begin{aligned}&\ddot{F}(t) \geq 2(\delta_2 - 1)\|\nabla u\|^2 - 2I_{\delta_2}(u) \geq 2(\delta_2 - 1)r^2(\delta_2) > 0,\\ &\dot{F}(t) \geq 2(\delta_2 - 1)r^2(\delta_2)t + \dot{F}(0), .10\end{aligned} \tag{4}$$

which shows that there exists a $t_0 \geq 0$ such that $\dot{F}(t_0) > 0$ and $F(t) \geq \dot{F}(t_0)(t - t_0) + F(t_0)$, $t \geq t_0$. So for sufficientl large t we have $(p-1)\lambda_1 F(t) > 2(p+1)E(0)$ and (4.9) holds. Finally (4.9) gives (see [5], p. 293)

$$F(t)\ddot{F}(t) - \frac{p+3}{4}(\dot{F}(t))^2 \geq (p+3)(\|u\|^2\|u_t\|^2 - (u, u_t)^2) \geq 0,$$

$$(F^{-\alpha}(t))'' = -\frac{\alpha}{F^{\alpha+2}(t)}(F(t)\ddot{F}(t) - (\alpha+1)(\dot{F}(t))^2) \leq 0, \quad \alpha = \frac{p-1}{4}$$

and (4.6) for some $T > 0$. □

Corollary 4.8. *Under the conditions of Theorem* 4.7 *we further have*

$$\lim_{t\to T} \|\nabla u(t)\| = +\infty \tag{4.11}$$

and

$$\lim_{t\to T} \|u(t)\|_r = +\infty, \tag{4.12}$$

where $2 < r \leq \infty$ if $n = 1$; $2 < r < \infty$ if $n = 2$; $2 < r \leq \frac{2n}{n-2}$ if $n \geq 3$.

Proof. (4.11) follows from (4.6) and $\|\nabla u\|^2 \geq \lambda_1\|u\|^2$. From Sobolev embedding theorem and $\|\nabla u(t)\| < C(t)$ for $0 \leq t < T$ we get $\|u\|_r < C(t)$ for $0 \leq t < T$, where r is define above. And by Hölder inequality we get

$$\|u\| \leq |\Omega|^{\frac{r-2}{2r}}\|u\|_r. \tag{4.13}$$

Hence (4.12) follows from (4.13) and (4.6). □

Note that from

$$\frac{1}{2}\|u_1\|^2+\frac{p-1}{2(p+1)}\|\nabla u_0\|^2+\frac{1}{p+1}I(u_0)\leq\frac{1}{2}\|u_1\|^2+J(u_0)=E(0)$$

we see that if $0<E(0)<d$, then $I(u_0)=0$, $\|\nabla u_0\|\neq 0$ is impossible; if $E(0)=0$, then $I(u_0)>0$ or $I(u_0)=0$, $\|\nabla u_0\|\neq 0$ is impossible; if $E(0)<0$, then $I(u_0)\geq 0$ is impossible. Hence all possible cases already have been discussed in Theorem 4.1 and Theorem 4.7 and a threshold result already has been obtained as follows

Corollary 4.9. *Let $u_0(x)\in H_0^1(\Omega)$, $u_1(x)\in L^2(\Omega)$. Assume that $E(0)<d$. Then when $I(u_0)\geq 0$ problem (1.1)–(1.3) admits a global weak solution and when $I(u_0)<0$ the problem does not admit any global weak solution.*

Remark 4.1. If in the proof of Theorem 4.7 we do not use the family of sets V_δ and only use the set V as did in [11], then for the case $0<E(0)<d$ we can only obtain $\ddot{F}(t)>0$ and $\dot{F}(t)$ is increasing, but can not prove $\dot{F}(t)>0$ and $F(t)$ is increasing, thereby can not obtain $(p-1)\lambda_1F(t)>2(p+1)E(0)$ and (4.9) for sufficientl large t, unless we further assume $\dot{F}(0)=2(u_0,u_1)>0$. Therefor in the case $0<E(0)<d$, $I(u_0)<0$ in order to prove the finit time blow up of solutions for problem $(1.1)-(1,3)$ without the assumption $(u_0,u_1)>0$ the introducing of the family of set V_δ is necessary.

5. Problem (1.1)–(1.3) with Critical Initial Conditions $I(u_0)\geq 0$, $E(0)=d$

In this section we prove the global existence of solutions for problem (1.1)–(1.3) with critical initial conditions $I(u_0)\geq 0$, $E(0)=d$.

Theorem 5.1. *Let $u_0(x)\in H_0^1(\Omega)$, $u_1(x)\in L^2(\Omega)$. Assume that $E(0)=d$ and $I(u_0)\geq 0$. Then problem (1.1)–(1.3) admits a global weak solution $u(t)\in L^\infty(0,\infty;H_0^1(\Omega))$ with $u_t(t)\in L^\infty(0,\infty;L^2(\Omega))$ and $u(t)\in\overline{W}$ for $0\leq t<\infty$.*

Proof. We prove this theorem for the following two cases (i) and (ii).

(i) $\|\nabla u_0\|\neq 0$.

Let $\lambda_m=1-\dfrac{1}{m}$ and $u_{0m}=\lambda_m u_0$, $m=2,3,\ldots$. Consider the initial conditions

$$u(x,0)=u_{0m}(x),\quad u_t(x,0)=u_1(x) \tag{5.1}$$

and corresponding problem (1.1), (5.1), (1.3). From $I(u_0)\geq 0$ and Lemma 2.1 we have $\bar{\lambda}=\bar{\lambda}(u_0)\geq 1$. Hence $I(u_{0m})>0$,

$$\begin{aligned}J(u_{0m})&\geq\frac{1}{2}\|\nabla u_{0m}\|^2-\frac{1}{p+1}\left(a\|u_{0m}\|_{p+1}^{p+1}-b\|u_{0m}\|_{q+1}^{q+1}\right)\\&=\left(\frac{1}{2}-\frac{1}{p+1}\right)\|\nabla u_{0m}\|^2+\frac{1}{p+1}I(u_{0m})>0\end{aligned}$$

and $J(u_{0m})=J(\lambda_m u_0)<J(u_0)$. And

$$0<E_m(0)\equiv\frac{1}{2}\|u_1\|^2+J(u_{0m})<\frac{1}{2}\|u_1\|^2+J(u_0)=E(0)=d.$$

So it follows from Theorem 4.1 that for each m problem (1.1), (5.1), (1.3) admits a global weak solution $u_m(t) \in L^\infty(0,\infty; H_0^1(\Omega))$ with $u_{mt}(t) \in L^\infty(0,\infty; L^2(\Omega))$ and $u_m(t) \in W$ for $0 \le t < \infty$ satisfying

$$(u_{mt}, v) + \int_0^t (\nabla u_m, \nabla v) d\tau = \int_0^t (f(u_m), v) d\tau + (u_1, v),$$
$$\forall\, v \in H_0^1(\Omega), \quad 0 \le t < \infty \tag{5.2}$$

and

$$\frac{1}{2}\|u_{mt}\|^2 + J(u_m) = E_m(0) < d. \tag{5.3}$$

From (5.3) and

$$J(u_m) \ge \left(\frac{1}{2} - \frac{1}{p+1}\right)\|\nabla u_m\|^2 + \frac{1}{p+1} I(u_m) \ge \frac{p-1}{2(p+1)}\|\nabla u_m\|^2$$

we can obtain

$$\|\nabla u_m\|^2 \le \frac{2(p+1)}{p-1} d, \quad 0 \le t < \infty,$$

$$\|u_m\|_{r+1}^2 \le C_r^2 \|\nabla u_m\|^2 \le C_r^2 \frac{2(p+1)}{p-1} d, \quad r = p, q, \ 0 \le t < \infty$$

and

$$\|u_{mt}\|^2 \le 2d, \quad 0 \le t < \infty.$$

Hence there exist a $u(x,t) \in \overline{W}$ and a subsequence $\{u_\nu\}$ of $\{u_m\}$ such that as $\nu \to \infty$

$$u_\nu \longrightarrow u \text{ in } L^\infty(0,\infty; H_0^1(\Omega)) \text{ weak star and } a.e. \text{ in } Q = \Omega \times [0,\infty),$$
$$u_{\nu t} \longrightarrow u_t \text{ in } L^\infty(0,\infty; L^2(\Omega)) \text{ weak star},$$
$$|u_\nu|^{r-1}u_\nu \longrightarrow |u|^{r-1}u \text{ in } L^\infty(0,\infty; L^{r'}(\Omega)) \text{ weak star}, \ r' = \frac{r+1}{r}, \quad r = p, q.$$

Letting $m - \nu \to \infty$ in (5.2) we get

$$(u_t, v) + \int_0^t (\nabla u, \nabla v) d\tau = \int_0^t (f(u), v) d\tau + (u_1, v),$$
$$\forall\, v \in H_0^1(\Omega), \ 0 \le t < \infty. \tag{5.4}$$

Eq. (5.4) implies that $u(x,t)$ satisfie equation (1.1) and $u_t(x,0) = u_1(x)$. On the other hand, clearly we have $u(x,0) = u_0(x)$. So $u(x,t)$ is a global weak solution of problem (1.1)–(1.3).

(ii) $\|\nabla u_0\| = 0$.

Note that $\|\nabla u_0\| = 0$ implies $J(u_0) = 0$ and $\frac{1}{2}\|u_1\|^2 = E(0) = d$. Let $\lambda_m = 1 - \frac{1}{m}$, $u_{1m}(x) = \lambda_m u_1(x)$, $m = 2, 3, \cdots$. Consider the initial conditions

$$u(x,0) = u_0(x), \quad u_t(x,0) = u_{1m}(x) \tag{5.5}$$

and corresponding problem (1.1), (5.5), (1.3). From $\|\nabla u_0\| = 0$,

$$0 < E_m(0) = \frac{1}{2}\|u_{1m}\|^2 + J(u_0) = \frac{1}{2}\|\lambda_m u_1\|^2 < E(0) = d$$

and Theorem 4.1 it follows that for each m problem (1.1), (5.5), (1.3) admits a global weak solution $u_m(t) \in L^\infty(0,\infty; H_0^1(\Omega))$ with $u_{mt}(t) \in L^\infty(0,\infty; L^2(\Omega))$ and $u_m(t) \in W$ for $0 \le t < \infty$ satisfying (5.2) and (5.3). The remainder of proof is the same as that in the part (i) of proof of this theorem. □

6. Reaction–Diffusion Equations

In this section we discuss the invariant sets, vacuum isolating of solutions, global existence and blow up of solutions and the problem with critical initial conditions for problem (1.4), (1.5), (1.6). First as did in [11] we assume the weak solution $u(x,t)$ of problem (1.4), (1.5), (1.3) satisfie the following condition

$$\int_0^t \|u_t\|^2 \mathrm{d}\tau + J(u) \le J(u_0), \quad \forall\, t \in [0,T), \tag{6.1}$$

where T is the existence time of solution $u(x,t)$. Exam carefully the proof of Theorem 3.1 – Corollary 3.10, Theorem 4.1 – Corollary 4.6 and Theorem 5.1 we see that if in these theorems the assumptions "$u_1(x) \in L^2(\Omega)$" and "or $\|\nabla u_0\| = 0$" are taken out, $E(0)$ is replaced by $J(u_0)$ and the energy equality $E(t) = E(0)$ is replaced by (6.1), and the conclusion $u_t(t) \in L^\infty\left(0,\infty; L^2(\Omega)\right)$ is replaced by $u_t(t) \in L^2\left(0,\infty; L^2(\Omega)\right)$, then the conclusions of these theorems also hold. Next we prove the blow up of solutions for problem (1.4), (1.5), (1.3).

Theorem 6.1. *Let $u_0(x) \in H_0^1(\Omega)$. Assume that $J(u_0) < d$ and $I(u_0) < 0$. Then the solution of problem (1.4), (1.5), (1.3) must blow up in finite time i.e. there exists a $T > 0$ such that*

$$\lim_{t\to T} \int_0^t \|u\|^2 \mathrm{d}\tau = +\infty. \tag{6.2}$$

Proof. Let $u(t)$ be any solution of problem (1.4), (1.5), (1.3), with $J(u_0) < d$, $I(u_0) < 0$,

$$F_1(t) = \int_0^t \|u\|^2 \mathrm{d}\tau.$$

Then $\dot{F}_1(t) = \|u\|^2$,

$$\ddot{F}_1(t) = 2(u_t, u) = 2((f(u),u) - \|\nabla u\|^2) = -2I(u). \tag{6.3}$$

From (6.3), (6.1) and

$$(f(u),u) \ge (p+1)\left(\frac{a}{p+1}\|u\|_{p+1}^{p+1} - \frac{b}{q+1}\|u\|_{q+1}^{q+1}\right)$$

we can obtain

$$\begin{aligned}\ddot{F}_1(t) &\geq 2(p+1)\int_0^t \|u_t\|^2 \mathrm{d}\tau + (p-1)\|\nabla u\|^2 - 2(p+1)J(u_0)\\ &\geq 2(p+1)\int_0^t \|u_t\|^2 \mathrm{d}\tau + (p-1)\lambda_1 \dot{F}_1(t) - 2(p+1)J(u_0)\end{aligned}$$

and

$$\begin{aligned}F_1\ddot{F}_1 - \frac{p+1}{2}(\dot{F}_1)^2 \geq\ & 2(p+1)\left[\int_0^t \|u\|^2 \mathrm{d}\tau \int_0^t \|u_t\|^2 \mathrm{d}\tau - \left(\int_0^t (u, u_t)\mathrm{d}\tau\right)^2\right]\\ &+ (p-1)\lambda_1 F_1 \dot{F}_1 - (p+1)\|u_0\|^2 \dot{F}_1\\ &- 2(p+1)J(u_0)F_1 + \frac{p+1}{2}\|u_0\|^2,\end{aligned}$$

Hölder inequality gives

$$\begin{aligned}&F_1\ddot{F}_1 - \frac{p+1}{2}(\dot{F}_1)^2\\ \geq&(p-1)\lambda_1 F_1\dot{F}_1 - (p+1)\|u_0\|^2\dot{F}_1 - 2(p+1)J(u_0)F_1 + \frac{p+1}{2}\|u_0\|^2. \qquad (6.4)\end{aligned}$$

(i) If $J(u_0) \leq 0$, then

$$F_1\ddot{F}_1 - \frac{p+1}{2}(\dot{F}_1)^2 \geq (p-1)\lambda_1 F_1\dot{F}_1 - (p+1)\|u_0\|^2\dot{F}_1.$$

First we prove $I(u) < 0$ for $t > 0$. If it is false, then must have $t_0 > 0$ such that $I(u(t_0)) = 0$. Let $t_0 > 0$ be the firs time such that $I(u(t_0)) = 0$, then $I(u(t)) < 0$ for $0 \leq t < t_0$. By Lemma 2.3 we have $\|\nabla u\| > r(1)$ for $0 < t < t_0$. Hence we have $\|\nabla u(t_0)\| \geq r(1)$ and $J(u(t_0)) \geq d$ which contradicts (6.1). So by (6.3) we get $\ddot{F}_1(t) > 0$ for $t > 0$. From this and $\dot{F}_1(0) = \|u_0\|^2 \geq 0$ it follows that there exists a $t_0 \geq 0$ such that $\dot{F}_1(t_0) > 0$ and

$$F_1(t) \geq \dot{F}_1(t_0)(t-t_0) + F_1(t_0) \geq \dot{F}_1(t_0)(t-t_0), t \geq t_0.$$

Hence for sufficientl large t we have $(p-1)\lambda_1 F_1 > (p+1)\|u_0\|^2$ and

$$F_1\ddot{F}_1 - \frac{p+1}{2}(\dot{F}_1)^2 > 0. \qquad (6.5)$$

(ii) If $0 < J(u_0) < d$, then by an argument similar to that in the proof of Theorem 4.7 we can obtain $I_{\delta_2}(u) \leq 0$ and $\|\nabla u\| \geq r(\delta_2)$ for $t \geq 0$, where $\delta_2 > 1$ is the larger real root (if $d(0) \leq J(u_0) < d$) or the unique real root (if $0 < J(u_0) < d(0)$) of equation $d(\delta) = J(u_0)$. Thus (6.3) gives

$$\begin{aligned}&\ddot{F}_1(t) = -2I(u) = 2(\delta_2 - 1)\|\nabla u\|^2 - 2I_{\delta_2}(u) \geq 2(\delta_2-1)r^2(\delta_2) > 0, \quad t \geq 0,\\ &\dot{F}_1(t) \geq 2(\delta_2-1)r^2(\delta_2)t + \dot{F}_1(0) \geq 2(\delta_2-1)r^2(\delta_2)t, \quad t \geq 0,\\ &F_1(t) \geq (\delta_2-1)r^2(\delta_2)t^2 + F_1(0) = (\delta_2-1)r^2(\delta_2)t^2, \quad t \geq 0.\end{aligned}$$

Hence for sufficientl large t we have

$$\frac{1}{2}(p-1)\lambda_1 F_1(t) > (p+1)\|u_0\|^2,$$
$$\frac{1}{2}(p-1)\lambda_1 \dot{F}_1(t) > 2(p+1)J(u_0).$$

Therefor by (6.4) we again obtain (6.5) for sufficientl large t. The remainder of proof is the same as that in [5]. □

Corollary 6.2. *Under the conditions of Theorem* 6.1 *we further have*

$$\overline{\lim_{t\to T}}\|u\|_r = +\infty, \tag{6.6}$$

and

$$\overline{\lim_{t\to T}}\|\nabla u\| = +\infty. \tag{6.7}$$

where $2 \le r \le \infty$ *if* $n = 1$; $2 \le r < \infty$ *if* $n = 2$; $2 \le r \le \dfrac{2n}{n-2}$ *if* $n \ge 3$.

Proof. From (6.2) we obtain

$$\overline{\lim_{t\to T}}\|u\| = +\infty.$$

The remainder of proof is similar to that in the proof of Corollary 4.8. □

Corollary 6.3. *Let* $u_0(x) \in H_0^1(\Omega)$. *Assume that* $J(u_0) < d$. *Then when* $I(u_0) \ge 0$ *problem* (1.4), (1.5), (1.3) *admits a global weak solution, when* $I(u_0) < 0$ *the problem does not admit any global weak solution.*

Remark 6.1. From (6.4) we see that in the case $0 < J(u_0) < d$ in order to prove (6.5) we must prove

$$(p-1)\lambda_1 F_1 \dot{F}_1 \ge (p+1)\|u_0\|^2 \dot{F}_1 + 2(p+1)J(u_0)F_1 \tag{6.8}$$

at least for sufficientl large t. Since $\dot{F}_1(t) > 0$ for $t > 0$, (6.8) yields

$$(p-1)\lambda_1 F_1 \dot{F}_1 > 2(p+1)J(u_0)F_1$$

and

$$\dot{F}_1(t) > \frac{2(p+1)}{(p-1)\lambda_1}J(u_0) \tag{6.9}$$

for sufficientl large t. However if we do not use the family of sets V_δ and only use the set V as did in [11], then by (6.4) we can only obtain $\ddot{F}_1(t) > 0$ and $\dot{F}_1(t)$ is increasing, but can not obtain (6.9) even if for sufficientl large t. Therefore in the case $0 < J(u_0) < d$, $I(u_0) < 0$ in order to prove the finit time blow up of solutions for problem (1.4), (1.5), (1.3) the introducing of the family of sets V_δ is necessary.

References

[1] J.M. Ball, Remarks on blow–up and nonexistence theorems for nonlinear evolution equations, *Quart. J. Math.* **28** (1977), 473–486.

[2] V.A. Galaktionov and S.I. Pohozaev, Blow–up and critical exponents for nonlinear hyperbolic equations, *Nonlinear Anal.* **53** (2003), 453–466.

[3] V. Geogev and G. Todorova, Existence of a solution of the wave equation with non-linear damping and source terms, *J. Differential Equations* **109**(1994), 295–308.

[4] R.T. Glassay, Blow–up theorems for nonlinear wave equations, *Math. Z.* **32** (1973), 183–203.

[5] M. Grillakis, Regularity for the wave equation with critical nonlinearity, *Ann. Math.* **132** (1990), 485–509.

[6] Hiroshi Uesaka, Oscillation or nonosillation property for semiliear wave equations, *J. Comput. Appl. Math.* **164-165** (2004), 723–730.

[7] H.A. Levine, Instablity and nonexistence of global solutions to nonlinear wave equations of the form $Pu = Au + F(u)$, *Trans. of Math, AMS,* **92** (1974), 1–21.

[8] H.A. Levine, Some additional remarks on the nonexistence of global solutions to nonlinear wave equations, *SIAM J. Math. Anal*. **5** (1974), 138–146.

[9] J.L. Lions, *Quelques méthodes de résolution des problèmes aux limites non linéaires*, Dunod Gauthier-Villars, Paris, 1969.

[10] Liu Yacheng, On potential wells and vacuum isolating of solutions for semilinear wave equations, *J. Differential Equations* **192** (2003),155–169.

[11] L.E. Payne and D.H. Sattinger, Sadle points and instability of nonlinear hyperbolic equations, *Israel. J. Math.* **22** (1975), 273–303.

[12] D.H. Sattinger, On global solution of nonlinear hyperbolic equations, *Arch. Rat. Mech. Anal.* **30** (1968), 148–172.

[13] J. Shatah and M. Struwe, Regularity results for nonlinear wave equations, *Ann. Math.* **138** (1993), 503–518.

[14] M. Tsutsumi, On solutions of semilinear differential equations in a Hilbert space, *Math. Japan* **17** (1972), 173–193.

In: Evolution Equations: Theories, Solutions and Functions ISBN 978-1-61324-090-8
Editor: Gaston M. N'Guerekata

Chapter 12

INTEGRAL SOLUTIONS FOR NONDENSELY DEFINED EVOLUTION EQUATIONS WITH NONLOCAL CONDITIONS

Zhenbin Fan* and Li Zhang
Department of Mathematics, Changshu Institute of Technology,
Suzhou, Jiangsu 215500, P. R. China

Abstract

This paper is concerned with the existence for nondensely define evolution equations with nonlocal conditions. Using the techniques of fi ed point theory, existence results are obtained, for integral solutions, without the compactness assumption on nonlocal items.

2000 Mathematics Subject Classification: 34G10, 47D06,

Keywords and phrases: Nonlocal condition, nondensely defined fi ed point, C_0-semigroup, integrated semigroup.

1. Introduction

In this work, we study the existence results for partial differential equations with nonlocal conditions in a real Banach space X of the following form:

$$\begin{aligned} &u'(t) = Au(t) + f(t, u(t)), \quad t \in (0, b], \\ &u(0) = g(u), \end{aligned} \tag{1.1}$$

where $A : D(A) \subset X \to X$ is a nondensely define operator, $f : [0, b] \times X \to X$ and $g : C(0, b; X) \to \overline{D(A)}$ are given X-valued functions.

The study of abstract nonlocal semilinear initial value problems was initiated by Byszewski [7, 8, 9]. Since it is more precise for describing physical phenomena than the classical Cauchy problems, the nonlocal problems have been studied extensively. Ntouyas and

*E-mail address: fzbmath@yahoo.com.cn

Tsamatos [25, 26], Byszewski and Akca [10], Liang, Liu and Xiao [21], Xue [29] and Fan et al. [15] have studied the case when A is a linear, densely define operator on X which generates a C_0-semigroup. Ezzinbi and Fu, etc. [14, 17] have considered the case when A is the infinitesima generator of an analytic semigroup on X. They obtained the existence and regularity results for some neutral partial differential equations with nonlocal conditions. G. M. Mophou, G. M. N'Guerekata [23, 24] have studied and obtained the existence and uniqueness of mild solutions for fractional semilinear differential equation with nonlocal conditions in the α-norm. Aizicovici [2, 3, 4], Xue [30, 31] and Garcia-Falset [19] have discussed the case when $-A$ is a nonlinear m-accretive operator on X which generates a nonlinear contraction semiguoup. Especially, in [2, 4], the authors mainly studied the equation (1.1) with nonlinear operator and multivalued perturbations. In [31], the author studied the equation (1.1) when the nonlinear operator generates a noncompact semigroup.

However, as indicated in [11], we sometimes need to study nondensely define operators. It occurs in many situations due to restrictions on the space where the equation is considered or due to boundary conditions (see examples in [11]). Recently, Ezzinbi and Liu [12], Fu [18] and Fan [16] have investigated the existence and regularity of integral solutions for nondensely define evolution equations with nonlocal conditions. In [12, 18], they mainly supposed that the nonlocal item g is a compact operator. In [16], the author studied the case when the nonlocal item g is Lipschitz continuous or continuous on $C(0, b; X)$, respectively. Other contributions on the nonlocal problems see [5, 6, 13, 20, 22, 28].

The purpose of this paper is to continue the study of these authors. We get the existence results for integral solutions of nondensely define equation (1.1) when the nonlocal item g has no compactness assumption. The organization of this work is as follows. In section 2, we recall some definition and facts about nondensely define operators and integral solutions. In section 3, we give the existence results of equation (1.1) when the nonlocal item g has no compactness assumption. Special types of nonlocal item g are also considered.

2. Preliminaries

Let $(X, \|\cdot\|)$ be a real Banach space. We denote by $C(0, b; X)$ the space of X-valued continuous functions on $[0, b]$ with the norm $\|u\| = \sup\{\|u(t)\|,\ t \in [0, b]\}$ and by $L^1(0, b; X)$ the space of X-valued Bochner integrable functions on $[0, b]$ with the norm $\|f\|_{L^1} = \int_0^b \|f(t)\|\, dt$.

Let us recall the following definition and facts.

Throughout this work, we suppose that

(HA) The linear operator $A : D(A) \subset X \to X$ satisfie the Hille-Yosida condition, i.e., there exist $\overline{M} \geq 0$ and $\omega \in R$ such that $(\omega, +\infty) \subset \rho(A)$ and

$$\sup\{(\lambda - \omega)^n \|R(\lambda, A)^n\| : n \in N,\ \lambda > \omega\} \leq \overline{M},$$

where $R(\lambda, A) = (\lambda I - A)^{-1}$, $\rho(A)$ is the resolvent set of A.

Definition 2.1. A function $u : [0, b] \to X$ is said to be an integral solution of equation (1.1) on $[0, b]$ if the following conditions hold:

(i) $u : [0, b] \to X$ is continuous;

(ii) $\int_0^t u(s)\,\mathrm{d}s \in D(A)$ for $t \in [0,b]$;

(iii) $u(t) = g(u) + A\int_0^t u(s)\,\mathrm{d}s + \int_0^t f(s,u(s))\,\mathrm{d}s$ for $t \in [0,b]$.

Remark 2.2. From the closedness property of A, one can see that if u is an integral solution of equation (1.1) on $[0,b]$, then for all $t \in [0,b]$, $u(t) \in \overline{D(A)}$. In particular, $u(0) \in \overline{D(A)}$.

Let A_0 be the part of A in $\overline{D(A)}$ define by

$D(A_0) = \{x \in D(A) : \; Ax \in \overline{D(A)}\}$,

$A_0 x = Ax$.

Then A_0 generates a C_0-semigroup $\{T(t)\}_{t\geq 0}$ on $\overline{D(A)}$ (see [27]) and the integral solution in Definitio 2.1 is given by

$$u(t) = T(t)g(u) + \lim_{\lambda\to+\infty}\int_0^t T(t-s)B(\lambda)f(s,u(s))\,\mathrm{d}s,\; t \geq 0,$$

where $B(\lambda) = \lambda R(\lambda, A)$. For more details about nondensely define operators and integrated semigroups we refer to [1, 11, 12, 18].

3. Main Results

In this section, we give the existence of integral solutions for nonlocal problem (1.1) when the nonlocal item g has no compactness assumption.

Let r be finit positive constant. We consider the sets $B_r = \{x \in X : \|x\| \leq r\}$, $W_r = \{u \in C(0,b;X) : u(t) \in B_r, \forall\, t \in [0,b]\}$.

Defin the operator $Q : C(0,b;X) \to C(0,b;X)$ by

$$(Qu)(t) = T(t)g(u) + \lim_{\lambda\to+\infty}\int_0^t T(t-s)B(\lambda)f(s,u(s))\,\mathrm{d}s$$

for all $t \in [0,b]$.

It is easy to see that the fi ed point of Q is the integral solution of nonlocal Cauchy problem (1.1).

We give the following assumptions:

(HA_0): The operator $T(t)$ generated by A_0 is compact in $\overline{D(A)}$ when $t > 0$ and $M = \sup_{t\in[0,b]}\|T(t)\|$;

(Hf): $f : [0,b] \times X \to X$, for a.e. $t \in [0,b]$, the function $f(t,\cdot) : X \to X$ is continuous and for all $x \in X$, the function $f(\cdot,x) : [0,b] \to X$ is measurable. Moreover, for any $l > 0$, there exists a function $\rho_l \in L^1(0,b;R)$ such that

$$\|f(t,x)\| \leq \rho_l(t)$$

for a.e. $t \in [0,b]$ and all $x \in B_l$;

(Hg): $g : C(0,b;X) \to \overline{D(A)}$ is continuous and

$$\sup_{u\in W_r} M\|g(u)\| + M\overline{M}\|\rho_r\|_{L^1} \leq r.$$

(H): The set $g(\overline{conv}QW_r)$ is pre-compact, where $\overline{conv}B$ denotes the convex closed hull of set $B \subseteq C(0,b;X)$.

It is easy to see that condition (H) is weaker than the compactness and convexity of g. The same hypothesis can be seen from [29], where the author considered the existence of mild solutions for semilinear nonlocal problems when A is a linear, densely define operator on X which generates a C_0-semigroup.

Lemma 3.1. *Assume that the conditions (HA), (HA_0), (Hf) and (Hg) are satisfied. Then the operator $Q : W_r \to W_r$ is a continuous operator. Moreover, for each $t \in (0, b]$, $QW_r(t)$ is relatively compact and QW_r is equicontinuous on $(0, b]$.*

Proof. Firstly, we claim that $QW_r \subseteq W_r$. In fact, for any $u \in W_r$, by (Hf), (Hg) and

$$\|B(\lambda)\| \leq \frac{\lambda \overline{M}}{\lambda - \omega} \to \overline{M}, \quad (\lambda \to +\infty),$$

we have

$$\begin{aligned}\|(Qu)(t)\| &\leq \|T(t)g(u)\| + \|\lim_{\lambda\to+\infty} \int_0^t T(t-s)B(\lambda)f(s,u(s))\,\mathrm{d}s\| \\ &\leq \sup_{u\in W_r} M\|g(u)\| + M\overline{M}\|\rho_r\|_{L^1} \\ &\leq r,\end{aligned}$$

i.e., $QW_r \subseteq W_r$.

Secondly, we prove that the mapping Q is continuous on $C(0, b; X)$. For this purpose, let $\{u_n\}_{n=1}^{+\infty}$ be a sequence in $C(0, b; X)$ with $\lim_{n\to\infty} u_n = u$ in $C(0, b; X)$. By the continuity of f with respect to the second argument, we deduce that for each $s \in [0, b]$, $f(s, u_n(s))$ converges to $f(s, u(s))$ in X; and we have

$$\|Qu_n - Qu\| \leq M\|g(u_n) - g(u)\| + M\overline{M} \int_0^b \|f(s, u_n(s)) - f(s, u(s))\|\,\mathrm{d}s.$$

Then by the continuity of g, and using the dominated convergence theorem, we get $\lim_{n\to\infty} Qu_n = Qu$ in $C(0, b; X)$, which implies that the mapping Q is continuous on $C(0, b; X)$.

Let $t \in (0, b], \varepsilon > 0$, $u \in QW_r$. By the definitio of integral solution for equation (1.1) and hypothesis (Hf), there exists $h \in (0, t)$ such that

$$\begin{aligned}\|u(t) - T(h)u(t-h)\| &= \|\lim_{\lambda\to+\infty} \int_{t-h}^t T(t-s)B(\lambda)f(s,u(s))\,\mathrm{d}s\| \\ &\leq M\overline{M} \int_{t-h}^t \rho_r(s)\,\mathrm{d}s \\ &\leq \varepsilon\end{aligned}$$

for all $u \in QW_r$.

Combining the above inequality with the compactness of the operator $T(h)$, one has $QW_r(t)$ is relatively compact.

Now, for $t \in (0, b]$, since $QW_r(t)$ is relatively compact, it is pre-compact. Thus, there exists a finit family $\{v_1, v_2, \ldots, v_j\} \subset QW_r(t)$ such that for any $u \in QW_r$, there exists $i \in J = \{1, 2, \ldots, j\}$ with

$$(M+1)\|u(t) - v_i\| \leq \varepsilon/3.$$

On the other hand, we can choose $h \in (0, t)$ such that

$$M\overline{M} \int_t^{t+h} \|f(s, u(s))\| \, \mathrm{d}s \leq \varepsilon/3, \quad \|T(h)v_i - v_i\| \leq \varepsilon/3,$$

for any $u \in QW_r$, $i \in J$. Therefore,

$$\begin{aligned}
&\|u(t+h) - u(t)\| \\
\leq &\|u(t+h) - T(h)u(t)\| + \|T(h)u(t) - u(t)\| \\
\leq &M\overline{M} \int_t^{t+h} \|f(s, u(s))\| \, \mathrm{d}s + \|T(h)v_i - v_i\| + (M+1)\|u(t) - v_i\| \\
\leq &\varepsilon/3 + \varepsilon/3 + \varepsilon/3 \\
\leq &\varepsilon
\end{aligned}$$

for any $u \in QW_r$, i.e., $QW_r(t)$ is equicontinuous for $t \in (0, b]$. □

Theorem 3.2. *Assume that the conditions (HA), (HA_0), (Hf), (Hg) and (H) are satisfied. Then the nonlocal Cauchy problem (1.1) has at least one integral solution on $[0, b]$.*

Proof. Let $W = \overline{conv}QW_r$, we get that W is a bounded closed and convex subset of $C(0, b; X)$ and $QW \subseteq W$. Moreover, we know that $g(W) = g(\overline{conv}QW_r)$ is pre-compact due to the condition (H).

Now, for any $u \in W$ and $t \in (0, b]$, we have

$$\|Qu(t) - Qu(0)\| \leq \|(T(t) - I)g(u)\| + M\overline{M} \int_0^t \rho_r(s) \, \mathrm{d}s.$$

Since $g(W)$ is pre-compact, we have $\|Qu(t) - Qu(0)\| \to 0$ uniformly on W as $t \to 0$. Thus, QW is equicontinuous at $t = 0$ and $QW(0) = g(W)$ is relatively compact which, combining with Lemma 3.1, implies that $Q : W \to W$ is a compact continuous operator. Therefore, by Schauder's fi ed point theorem, we know that the operator Q has at least one fi ed point on W, which is just the integral solution of nonlocal Cauchy problem (1.1). □

Next, we will give several special types of nonlocal item g, which satisfie the condition (H) under the the hypotheses (HA), (HA_0), (Hf) and (Hg).

Assume that

(Hg_1): $g_1 : L^1(0, b; X) \to \overline{D(A)}$ is continuous. And the initial value of equation (1.1) satisfie $g(u) = g_1(u)$.

(Hg_2): For fi ed $\tau \in (0, b)$ and $u \in C(0, b; X)$, let $D = \{u_\tau \in C(0, b; X) : u_\tau(t) = u(t + \tau)$ for $t \in [0, b - \tau]$, $u_\tau(t) = u(b)$ for $t \in (b - \tau, b]\}$ and $g_2 : D \to X$ is continuous. The initial value of equation (1.1) satisfie $g(u) = g_2(u_\tau)$.

(Hg_3): Suppose that $g_3(u) = \sum_{i=1}^{p} c_i u(t_i)$, where c_i are given constants, and $0 < t_1 < t_2 < \cdots < t_p \leq b$. The initial value of equation (1.1) satisfie $u(0) = g_3(u)$.

Corollary 3.3. *Assume that the conditions* (HA), (HA_0), (Hf) *and* (Hg) *are satisfied with* $g(u) = g_1(u)$. *Then the nonlocal Cauchy problem (1.1) has at least one integral solution on* $[0, b]$.

Proof. According to Theorem 3.2, we should only prove that the hypothesis (H) is satisfied

For arbitrary $\epsilon > 0$, there exists $\delta > 0$ such that $\int_0^\delta \|u(s)\|\, ds < \epsilon$ for all $u \in QW_r$. Set $W = \{v \in L^1(0, b; X) : v(t) = u(\delta)$ when $t \in [0, \delta)$ and $v(t) = u(t)$ when $t \in [\delta, b]$ for some $u \in QW_r\}$. From Lemma 3.1 we know that $(QW_r)_\delta = \{u \in C(\delta, b; X) : u(t) = v(t)\ for\ t \in [\delta, b],\ v \in QW_r\}$ is pre-compact in $C(\delta, b; X)$, which implies that $(QW_r)_\delta$ is pre-compact in $L^1(\delta, b; X)$. So W is pre-compact in $L^1(0, b; X)$. Then QW_r is pre-compact in $L^1(0, b; X)$ as it has an ϵ-net W. By the hypothesis $g = g_1$ is continuous from $L^1(0, b; X)$ to $\overline{D(A)}$ and $\overline{conv}QW_r \subseteq (L)\overline{conv}QW_r$, we know that the condition (H) is satisfied Thus the nonlocal Cauchy problem (1.1) has at least one integral solution on $[0, b]$. □

Remark 3.4. In [2], S. Aizicovici and M. McKibben also supposed the condition (Hg_1). By using the fi ed point theorem on $L^1(0, b; X)$, they obtained the existence of integral solutions for fully nonlinear nonlocal problems with multivalued perturbations. Here, we use Schauder's fi ed point theorem on $C(0, b; X)$ rather than on $L^1(0, b; X)$ and also get the integral solutions for nondensely define equation (1.1).

Corollary 3.5. *Assume that the conditions* (HA), (HA_0), (Hf) *and* (Hg) *are satisfied with* $g(u) = g_2(u)$. *Then the nonlocal Cauchy problem* (1.1) *has at least one integral solution on* $[0, b]$.

Proof. It is easy to see that $g((QW_r)_\delta) = g(QW_r)$ when $\delta \in (0, \tau)$ and $g(u) = g_2(u_\tau)$. Since $(QW_r)_\delta$ is pre-compact due to Lemma 3.1, we get that the condition (H) is satisfied Thus the nonlocal Cauchy problem (1.1) has at least one integral solution on $[0, b]$. □

Corollary 3.6. *Assume that the conditions* (HA), (HA_0) *and* (Hf) *are satisfied and* $g(u) = g_3(u)$. *Then the nonlocal Cauchy problem* (1.1) *has at least one integral solution on* $[0, b]$ *when the inequality* $M(\sum_{i=1}^{p} |c_i| r + \overline{M}\rho_r) \le r$ *holds.*

Proof. Clearly, g_3 satisfie the hypothesis (Hg) and $g(\overline{conv}QW_r) = g(\overline{conv}(QW_r)_\delta)$ for $\delta \in (0, t_1)$. Since $(QW_r)_\delta$ is pre-compact, we know that the condition (H) is satisfied Thus the nonlocal Cauchy problem (1.1) has at least one integral solution on $[0, b]$. □

Remark 3.7. The condition (Hg_3) has been discussed in some previous papers [3, 16, 21, 29, 30] in different frames. We remark that in [29, 30], the author obtained the existence results under the condition (Hg_3) when A generates a C_0-semigroup or nonlinear contraction semigroup, respectively. In [16], the author mainly used the approximate technique to get the similar result to the above Corollary 3.6 (see Corollary 3.1 in [16]). However, using Theorem 3.2 in this work, we can not only recapture the Corollary 3.1 in [16], but also obtain two new existence results, i.e., Corollary 3.3 and Corollary 3.5 above.

4. An Example

Example 4.1. Consider the following partial equation:

$$\begin{aligned} &\frac{\partial}{\partial t}w(t,x) = \Delta w(t,x) + F(t, w(t,x)), \quad 0 \le t \le b,\ x \in \Omega, \\ &w(t,x) = 0, \quad 0 \le t \le b,\ x \in \partial\Omega, \\ &w(0,x) + c(x)\int_0^b w(t,x)\,\mathrm{d}t = u_0(x), \quad x \in \Omega, \end{aligned} \tag{4.1}$$

where Ω is a bounded open set in R^n with regular boundary $\partial\Omega$, $c \in C(\overline{\Omega}; R^n)$, $u_0 \in C(\overline{\Omega}; R^n)$, $\Delta = \sum_{k=1}^{n} \frac{\partial^2}{\partial x_k^2}$, $F : [0,b] \times \overline{\Omega} \to R^n$ and $c(x) = 0$ for $x \in \partial\Omega$.

We choose $X = C(\overline{\Omega}; R^n)$, and we consider the operator $A : D(A) \subseteq X \to X$ define by

$D(A) = \{z \in X : \Delta z \in X \ and \ z = 0 \ on \ \partial\Omega\}$,
$Az = \Delta z$.

Now, we have

$$\overline{D(A)} = C_0(\overline{\Omega}; R^n) = \{z \in X :\ z = 0 \ on \ \partial\Omega\} \neq X,$$

$$(0, +\infty) \subseteq \rho(A),$$

$$\|R(\lambda, A)\| \le \frac{1}{\lambda}, \quad for \ \lambda > 0.$$

This implies that operator A satisfie the conditions (HA) and (HA_0) with $M = \overline{M} = 1$.

We assume that:

(1) $f : [0,b] \times X \to X$ is a continuous function define by

$$f(t,z)(x) = F(t, z(x)),\ x \in \overline{\Omega}.$$

Moreover, for each $r > 0$, there exists an integrable function $\theta_r : [0,b] \to R \ (r > 0)$ such that $\|f(t,z)\| \le \theta_r(t)$ for $t \in [0,b]$ and all $\|z\| \le r$;

(2) $g : C(0,b;X) \to X$ is a continuous function define by

$$g(u)(x) = u_0(x) - c(x)\int_0^b u(t)(x)\,\mathrm{d}t,\ x \in \overline{\Omega},$$

where $u(t)(x) = w(t,x),\ t \ge 0,\ x \in \overline{\Omega}$.

Note that the nonlocal item g define as above is continuous in $(C(0,b;X), \|\cdot\|_{L^1})$. In fact, for $u_n, u \in C(0,b;X),\ n \ge 1$ and $\|u_n - u\|_{L^1} \to 0$, we have

$$\begin{aligned} \|g(u_n) - g(u)\|_X &= \sup_{x\in\overline{\Omega}} |g(u_n)(x) - g(u)(x)| \\ &= \sup_{x\in\overline{\Omega}} |c(x)\int_0^b [u_n(t)(x) - u(t)(x)]\,\mathrm{d}t| \\ &\le \|c\| \int_0^b \sup_{x\in\overline{\Omega}} |u_n(t)(x) - u(t)(x)|\,\mathrm{d}t \\ &= \|c\|\|u_n - u\|_{L^1} \to 0, \end{aligned}$$

as $n \to \infty$. Moreover, if $u_0 \in C_0(\overline{\Omega}; R^n)$, we can verify that the range of g is exactly contained in $\overline{D(A)}$ since $c \in C_0(\overline{\Omega}; R^n)$. Thus, under these assumptions, the problem (4.1) can be reformulated as the abstract problem (1.1), and conditions (Hf), (Hg_1) are satisfie with $\rho_r = \theta_r$. If also the following inequality

$$\|u_0\| + b\|c\|r + \|\theta_r\|_{L^1} \leq r$$

holds for some $r > 0$. Then according to Corollary 3.3, the problem (4.1) has at least one integral solution in $C(0, b; X)$ for every $u_0 \in C_0(\overline{\Omega}; R^n)$.

Note that we can not expect to get the mild solution for problem (4.1) by the Theorem 3.2 in [16], for the nonlocal item g define as above satisfie the condition (Hg_1), but does not satisfy the condition $(H2')$ in [16].

References

[1] M. Adimy, H. Bouzahir, K. Ezzinbi, Local existence and stability for some partial functional differential equations with infinit delay, *Nonlinear Anal.* **48** (2002), 323-348.

[2] S. Aizicovici, M. McKibben, Existence results for a class of abstract nonlocal Cauchy problems, *Nonlinear Anal.* **39** (2000) 649–668.

[3] S. Aizicovici, H. Lee, Nonlinear nonlocal Cauchy problems in Banach spaces, *Appl. Math. Lett.* **18** (2005), 401–407.

[4] S. Aizicovici, V. Staicu, Multivalued evolution equations with nonlocal initial conditions in Banach spaces, *NoDEA Nonlinear Differential Equations Appl.* **14** (2007), 361–376.

[5] M. Benchohra, S. Ntouyas, Nonlocal Cauchy problems for neutral functional differential and integrodifferential inclusions in Banach spaces, *J. Math. Anal. Appl.* **258** (2001), 573–590.

[6] M. Benchohra, E. P. Gatsori, J. Henderson, S.K. Ntouyas, Nondensely define evolution impulsive differential inclusions with nonlocal conditions, *J. Math. Anal. Appl.* **286** (2003), 307–325.

[7] L. Byszewski, Theorems about the existence and uniqueness of solutions of a semilinear evolution nonlocal Cauchy problem, *J. Math. Anal. Appl.* **162** (1991), 494–505.

[8] L. Byszewski, Existence and uniqueness of solutions of semilinear evolution nonlocal Cauchy problem, *Zesz. Nauk. Pol. Rzes. Mat. Fiz.* **18** (1993), 109–112.

[9] L. Byszewski, V. Lakshmikantham, Theorem about the existence and uniqueness of solutions of a nonlocal Cauchy problem in a Banach space, *Appl. Anal.* **40** (1990), 11–19.

[10] L. Byszewski, H. Akca, Existence of solutions of a semilinear functional-differential evolution nonlocal problem, *Nonlinear Anal.* **34** (1998), 65–72.

[11] G. Da Prato, E. Sinestrari, Differential operators with non-dense domain, *Ann. Scuola Norm. Sup. Pisa Sci.* **14** (1987) 285–344.

[12] K. Ezzinbi, J. H. Liu, Nondensely define evolution equations with nonlocal conditions, *Math. Comp. Model.* **36** (2002), 1027–1038.

[13] K. Ezzinbi, X. Fu, Existence and regularity of solutions for some neutral partial differential equations with nonlocal conditions, *Nonlinear Anal.* **57** (2004), 1029–1041.

[14] K. Ezzinbi, X. Fu, K. Hilal, Existence and regularity in the α-norm for some neutral partial differential equations with nonlocal conditions, *Nonlinear Anal.* **67** (2007), 1613–1622.

[15] Z. Fan, Q. Dong, G. Li, Semilinear differential equations with nonlocal conditions in Banach spaces, Inter. J. Nonlinear Sci. **2(3)** (2006), 131–139.

[16] Z. Fan, Existence of nondensely define evolution equations with nonlocal conditions, *Nonlinear Anal.* **70** (2009), 3829–3836.

[17] X. Fu, K. Ezzinbi, Existence of solutions for neutral functional differential evolution equations with nonlocal conditions, *Nonlinear Anal.* **54** (2003), 215–227.

[18] X. Fu, On solutions of neutral nonlocal evolution equations with nondense domain, *J. Math. Anal. Appl.* **299** (2004), 392–410.

[19] J. Garcia–Falset, Existence results and asymptotic behavior for nonlocal abstract Cauchy problems, *J. Math. Anal. Appl.* **338** (2008) 639–652.

[20] J. Liang, J. Van Casteren, T. J. Xiao, Nonlocal Cauchy problems for semilinear evolution equations, *Nonlinear Anal.* **50A** (2002) 173–189.

[21] J. Liang, J. Liu, T. J. Xiao, Nonlocal Cauchy problems governed by compact operator families, *Nonlinear Anal.* **57** (2004) 183–189.

[22] Y. Lin, J. Liu, Semilinear integrodifferential equations with nonlocal Cauchy problems, *Nonlinear Anal.* **26** (1996), 1023–1033.

[23] G. M. Mophou, G. M. N'Guerekata, Mild solutions for semilinear fractional differential equations, *Elec. J. Diff. Equ.* **21** (2009) 1–9.

[24] G. M. Mophou, G. M. N'Guerekata, Existence of mild solution for some fractional differential equations with nonlocal conditions, *Semigroup Forum,* **79** (2009), 315–322.

[25] S. Ntouyas, P. Tsamatos, Global existence for semilinear evolution equations with nonlocal conditions, *J. Math. Anal. Appl.* **210** (1997), 679–687.

[26] S. Ntouyas, P. Tsamatos, Global existence for semilinear integrodifferential equations with delay and nonlocal conditions, *Anal. Appl.* **64** (1997), 99–105.

[27] A. Pazy, *Semigroups of Linear Operators and Applications to Partial Differential Equations*, Springer–Verlag, New York, 1983.

[28] T. J. Xiao, J. Liang, Existence of classical solutions to nonautonomous nonlocal parabolic problems, *Nonlinear Anal.* **63** (2005), 225–232.

[29] X. Xue, Existence of semilinear differential equations with nonlocal initial conditions, *Acta Math. Sinica.* **23(6)** (2007) 983–988.

[30] X. Xue, Nonlinear differential equations with nonlocal conditions in Banach spaces, *Nonlinear Anal.* **63** (2005), 575–586.

[31] X. Xue, Nonlocal nonlinear differential equations with a measure of noncompactness in Banach spaces, *Nonlinear Anal.* **70(7)** (2009), 2593–2601.

In: Evolution Equations: Theories, Solutions and Functions ISBN 978-1-61324-090-8
Editor: Gaston M. N'Guerekata

Chapter 13

POSITIVE SOLUTIONS OF A NONLINEAR ELLIPTIC EQUATION ON UNBOUNDED NON-SMOOTH DOMAINS

***Badreddine Gouaid*[1,*] *and Lotfi Riahi*[2,†]**
[1]Department of Mathematics, Preparatory Institute for Engineering Studies, La Marsa, Tunis, Tunisia
[2]Department of Mathematics, National Institute of Applied Sciences and Technology, Charguia 1, 1080, Tunis, Tunisia

Abstract

Under general conditions on the function f, we prove the existence of positive continuous solutions for the nonlinear elliptic equation $\Delta u(x) + f(x, u(x)) = 0$ with zero Dirichlet boundary condition on unbounded non-smooth domain D in $\mathbb{R}^n$, $n \geq 2$. In particular, we extend known results to more general classes of functions and domains and we provide the optimal class of nonnegative potentials V, which guarantees the existence of such solutions for the semi-linear equation $\Delta u(x) + V(x)u^p(x) = 0$, $p > 1$ studied by several authors.

2000 Mathematics Subject Classification: 35J60, 35K55.

Keywords and phrases: Nonlinear elliptic equation, Dirichlet boundary condition, positive solution, Geen function, Kato class, non-smooth domain, fi ed point theorem.

1. Introduction

Over the last three decades, the study of the nonlinear elliptic equation $\Delta u + f(., u) = 0$ with Dirichlet boundary condition, where Δ is the Laplacian operator, has attracted a

*E-mail address: gouaidbadreddine@yahoo.fr
†E-mail address: Lotfi.Riahi@fst.rnu.tn ltfrh_2006@yahoo.com

number of mathematicians (see [1, 2, 7-11, 15, 17-19] and the references therein). In this paper we will especially interest in the following nonlinear elliptic problem

$$(\mathcal{E})\begin{cases}\Delta u(x)+f(x,u(x))=0, \text{ on D},\\ u(x)>0, \quad x\in D,\\ u(x)=0, \quad x\in\partial D,\end{cases}$$

where D is either an unbounded nontangentially accessible (NTA for short) domain (see definitio in [4]) with a compact boundary ∂D in $\mathbb{R}^n$, $n\geq 3$, or an unbounded Jordan domain in $\mathbb{R}^2$, i.e. an unbounded domain with a compact nonempty boundary ∂D consisting of finitel many disjoint closed Jordan curves in $\mathbb{R}^2$, and f satisfie some general conditions that will be detailed later. We recall that this problem is studied by Zhao [18] in the special case of unbounded Lipschitz domain with a compact boundary in $\mathbb{R}^n$, $n\geq 3$, and by Ufektepe and Zhao [17] on unbounded Jordan domain in $\mathbb{R}^2$. By Probabilistic method using Brownian path integration and potential theory they proved existence theorems for the problem $(\mathcal{E})$ when f satisfie some general conditions related to the well known Kato class at infinit K_n^∞. In the special case, when D is a Dini-smooth Jordan domain in the euclidian plane, the same result is obtained, basing on the $3G$-Theorem proved in [16], for a more general class denoted by K in [10].

Our aim in this paper is to prove an existence result for the problem $(\mathcal{E})$ when the function f satisfie general conditions related to a new functional class of potentials denoted by $\mathcal{K}_n^\infty(D)$ more general than the Kato class K_n^∞. In the special case of Dini-smooth Jordan domains in the euclidien plane, our class $\mathcal{K}_n^\infty(D)$ coincides with the class K used in [10]. We also prove that $\mathcal{K}_n^\infty(D)$ is the optimal class insuring the existence result for the semilinear problem $\Delta u(x)+V(x)u^p(x)=0$, $p>1$ with nonnegative potential V. Interesting examples of functions in $\mathcal{K}_n^\infty(D)\setminus K_n^\infty$ are also given.

Throughout the paper the letter C means a positive constant which may vary in value from line to line but not in dependence. For simplicity and without loss of generality we will assume that $D\subset B^c(0,1)$, the complement of the unit ball $B(0,1)$ in $\mathbb{R}^n$, $n\geq 2$. Let G be the Laplacian Green function with Dirichlet boundary condition on the domain D. Since the Kelvin transformation: $x\to x^*=x/|x|^2$ from $D\cup\{\infty\}$ onto $D^*:=\{x^*\in B(0,1)\;:\;x\in D\cup\{\infty\}\}$ preserves harmonicity (see [5] p. 36) and D^* is a bounded regular domain, we clearly have

$$G(x,y)=\begin{cases}G^*(x^*,y^*) \quad \text{if} \quad n=2,\\ \dfrac{G^*(x^*,y^*)}{|x|^{n-2}|y|^{n-2}} \quad \text{if} \quad n\geq 3,\end{cases} \tag{1.1}$$

where G^* is the Laplacian Green function with Dirichlet boundary condition on the bounded domain D^*. For $x\in D$, let

$$h(x)=\begin{cases}2\pi G^*(x^*,0)=2\pi\lim\limits_{|z|\to\infty}G(x,z) \quad \text{if} \quad n=2,\\ \dfrac{4\pi^{n/2}}{\Gamma(n/2-1)}\dfrac{G^*(x^*,0)}{|x|^{n-2}}=\dfrac{4\pi^{n/2}}{\Gamma(n/2-1)}\lim\limits_{|z|\to\infty}|z|^{n-2}G(x,z) \quad \text{if} \quad n\geq 3.\end{cases} \tag{1.2}$$

Clearly, h is a positive harmonic function on D satisfying:

$$\begin{cases} \lim_{x \to z} h(x) = 0, \text{ for all } z \in \partial D, \\ \lim_{|x| \to \infty} \frac{h(x)}{\ln |x|} = 1 \quad \text{if} \quad n = 2, \\ \lim_{|x| \to \infty} h(x) = 1 \quad \text{if} \quad n \geq 3. \end{cases} \tag{1.3}$$

To present our results we firs introduce the new functional class $\mathcal{K}_n^\infty(D)$.

Definition 1.1. A Borel measurable function V on D is said to be in the class $\mathcal{K}_n^\infty(D)$ if it satisfies

$$\lim_{M \to \infty} \sup_{x \in D} \int_{D \cap (|y| \geq M)} \frac{h(y)}{h(x)} G(x, y) |V(y)| dy = 0, \tag{1.4}$$

and for all $x_0 \in \overline{D}$

$$\lim_{r \to 0} \sup_{x \in D} \int_{D \cap (|y - x_0| < r)} \frac{h(y)}{h(x)} G(x, y) |V(y)| dy = 0. \tag{1.5}$$

REMARKS.

1. Note that if $V \in \mathcal{K}_n^\infty(D)$, then

$$\|V\| := \sup_{x \in \overline{D}} \int_D \frac{h(y)}{h(x)} G(x, y) |V(y)| dy < \infty.$$

In particular $V \in L^1_{\text{loc}}(D)$.

2. If $V \in \mathcal{K}_n^\infty(D)$, then, by (1.2) and (1.3), there is $M > 0$ such that

$$\int_{|y| \geq M} \frac{|V(y)|}{|y|^{n-2}} dy < \infty \quad \text{if } n \geq 3, \tag{1.6}$$

and

$$\int_{|y| \geq M} |V(y)| \ln |y| dy < \infty \quad \text{if } n = 2. \tag{1.7}$$

The function f is assumed to satisfy the following conditions:

(H_1) $f : D \times]0, \infty[\to \mathbb{R}$ is Borel measurable, continuous with respect to the second variable, and there is a positive nondecreasing function $F : D \times]0, \infty[\to \mathbb{R}$ with $\lim_{t \to 0} F(x, t) = 0$ such that for all $(x, t) \in D \times]0, \infty[$,

$$|f(x, t)| \leq t F(x, t),$$

(H_2) $F(., h(.)) \in \mathcal{K}_n^\infty(D)$.

As important examples of functions f satisfying (H_1) are $f(x,t) = V(x)g(t)$, where g is a positive nondecreasing function on $(0,\infty)$ such that $\lim_{t\to 0} \frac{g(t)}{t} = 0$.
If we take $g(t) = t^p$, $p > 1$, we get the semi-linear elliptic equation $\Delta u(x) + V(x)u^p(x) = 0$ studied by several authors (see [7-9, 11, 15] and the references therein).

The results of this paper are the followings.

Theorem 1.2. *Let f be a function satisfying the conditions (H_1) and (H_2). Then, there exists a $\lambda_0 > 0$ such that for all $\lambda \in]0, \lambda_0]$, there is a solution u of the problem $(\mathcal{E})$ satisfying*

$$\frac{u}{h} \in C(\overline{D}), \quad \frac{\lambda}{2} \leq \frac{u}{h} \leq \frac{3\lambda}{2} \quad \text{and} \quad \lim_{|x|\to\infty} \frac{u(x)}{h(x)} = \lambda.$$

Corollary 1.3. *Let V be a Borel measurable function on D and $p > 1$ such that $Vh^{p-1} \in \mathcal{K}_n^\infty(D)$. Then, there exists a $\lambda_0 > 0$ such that for all $\lambda \in]0, \lambda_0]$, there is a solution u of the semi-linear problem*

$$(\mathcal{E}') \begin{cases} \Delta u(x) + V(x)u^p(x) = 0, & \text{on } D, \\ u(x) > 0, & x \in D, \\ u(x) = 0, & x \in \partial D, \end{cases}$$

satisfying

$$\frac{u}{h} \in C(\overline{D}), \quad \frac{\lambda}{2} \leq \frac{u}{h} \leq \frac{3\lambda}{2} \quad \text{and} \quad \lim_{|x|\to\infty} \frac{u(x)}{h(x)} = \lambda.$$

Theorem 1.4. *Assume that V is a nonnegative Borel measurable function and the results of Corollary 1.3 hold, then $Vh^{p-1} \in \mathcal{K}_n^\infty(D)$.*

2. Preliminaries and Known Results

In this section we need to recall and prove some preliminary results which are the basic tools in this work. For $x \in D$, let $\varphi(x) = 1 \wedge h(x)$ if $n = 2$ and $\varphi(x) = |x|^{2-n} \wedge h(x)$ if $n \geq 3$. By (1.1) and the $3G$-Theorems (Theorem 3.2 in [4] proved for bounded NTA-domains in $\mathbb{R}^n$, $n \geq 3$ and Theorem 3.1 in [13] proved for bounded Jordan domains in the plane), we have the following.

Theorem 2.1 (3G-Theorem). *There exists a constant $C_0 = C_0(D) > 0$ such that for all x, y and $z \in D$,*

$$\frac{G(x,y)G(y,z)}{G(x,z)} \leq C_0\Big[\frac{\varphi(y)}{\varphi(x)}G(x,y) + \frac{\varphi(y)}{\varphi(z)}G(y,z)\Big].$$

We also have the following inequalities.

Lemma 2.2. *There exists a constant $C = C(D) > 0$ such that for all x and $y \in D$,*

$$\frac{\varphi(y)}{\varphi(x)}G(x,y) \leq \begin{cases} C\dfrac{|x|^{n-2}}{|y|^{n-2}|x-y|^{n-2}} & \text{if } n \geq 3, \\ C\ln(1 + \dfrac{|x||y|}{|x-y|}) & \text{if } n = 2. \end{cases}$$

Proof. Let $\varphi^* = 1 \wedge G^*(.,0)$. In the case $n \geq 3$, D^* is a bounded NTA-domain, and so its Green function G^* satisfie the $\mathcal{B}$-approximation (Theorem 3.2 in [4]). Thus, following the proof of Lemma 3.2 in [14] for bounded Lipschitz domains, we get

$$\frac{\varphi^*(y^*)}{\varphi^*(x^*)} G^*(x^*, y^*) \leq \frac{C}{|x^* - y^*|^{n-2}}. \tag{2.8}$$

Combining (2.1), (1.1) and (1.2) and recalling that $\varphi(x) = |x|^{2-n} \wedge h(x)$, we obtain the inequality stated in the Lemma. For $n = 2$, the inequality holds from Lemmas 3.5 and 3.6 in [13], (1.1), (1.2) and the fact that $\varphi(x) = 1 \wedge h(x)$. □

Theorem 2.3 (Harnack inequality). *For any compact subset $E \subset D$, there exists a constant $C = C(D, E) > 0$ such that for all nonnegative harmonic functions u on D, we have*

$$\sup_{x \in E} u(x) \leq C \inf_{x \in E} u(x).$$

The Boundary Harnack principle (Theorem 5.1 in [6]) is valid for NTA-domains in $\mathbb{R}^n$, $n \geq 2$. In the euclidian plane, by the extended Riemann mapping theorem (Proposition 6.16 in [3]), we know the existence of an extended conformal mapping from the Jordan domain D onto a C^∞-domain. Since conformal mapping preserves harmonicity, by Theorem 5.1 in [6] the boundary Harnack principle is also valid for Jordan domains in $\mathbb{R}^2$. In fact, we have the following.

Theorem 2.4 (Boundary Harnack principle). *There exist constants $r_0 > 0$ and $C > 0$ depending only on D such that for any $x_0 \in \partial D$, $0 < r < r_0$ and any two positive harmonic functions u and v on $D \cap B(x_0, 2r)$ continuously vanishing on $\partial D \cap B(x_0, 2r)$, we have*

$$\frac{u(x)}{v(x)} \leq C \frac{u(y)}{v(y)},$$

for all x and $y \in D \cap B(x_0, r)$.

Corollary 2.5. *Let V be in the class $\mathcal{K}_n^\infty(D)$. Then for all $M > 0$,*

$$\int_{\overline{D} \cap (|y| \leq M)} h^2(y)|V(y)|dy < \infty.$$

Proof. Let $x \in D$ be fi ed with $|x| \geq 2M$. We have,

$$\int_D h(y)G(x, y)|V(y)|dy \leq \|V\| h(x) < \infty.$$

Moreover, by the boundary Harnack principle (Theorem 2.4), there is $r > 0$ and $C > 0$ such that

$$C^{-1} h(y) \leq G(x, y) \leq Ch(y),$$

for all $y \in E$, where $E = \{y \in \overline{D} : d(y, \partial D) \leq r\}$. Hence

$$\int_E h^2(y)|V(y)|dy < \infty. \tag{2.9}$$

Since $V \in L^1_{\mathrm{loc}}(D)$, h is continuous and $E' := \{y \in D : d(y, \partial D) \geq r\} \cap \{y \in D : |y| \leq M\}$ is a compact subset in D, we have

$$\int_{E'} h^2(y)|V(y)|dy < \infty. \tag{2.10}$$

Combining (2.2) and (2.3), we obtain

$$\int_{\overline{D} \cap (|y| \leq M)} h^2(y)|V(y)|dy < \infty.$$

□

3. The Class $\mathcal{K}_n^\infty(D)$

In this section, we study some properties of the class $\mathcal{K}_n^\infty(D)$. We prove that $\mathcal{K}_n^\infty(D)$ is more general than the Kato class K_n^∞ and we provide interesting examples of potentials V in the class $\mathcal{K}_n^\infty(D)$ which do not belong to K_n^∞ for domains with smooth boundaries. In the special case of Dini-smooth Jordan domains in the plane $\mathcal{K}_2^\infty(D)$ coincides with the class K considered in [10]. For simplicity, we will use the notation $a \sim b$, for two positive quantities a and b, to mean that $C^{-1}a \leq b \leq Ca$ for some constant $C = C(D, n) > 0$, $a \wedge b$ to mean $\min(a, b)$ and $a \vee b$ to mean $\max(a, b)$. It is firs interesting to know that in the definitio of the class $\mathcal{K}_n^\infty(D)$ we may replace h by φ. This is the object of the following proposition.

Proposition 3.1. *Let V be a Borel measurable function on D. Then V is in the class $\mathcal{K}_n^\infty(D)$ if and only if it satisfies*

$$\lim_{M \to \infty} \sup_{x \in D} \int_{D \cap (|y| \geq M)} \frac{\varphi(y)}{\varphi(x)} G(x, y)|V(y)|dy = 0, \tag{3.11}$$

and for all $x_0 \in \overline{D}$,

$$\lim_{r \to 0} \sup_{x \in D} \int_{D \cap (|y - x_0| < r)} \frac{\varphi(y)}{\varphi(x)} G(x, y)|V(y)|dy = 0. \tag{3.12}$$

Proof. Assume that V satisfie (3.1) and (3.2). The function h is harmonic and so by Theorem 2.1 in [12], there is an increasing sequence $(f_n)_n$ of nonnegative Borel measurable functions on D such that $h(x) = \sup_n \int_D G(x, y) f_n(y)dy$. By using the 3$G$-Theorem and the Fubini's theorem, we obtain

$$\sup_{x \in D} \int_{D \cap (|y| \geq M)} \frac{h(y)}{h(x)} G(x, y)|V(y)|dy \leq 2C_0 \sup_{x \in D} \int_{D \cap (|y| \geq M)} \frac{\varphi(y)}{\varphi(x)} G(x, y)|V(y)|dy,$$

and

$$\sup_{x \in D} \int_{D \cap (|y - x_0| < r)} \frac{h(y)}{h(x)} G(x, y)|V(y)|dy \leq 2C_0 \sup_{x \in D} \int_{D \cap (|y - x_0| < r)} \frac{\varphi(y)}{\varphi(x)} G(x, y)|V(y)|dy.$$

It follows that V satisfie (1.4) and (1.5) and so $V \in \mathcal{K}_n^\infty(D)$.
Conversely, assume that V in the class $\mathcal{K}_n^\infty(D)$. We will prove that V satisfie (3.1) and (3.2). We firs treat the case $n \geq 3$. Let $R > 0$ sufficentl great so that $B^c(0,R) \subset D$. The function V satisfie (1.4) and (1.5) and so, by recalling that $\varphi(x) = |x|^{2-n} \wedge h(x)$, we have,

$$\lim_{M\to\infty} \sup_{x\in D, |x|\leq R} \int_{D\cap(|y|\geq M)} \frac{\varphi(y)}{\varphi(x)} G(x,y)|V(y)|dy = 0, \tag{3.13}$$

and for all $x_0 \in \overline{D}$,

$$\lim_{r\to 0} \sup_{x\in D, |x|\leq R} \int_{D\cap(|y-x_0|<r)} \frac{\varphi(y)}{\varphi(x)} G(x,y)|V(y)|dy = 0. \tag{3.14}$$

Moreover, for $|x| \geq R$ and $|y| \geq M$, with $M > 0$ sufficentl great, we have $h(x) \sim 1$, $\varphi(x) \sim |x|^{2-n}$ and $G(x,y) \sim |x-y|^{2-n}$. Thus

$$\begin{aligned}
&\sup_{|x|\geq R} \int_{D\cap(|y|\geq M)} \frac{\varphi(y)}{\varphi(x)} G(x,y)|V(y)|dy \\
\leq\ & C \sup_{|x|\geq R} \int_{|y|\geq M} \frac{|x|^{n-2}}{|y|^{n-2}|x-y|^{n-2}} |V(y)|dy \\
\leq\ & 2^{n-2} C\Big(\sup_{|x|\geq R} \int_{|y|\geq M \vee |x-y|} \frac{|V(y)|}{|x-y|^{n-2}} dy + \int_{|x-y|\geq|y|\geq M} \frac{|V(y)|}{|y|^{n-2}} dy \Big) \\
\leq\ & 2^{n-2} C\Big(\sup_{|x|\geq R} \int_{|y|\geq M} \frac{h(y)}{h(x)} G(x,y)|V(y)|dy + \int_{|y|\geq M} \frac{|V(y)|}{|y|^{n-2}} dy \Big).
\end{aligned} \tag{3.15}$$

The last member in (3.5) tends to 0 as $M \to \infty$ by (1.6) and so (3.3) and (3.5) imply that V satisfie (3.1). Moreover for $x_0 \in \overline{D}$, by the Harnack inequality (if $x_0 \in D$) and the boundary Harnack principle (if $x_0 \in \partial D$), there is $C > 0$ such that for all $|x| \geq R$ and all $y \in D$ with $|y - x_0| \leq r$,

$$\frac{G(x,y)}{\varphi(x)} \leq Ch(y).$$

Thus

$$\sup_{|x|\geq R} \int_{D\cap(|y-x_0|<r)} \frac{\varphi(y)}{\varphi(x)} G(x,y)|V(y)|dy \leq C \int_{D\cap(|y-x_0|<r)} h^2(y)|V(y)|dy. \tag{3.16}$$

The last member in (3.6) tends to 0 as $r \to 0$ by Corollary 2.5, and so (3.4) and (3.6) imply that V satisfie (3.2).
Now assume that V in $\mathcal{K}_2^\infty(D)$. As in the case $n \geq 3$, by recalling that $\varphi(x) = 1 \wedge h(x)$, it follows from (1.4) and (1.5) that V satisfie (3.3) and (3.4). Moreover, for $|x| \geq R$ and $|y| \geq M$, with $M > 0$ sufficentl great, we have $h(x) \sim \ln|x|$, $\varphi(x) \sim 1$ and

$G(x,y) \sim \ln(\frac{|x||y|}{|x-y|})$. Thus

$$\begin{aligned}
&\sup_{|x|\geq R} \int_{D\cap(|y|\geq M)} \frac{\varphi(y)}{\varphi(x)} G(x,y)|V(y)|dy \\
\leq\ & C \sup_{|x|\geq R} \int_{|y|\geq M} \ln(\frac{|x||y|}{|x-y|})|V(y)|dy \\
\leq\ & C\Big(\sup_{|x|\geq R} \int_{(|y|\geq M)\cap(|x-y|\leq 1)} \ldots dy + \sup_{|x|\geq R} \int_{(|y|\geq M)\cap(|x-y|\geq 1)} \ldots dy\Big) \\
\leq\ & C\Big(\sup_{|x|\geq R} \int_{|y|\geq M} \frac{\ln|y|}{\ln|x|} \ln(\frac{|x||y|}{|x-y|})|V(y)|dy + \int_{|y|\geq M} |V(y)| \ln|y| dy\Big) \\
\leq\ & C\Big(\sup_{|x|\geq R} \int_{|y|\geq M} \frac{h(y)}{h(x)} G(x,y)|V(y)|dy + \int_{|y|\geq M} |V(y)| \ln|y| dy\Big). \qquad (3.17)
\end{aligned}$$

The last member in (3.7) tends to 0 as $M \to \infty$ by (1.7) and so (3.3) and (3.7) imply that V satisfie (3.1). As in the case $n \geq 3$, V satisfie (3.2). □

REMARK. From Proposition 3.1, we deduce that, in the special case of Dini-smooth Jordan doamins in $\mathbb{R}^2$, the class $\mathcal{K}_2^\infty(D)$ coincides with the class K used in [10].

In what follows we will prove that $\mathcal{K}_n^\infty(D)$ is more general than the Kato class K_n^∞. For reader's convenience we recall the definitio of the Kato class K_n^∞ (see [17] and [18]).

Definition 3.2. Let V be a Borel measurable function on D.
We say that V is in the Kato class K_n^∞, $n \geq 3$, if it satisfies

$$\lim_{r\to 0} \sup_{x\in\mathbb{R}^n} \int_{D\cap(|x-y|<r)} \frac{|V(y)|}{|x-y|^{n-2}} dy = 0 \text{ and } \lim_{M\to\infty} \sup_{x\in\mathbb{R}^n} \int_{D\cap(|y|\geq M)} \frac{|V(y)|}{|x-y|^{n-2}} dy = 0.$$

We say that V is in the Kato class K_2^∞ if it satisfies

$$\lim_{r\to 0} \sup_{x\in\mathbb{R}^2} \int_{D\cap(|x-y|<r)} \ln(\frac{1}{|x-y|})|V(y)|dy = 0 \text{ and } \int_D |V(y)| \ln|y| dy < \infty.$$

REMARKS. The following properties of the Kato class hold by simple computations.
1. If $V \in K_n^\infty$, $n \geq 3$, then for all $x_0 \in \mathbb{R}^n$,

$$\begin{aligned}
&\lim_{r\to 0} \sup_{x\in\mathbb{R}^n} \int_{D\cap(|y-x_0|<r)} \frac{|V(y)|}{|x-y|^{n-2}} dy = 0, \text{ and} \\
&\sup_{x\in\mathbb{R}^n} \int_D \frac{|V(y)|}{|x-y|^{n-2}} dy < \infty. \qquad (3.18)
\end{aligned}$$

2. If $V \in K_2^\infty$, then for all $x_0 \in \mathbb{R}^2$,

$$\begin{aligned}
&\lim_{r\to 0} \sup_{x\in\mathbb{R}^2} \int_{D\cap(|y-x_0|<r)} \Big[1 + \ln(1 + \frac{1}{|x-y|})\Big] |V(y)|dy = 0, \text{ and} \\
&\int_D |V(y)| \ln(1+|y|) dy < \infty. \qquad (3.19)
\end{aligned}$$

Proposition 3.3. *We have the following inclusion:* $K_n^\infty \subseteq \mathcal{K}_n^\infty(D)$.

Proof. Let V be in the Kato class K_n^∞, $n \geq 3$ and $x_0 \in \overline{D}$. By Lemma 2.2 and the inequality $|x| \leq |x-y| + |y|$, we have

$$\begin{aligned}
&\sup_{x\in D}\int_{D\cap(|y-x_0|<r)} \frac{\varphi(y)}{\varphi(x)} G(x,y)|V(y)|dy \\
\leq\ & C\sup_{x\in\mathbb{R}^n}\int_{D\cap(|y-x_0|<r)} \frac{|x|^{n-2}}{|y|^{n-2}|x-y|^{n-2}}|V(y)|dy \\
=\ & C\sup_{x\in\mathbb{R}^n}\Big(\int_{D\cap(|y-x_0|<r)\cap(|x|\leq 2|y|)} \dots dy + \int_{D\cap(|y-x_0|<r)\cap(|x|\geq 2|y|)} \dots dy\Big) \\
\leq\ & 2^{n-2}C\sup_{x\in\mathbb{R}^n}\Big(\int_{D\cap(|y-x_0|<r)} \frac{|V(y)|}{|x-y|^{n-2}}dy + \int_{D\cap(|y-x_0|<r)} \frac{|V(y)|}{|y|^{n-2}}dy\Big) \\
\leq\ & 2^{n-1}C\sup_{x\in\mathbb{R}^n}\int_{D\cap(|y-x_0|<r)} \frac{|V(y)|}{|x-y|^{n-2}}dy.
\end{aligned}$$

The last member tends to 0 as $r \to 0$ by (3.8). In the same way,

$$\sup_{x\in D}\int_{D\cap(|y|\geq M)} \frac{\varphi(y)}{\varphi(x)} G(x,y)|V(y)|dy \leq 2^{n-1}C\sup_{x\in\mathbb{R}^n}\int_{D\cap(|y|\geq M)} \frac{|V(y)|}{|x-y|^{n-2}}dy,$$

and the right member in the inequality tends to 0 as $M \to \infty$ by Definitio 3.2. Thus V is in the class $\mathcal{K}_n^\infty(D)$, $n \geq 3$.
Now let V be in the Kato class K_2^∞ and $x_0 \in \overline{D}$. By Lemma 2.2 and the inequality $|x| \leq |x-y| + |y|$, we have

$$\begin{aligned}
&\sup_{x\in D}\int_{D\cap(|y-x_0|<r)} \frac{\varphi(y)}{\varphi(x)} G(x,y)|V(y)|dy \\
\leq\ & C\sup_{x\in D}\int_{D\cap(|y-x_0|<r)} \ln(1+\frac{|x||y|}{|x-y|})|V(y)|dy \\
\leq\ & C\Big(\sup_{x\in\mathbb{R}^2}\int_{D\cap(|y-x_0|<r)} [1+\ln(1+\frac{1}{|x-y|})]|V(y)|dy \\
&+\int_{D\cap(|y-x_0|<r)} \ln(1+|y|)|V(y)|dy\Big).
\end{aligned}$$

The last member tends to 0 as $r \to 0$ by (3.9). In the same way,

$$\begin{aligned}
&\sup_{x\in D}\int_{D\cap(|y|\geq M)}\frac{\varphi(y)}{\varphi(x)}G(x,y)|V(y)|dy\\
\leq\;& C\Big(\sup_{x\in\mathbb{R}^2}\int_{D\cap(|y|\geq M)}[1+\ln(1+\frac{1}{|x-y|})]|V(y)|dy\\
&+\int_{D\cap(|y|\geq M)}\ln(1+|y|)|V(y)|dy\Big)\\
\leq\;& C\Big(\sup_{x\in\mathbb{R}^2}\int_{D\cap(|y|\geq M)\cap(|x-y|<r)}[1+\ln(1+\frac{1}{|x-y|})]|V(y)|dy\\
&+\sup_{x\in\mathbb{R}^2}\int_{D\cap(|y|\geq M)\cap(|x-y|\geq r)}[1+\ln(1+\frac{1}{|x-y|})]|V(y)|dy\\
&+\int_{D\cap(|y|\geq M)}\ln(1+|y|)|V(y)|dy\Big)\\
\leq\;& 3C\Big(\sup_{x\in\mathbb{R}^2}\int_{D\cap(|x-y|<r)}\ln(\frac{1}{|x-y|})|V(y)|dy+\ln\frac{1}{r}\int_{D\cap(|y|\geq M)}|V(y)|dy\\
&+\int_{D\cap(|y|\geq M)}\ln(1+|y|)|V(y)|dy\Big).
\end{aligned}$$

The last member tends to 0 as $M \to \infty$ and $r \to 0$ by Definitio 3.2 and (3.9). Thus $V \in \mathcal{K}_2^\infty(D)$. □

In what follows we will assume that D is of $C^{1,1}$ boundary and we will present examples of singular potentials in the class $\mathcal{K}_n^\infty(D)$ which do not belong to the Kato class K_n^∞. For $y \in D$, we denote by $\delta(y)$ the euclidien distance from y to the boundary ∂D of D. By (1.1) and the Green functions estimates for bounded $C^{1,1}$-domains given in [3, Theorem 6.23] for $n = 2$ and in [14, Corollary 4.5] for $n \geq 3$, we have the following.

Lemma 3.4. *Assume that D is of $C^{1,1}$ boundary, then*

$$G(x,y)\sim\ln\Big(1+\frac{\delta(x)\delta(y)}{|x-y|^2}|x||y|\Big),\quad \text{if}\quad n=2,$$

and

$$G(x,y)\sim\min\Big(1,\frac{\delta(x)\delta(y)}{|x-y|^2}|x||y|\Big)|x-y|^{2-n},\quad \text{if}\quad n\geq 3.$$

By (1.2) and Lemma 3.4, we have the following.

Lemma 3.5. *Assume that D is of $C^{1,1}$ boundary, then*

$$h(x)\sim\ln(1+\delta(x)|x|),\quad \text{if}\quad n=2,$$

and

$$h(x)\sim\min(1,\delta(x)|x|),\quad \text{if}\quad n\geq 3.$$

Proposition 3.6. *Assume that D is of $C^{1,1}$ boundary, then for $\alpha < 2 < \beta$ the function $V_{\alpha,\beta}(y) := (1\wedge\delta(y))^{-\alpha}|y|^{-\beta}$ is in the class $\mathcal{K}_n^\infty(D)$.*

Proof. Since for $\alpha < 0$, we have $0 \leq V_{\alpha,\beta} \leq V_{0,\beta}$, it suffice to prove the result for $\alpha \in [0, 2[$.

Case 1: $n \geq 3$.
By Lemma 2.2 and the inequality $|x| \leq |x-y|+|y|$, we have, for $M > 0$ sufficientl great,

$$\begin{aligned}
&\sup_{x\in D}\int_{D\cap(|y|\geq M)}\frac{\varphi(y)}{\varphi(x)}G(x,y)V_{\alpha,\beta}(y)dy\\
\leq\;& C\sup_{x\in D}\int_{|y|\geq M}\frac{|x|^{n-2}}{|x-y|^{n-2}|y|^{n-2+\beta}}dy\\
\leq\;& C\sup_{x\in D}\int_{|y|\geq M}\Big(\frac{1}{|x-y|}+\frac{1}{|y|}\Big)^{n-2}\frac{dy}{|y|^{\beta}}\\
\leq\;& C\sup_{x\in D}\Big(\int_{|y|\geq M\geq|x-y|}\dots dy+\int_{|x-y|\geq|y|\geq M}\dots dy+\int_{|y|\geq|x-y|\geq M}\dots dy\Big)\\
\leq\;& 2^{n-2}C\sup_{x\in D}\Big(\frac{1}{M^{\beta}}\int_{|x-y|\leq M}\frac{dy}{|x-y|^{n-2}}+\int_{|y|\geq M}\frac{dy}{|y|^{n-2+\beta}}+\int_{|x-y|\geq M}\frac{dy}{|x-y|^{n-2+\beta}}\Big)\\
\leq\;& \frac{C}{M^{\beta-2}},
\end{aligned}$$

where $C = C(D, n, \alpha) > 0$. Thus $V_{\alpha,\beta}$ satisfie (3.1). We will prove that $V_{\alpha,\beta}$ satisfie (3.2). Let firs $x_0 \in D$. By Lemma 2.2 and the inequality $|x| \leq |x-y| + |y|$, we have

$$\begin{aligned}
&\sup_{x\in D}\int_{D\cap(|y-x_0|<r)}\frac{\varphi(y)}{\varphi(x)}G(x,y)V_{\alpha,\beta}(y)dy\\
\leq\;& C\sup_{x\in D}\int_{D\cap(|y-x_0|<r)}\frac{|x|^{n-2}}{|x-y|^{n-2}}dy\\
\leq\;& C\sup_{x\in D}\int_{D\cap(|y-x_0|<r)}\Big(1+\frac{|y|}{|x-y|}\Big)^{n-2}dy\\
:=\;& C\sup_{x\in D}\Big(\int_{D\cap(|y-x_0|<r)\cap(|x-y|<r)}\dots dy+\int_{D\cap(|y-x_0|<r\leq|x-y|)}\dots dy\Big)\\
\leq\;& C\sup_{x\in D}\Big(\int_{|x-y|<r}\frac{dy}{|x-y|^{n-2}}+\int_{|y-x_0|<r}\frac{dy}{|y-x_0|^{n-2}}\Big)\\
\leq\;& Cr^2,
\end{aligned}$$

where $C = C(D, n, x_0, \alpha, \beta) > 0$. Thus $V_{\alpha,\beta}$ satisfie (3.2) for $x_0 \in D$. Now let $x_0 \in \partial D$. By Lemmas 3.4 and 3.5, we have

$$\begin{aligned}
&\sup_{x\in D}\int_{D\cap(|y-x_0|<r)}\frac{\varphi(y)}{\varphi(x)}G(x,y)V_{\alpha,\beta}(y)dy\\
\leq\;& C\sup_{x\in D}\int_{D\cap(|y-x_0|<r)}\frac{\delta(y)^{1-\alpha}\min(1,\frac{\delta(x)\delta(y)}{|x-y|^2}|x||y|)}{\min(|x|^{2-n},\delta(x)|x|)|x-y|^{n-2}}dy\\
\leq\;& C\Big(\sup_{\delta(x)\geq1}\int_{D\cap(|y-x_0|<r)}\dots dy+\sup_{\delta(x)\leq1}\int_{D\cap(|y-x_0|<r)}\dots dy\Big)\\
:=\;& C(J_1(r)+J_2(r)),
\end{aligned}\tag{3.20}$$

where $C = C(D, n, \beta) > 0$. Clearly,

$$J_1(r) \leq C\int_{D\cap(|y-x_0|<r)}\delta(y)^{2-\alpha}dy \leq C\int_{|y-x_0|<r}|y-x_0|^{2-\alpha}dy = Cr^{n+2-\alpha}. \tag{3.21}$$

We estimate $J_2(r)$. For $x \in D$, let $D_x = \{y \in D : |x-y| \leq \delta(x)/2\}$. If $y \in D_x$, then $|\delta(y) - \delta(x)| \leq |x-y| \leq \delta(x)/2$ and so $\delta(x)/2 \leq \delta(y) \leq 2\delta(x)$. We have

$$\begin{aligned} J_2(r) &\leq C\Big(\sup_{\delta(x)\leq 1}\int_{D_x\cap(|y-x_0|<r)} \frac{\delta(x)^{-\alpha}}{|x-y|^{n-2}}dy + \sup_{\delta(x)\leq 1}\int_{D\cap(|y-x_0|<r)\setminus D_x} \frac{\delta(y)^{2-\alpha}}{|x-y|^n}dy\Big) \\ &\leq C\Big(\sup_{\delta(x)\leq 2r} \delta(x)^{-\alpha}(r\wedge\delta(x))^2 + r^{1-\alpha/2}\sup_{\delta(x)\leq 1}\int_{|x-y|\leq 4R}\frac{dy}{|x-y|^{n-1+\alpha/2}}\Big) \\ &\leq C(r^{2-\alpha} + r^{1-\alpha/2}). \end{aligned} \tag{3.22}$$

Combining (3.10)-(3.12), we get that $V_{\alpha,\beta}$ satisfie (3.2) for $x_0 \in \partial D$.

Case 2: $n = 2$.
By Lemma 2.2 and the inequalities $|x| \leq |y| + |x-y|$ and $\ln(1+t) \leq t$, for all $t \geq 0$, we have

$$\begin{aligned} &\sup_{x\in D}\int_{D\cap(|y|\geq M)} \frac{\varphi(y)}{\varphi(x)}G(x,y)V_{\alpha,\beta}(y)dy \\ \leq\ & C\sup_{x\in D}\int_{|y|\geq M} |y|^{-\beta}\ln(1+\frac{|x||y|}{|x-y|})dy \\ \leq\ & C\Big(\int_{|y|\geq M}|y|^{-\beta}\ln(1+|y|)dy + \sup_{x\in D}\int_{|y|\geq M}|y|^{-\beta}\ln(1+\frac{1}{|x-y|})dy\Big) \\ \leq\ & C\Big(\int_M^\infty t^{1-\beta}\ln(1+t)dt + \sup_{x\in D}\int_{|y|\geq M}\frac{dy}{|y|^\beta|x-y|}\Big) \\ \leq\ & C\Big(\int_M^\infty t^{1-\beta}\ln(1+t)dt + \sup_{x\in D}\int_{(|y|\geq M)\cap(|x-y|\leq 1)}\frac{dy}{|y|^\beta|x-y|} \\ & + \sup_{x\in D}\int_{(|y|\geq M)\cap(|x-y|\geq 1)}\frac{dy}{|y|^\beta|x-y|}\Big) \\ \leq\ & C\Big(\int_M^\infty t^{1-\beta}\ln(1+t)dt + M^{-\beta}\sup_{x\in D}\int_{|x-y|\leq 1}\frac{dy}{|x-y|} + \int_{|y|\geq M}\frac{dy}{|y|^\beta}\Big) \\ \leq\ & C\Big(\int_M^\infty t^{1-\beta}\ln(1+t)dt + M^{-\beta} + M^{2-\beta}\Big), \end{aligned}$$

where $C = C(D,\alpha) > 0$. Thus $V_{\alpha,\beta}$ satisfie (3.1). We will prove that $V_{\alpha,\beta}$ satisfie (3.2). Let firs $x_0 \in D$. By Lemma 2.2 and the inequalities $|x| \leq |y| + |x-y|$ and $\ln(1+t) \leq t$, for all $t \geq 0$, we have

$$\begin{aligned} &\sup_{x\in D}\int_{D\cap(|y-x_0|<r)} \frac{\varphi(y)}{\varphi(x)}G(x,y)V_{\alpha,\beta}(y)dy \\ \leq\ & C\sup_{x\in D}\int_{D\cap(|y-x_0|<r)} \ln(1+\frac{|x||y|}{|x-y|})dy \\ \leq\ & C\sup_{x\in D}\int_{D\cap(|y-x_0|<r)}\frac{dy}{|x-y|} \\ \leq\ & C\Big(\sup_{x\in D}\int_{|x-y|\leq|y-x_0|<r}\frac{dy}{|x-y|} + \sup_{x\in D}\int_{(|y-x_0|<r\wedge|x-y|)}\frac{dy}{|x-y|}\Big) \end{aligned}$$

$$\begin{aligned}
&\leq C\Big(\sup_{x\in D}\int_{|x-y|<r}\frac{dy}{|x-y|}+\int_{|y-x_0|<r}\frac{dy}{|y-x_0|}\Big)\\
&\leq Cr,
\end{aligned}$$

where $C=C(D,x_0,\alpha,\beta)>0$. Thus $V_{\alpha,\beta}$ satisfie (3.1) for $x_0\in D$. Now let $x_0\in\partial D$. By Lemmas 3.4 and 3.5, we have

$$\begin{aligned}
&\sup_{x\in D}\int_{D\cap(|y-x_0|<r)}\frac{\varphi(y)}{\varphi(x)}G(x,y)V_{\alpha,\beta}(y)dy\\
\leq\ & C\sup_{x\in D}\int_{D\cap(|y-x_0|<r)}\frac{\delta(y)^{1-\alpha}\ln(1+\frac{\delta(x)\delta(y)}{|x-y|^2}|x||y|)}{\min(1,\delta(x)|x|)}dy\\
\leq\ & C\Big(\sup_{\delta(x)\geq 1}\int_{D\cap(|y-x_0|<r)}\dots dy+\sup_{\delta(x)\leq 1}\int_{D\cap(|y-x_0|<r)}\dots dy\Big)\\
:=\ & C(J_1(r)+J_2(r)),
\end{aligned}\tag{3.23}$$

where $C=C(D,x_0,\beta)>0$. Clearly,

$$J_1(r)\leq C\int_{D\cap(|y-x_0|<r)}\delta(y)^{2-\alpha}dy\leq C\int_{|y-x_0|<r}|y-x_0|^{2-\alpha}dy=Cr^{4-\alpha}.\tag{3.24}$$

We estimate $J_2(r)$. For $x\in D$, let $D_x=\{y\in D:|x-y|\leq\delta(x)/2\}$. We have

$$\begin{aligned}
J_2(r)\ &\leq\ C\sup_{\delta(x)\leq 1}\int_{D\cap(|y-x_0|<r)}\frac{\delta(y)^{1-\alpha}\ln(1+\frac{\delta(x)\delta(y)}{|x-y|^2})}{\delta(x)}dy\\
&\leq\ C\Big(\sup_{\delta(x)\leq 1}\int_{D_x\cap(|y-x_0|<r)}\dots dy+\sup_{\delta(x)\leq 1}\int_{D\cap(|y-x_0|<r)\setminus D_x}\dots dy\Big)\\
&:=\ C(J_{21}(r)+J_{22}(r)).
\end{aligned}\tag{3.25}$$

Using the inequalities

$$\ln(1+\frac{\delta(x)\delta(y)}{|x-y|^2})\leq\ln(1+\frac{(\delta(y)+|x-y|)\delta(y)}{|x-y|^2})\leq 2\ln(1+\frac{\delta(y)}{|x-y|}),$$

$\ln(1+t)\leq t$, for all $t\geq 0$ and $\frac{\delta(x)}{2}\leq\delta(y)\leq 2\delta(x)$ for $y\in D_x$, we have

$$\begin{aligned}
J_{21}(r)\ &\leq\ C\sup_{\delta(x)\leq 1}\int_{D_x\cap(|y-x_0|<r)}\frac{\delta(y)^{2-\alpha}}{\delta(x)|x-y|}dy\\
&\leq\ Cr^{2-\alpha}\sup_{\delta(x)\leq 1}\int_{|x-y|\leq\frac{\delta(x)}{2}}\frac{dy}{\delta(x)|x-y|}\\
&=\ Cr^{2-\alpha},
\end{aligned}\tag{3.26}$$

and

$$J_{22}(r)\ \leq\ C\sup_{\delta(x)\leq 1}\int_{D\cap(|y-x_0|<r)\setminus D_x}\frac{\delta(y)^{2-\alpha}}{|x-y|^2}dy$$

$$\leq C \sup_{\delta(x)\leq 1} \int_{D\cap(|y-x_0|<r)\setminus D_x} \frac{\delta(y)^{1-\alpha/2}}{|x-y|^{1+\alpha/2}} dy$$
$$\leq Cr^{1-\alpha/2} \sup_{\delta(x)\leq 1} \int_{|y-x_0|<r} \frac{dy}{|x-y|^{1+\alpha/2}}$$
$$\leq Cr^{1-\alpha/2}\Big(\sup_{\delta(x)\leq 1} \int_{|y-x_0|\leq r\wedge |x-y|} \dots dy + \sup_{\delta(x)\leq 1} \int_{|x-y|\leq |y-x_0|\leq r} \dots dy\Big)$$
$$\leq Cr^{1-\alpha/2}\Big(\int_{|y-x_0|\leq r} \frac{dy}{|y-x_0|^{1+\alpha/2}} + \sup_{\delta(x)\leq 1} \int_{|x-y|\leq r} \frac{dy}{|x-y|^{1+\alpha/2}}\Big)$$
$$\leq Cr^{2-\alpha}. \tag{3.27}$$

Combining (3.13)-(3.17), we get that $V_{\alpha,\beta}$ satisfie (3.1) for $x_0 \in \partial D$. □

Corollary 3.7. *For $1 \leq \alpha < 2$ and $\beta > 2$, we have $V_{\alpha,\beta} \in \mathcal{K}_n^\infty(D) \setminus K_n^\infty$.*

Proof. By Proposition 4.7 in [14], which is also true by the same proof for the dimension $n = 2$, we have $V_{\alpha,\beta} \notin L^1_{\text{loc}}(\overline{D})$ and so $V_{\alpha,\beta} \notin K_n^\infty$, for $1 \leq \alpha < 2$ and $\beta > 2$. □

4. Proofs of Theorem 1.2 and Theorem 1.4

To prove Theorem 1.2, we need the following result. Let $C_0(\overline{D})$ be the set of all continuous functions v on $\overline{D}$ satisfying $\lim_{|x|\to\infty} v(x) = 0$.

Proposition 4.1. *Let V_0 be a nonnegative measurable function in the class $\mathcal{K}_n^\infty(D)$. Then, the family of functions*

$$\mathcal{F} = \Big\{H_V : x \to \int_D \frac{h(y)}{h(x)} G(x,y)|V(y)|dy, \quad |V| \leq V_0\Big\}$$

is uniformly bounded and equicontinuous in $C_0(\overline{D})$, and consequently is relatively compact in $C_0(\overline{D})$.

Proof. We have $0 \leq H_V \leq H_{V_0} \leq \|V_0\|$ and so $\mathcal{F}$ is uniformly bounded. We will prove that $H_{V_0}(x) \to 0$ as $|x| \to \infty$. Let $\varepsilon > 0$. By (1.4), there is $M > 0$ sufficientl great such that

$$\sup_{x\in D} \int_{D\cap\{|y|\geq M\}} \frac{h(y)}{h(x)} G(x,y)V_0(y)dy < \varepsilon. \tag{4.28}$$

On the other hand by the boundary Harnack principle (Theorem 2.4) and the Harnack inequality (Theorem 2.3), there is $C > 0$ such that for all x and y with $|x| \geq 2M$ and $|y| \leq M$,

$$C^{-1}h(y) \leq G(x,y) \leq Ch(y).$$

Moreover by (1.3), there is $M' > 2M$ such that for $|x| \geq M'$, $h(x) \geq \frac{1}{2}$. Thus for all x and y with $|x| \geq M'$ and $|y| \leq M$,

$$\frac{G(x,y)}{h(x)} \leq 2Ch(y). \tag{4.29}$$

By recalling that from (1.2) and (1.3), $\lim_{|x|\to\infty} \frac{G(x,y)}{h(x)} = 0$, using (4.2), Corollary 2.5 and the dominated convergence theorem, we obtain

$$\lim_{|x|\to\infty} \int_{D\cap\{|y|\le M\}} \frac{h(y)}{h(x)} G(x,y)V_0(y)dy = 0. \tag{4.30}$$

Combining (4.1) and (4.3) we obtain $\lim_{|x|\to\infty} H_{V_0}(x) = 0$.

We will prove the equicontinuity. Let $x_0 \in \overline{D}$ and $\varepsilon > 0$. By (1.5), there exists $r > 0$ small so that

$$\sup_{x\in D} \int_{D\cap(|x_0-y|<r)} \frac{h(y)}{h(x)} G(x,y)V_0(y)dy < \varepsilon. \tag{4.31}$$

Let $\rho \in]0, r]$. For x and $x' \in \overline{D} \cap B(x_0, \rho/2)$, we have

$$\begin{aligned} |H_V(x) - H_V(x')| &\le \int_D h(y)\left|\frac{G(x,y)}{h(x)} - \frac{G(x',y)}{h(x')}\right| V_0(y)dy \\ &= \int_{D\cap B(x_0,\rho)} \dots dy + \int_{D\cap B^c(x_0,\rho)} \dots dy \\ &:= I(x,x') + J(x,x'). \end{aligned} \tag{4.32}$$

By (4.4), we have

$$\begin{aligned} I(x,x') &\le \int_{D\cap B(x_0,\rho)} h(y)\frac{G(x,y)}{h(x)} V_0(y)dy + \int_{D\cap B(x_0,\rho)} h(y)\frac{G(x',y)}{h(x')} V_0(y)dy \\ &\le 2\varepsilon. \end{aligned} \tag{4.33}$$

We will prove that $J(x,x') \to 0$ as $|x - x'| \to 0$.

Case 1: $x_0 \in D$.

We choose $\rho = r \wedge d(x_0, \partial D)$. For $y \in D \cap B^c(x_0, \rho)$, $\frac{G(.,y)}{h}$ is the quotient of two nonnegative harmonic functions on $B(x_0, 2\rho/3)$, and so by the Harnack inequality (Theorem 2.3), there is a constant $C > 0$ depending only on D, r and x_0 such that for all $z \in B(x_0, \rho/2)$ and all $y \in D \cap B^c(x_0, \rho)$,

$$\frac{G(z,y)}{h(z)} \le C\frac{G(x_0,y)}{h(x_0)}. \tag{4.34}$$

By (4.7), the continuity of $\frac{G(.,y)}{h}$ on $B(x_0, \rho/2)$ and the dominated convergence theorem it follows that $J(x,x') \to 0$ as $|x - x'| \to 0$.

Case 2: $x_0 \in \partial D$.

We choose $\rho = r$ and $x_1 \in B(x_0, \rho/2)$. For $y \in D \cap B^c(x_0, \rho)$, $\frac{G(.,y)}{h}$ is the quotient of two nonnegative harmonic functions on $D \cap B(x_0, 2\rho/3)$ continuously vanishing on $\partial D \cap B(x_0, 2\rho/3)$ and so by the boundary Harnack principle (Theorem 2.4), there is a constant $C > 0$ depending only on D, r and x_0 such that for $z \in B(x_0, \rho/2)$ and $y \in D \cap B^c(x_0, \rho)$,

$$\frac{G(z,y)}{h(z)} \leq C \frac{G(x_1,y)}{h(x_1)}. \tag{4.35}$$

Using (4.8), the continuity of $\dfrac{G(.,y)}{h}$ on $B(x_0, \rho/2)$ and the dominated convergence theorem it follows that $J(x, x') \to 0$ as $|x - x'| \to 0$. By combining (4.5)-(4.8), $\mathcal{F}$ is equicontinuous. The familly $\mathcal{F}$ is uniformly bounded in $C_0(\overline{D})$ and equicontinuous and so by the Ascoli-Arzelà theorem $\mathcal{F}$ is relatively compact in $C_0(\overline{D})$. □

Proof of Theorem 1.2. We will prove the existence of $\lambda_0 > 0$ such that for all $\lambda \in]0, \lambda_0[$, there exists a continuous function u on $\overline{D}$ satisfying:

$$u(x) = \lambda h(x) + \int_D G(x,y) f(y, u(y))dy \quad \text{and} \quad \frac{\lambda}{2} \leq \frac{u}{h} \leq \frac{3}{2}\lambda.$$

Oubviously if u satisfie the above integral equation, by applying Δ to both members, we get that u is a solution of the problem $(\mathcal{E})$. Let $C(\overline{D} \cup \{\infty\})$ be the set of all continuous functions v on D such that $\lim_{x \to z \in \partial D} v(x)$ and $\lim_{|x| \to \infty} v(x)$ exist and finite By writing $w = \dfrac{u}{h}$, it then suffice to prove the existence of a function $w \in C(\overline{D} \cup \{\infty\})$ satisfying:

$$w(x) = \lambda + \frac{1}{h(x)} \int_D G(x,y) f(y, h(y)w(y))dy \quad \text{and} \quad \frac{\lambda}{2} \leq w \leq \frac{3}{2}\lambda.$$

For $\lambda > 0$, let consider the set

$$S_\lambda = \Big\{ w \in C(\overline{D} \cup \{\infty\}) : \frac{\lambda}{2} \leq w \leq \frac{3}{2}\lambda \Big\}.$$

Clearly S_λ is a nonempty, bounded, closed and convex subset in the Banach space $(C(\overline{D} \cup \{\infty\}), \|.\|_\infty)$, where $\|w\|_\infty = \sup_{x \in \overline{D} \cup \{\infty\}} |w(x)|$. Let us defin the operator T on S_λ by

$$Tw(x) = \lambda + \frac{1}{h(x)} \int_D G(x,y) f(y, w(y)h(y))dy,$$

for all $x \in D$. By Proposition 4.1, for $w \in S_\lambda$, $Tw \in C(\overline{D} \cup \{\infty\})$. Moreover, by the assumptions (H_1) and (H_2) on f, we have

$$\begin{aligned}
\left| \frac{Tw(x)}{\lambda} - 1 \right| &\leq \frac{1}{\lambda h(x)} \int_D G(x,y) |f(y, w(y)h(y))| dy \\
&\leq \int_D \frac{w(y)}{\lambda} \frac{h(y)}{h(x)} G(x,y) F(y, w(y)h(y)) dy \\
&\leq \frac{3}{2} \int_D \frac{h(y)}{h(x)} G(x,y) F(y, \frac{3}{2}\lambda h(y)) dy \\
&\leq \frac{3}{2} \Big\| F(., \frac{3}{2}\lambda h(.)) \Big\|.
\end{aligned} \tag{4.36}$$

By Proposition 4.1, there exists $M > 0$ sufficientl great such that

$$\sup_{|x|\geq M}\int_D \frac{h(y)}{h(x)} G(x,y)F(y,h(y))dy \leq \frac{1}{6}, \tag{4.37}$$

and by the Dini-Theorem, there is $\lambda_0 > 0$ small so that for all $\lambda \in]0, \lambda_0[$,

$$\sup_{x\in\overline{D}\cap(|x|\leq M)}\int_D \frac{h(y)}{h(x)} G(x,y)F\Big(y, \frac{3}{2}\lambda h(y)\Big)dy \leq \frac{1}{6}. \tag{4.38}$$

Combining (4.10) and (4.11), we obtain, for all $\lambda \in]0, \lambda_0[$,

$$\Big\|F(., \frac{3}{2}\lambda h(.))\Big\| \leq \frac{1}{3}. \tag{4.39}$$

From (4.9) and (4.12), it follows that $\frac{\lambda}{2} \leq Tw \leq \frac{3}{2}\lambda$ and so $Tw \in S_\lambda$. Hence $T(S_\lambda) \subset S_\lambda$. We will prove that the operator T is continuous. For w and w' in S_λ and $x \in \overline{D} \cup \{\infty\}$, we have

$$|Tw(x) - Tw'(x)| \;\leq\; \frac{1}{h(x)}\int_D G(x,y)|f(y,w(y)h(y)) - f(y,w'(y)h(y))|dy.$$

Since $f(y,.)$ is continuous and $|f(y,w(y)h(y)) - f(y,w'(y)h(y))| \leq \frac{3\lambda}{2}h(y)F(y,h(y))$ with $\int_D h(y)G(x,y)F(y,h(y))dy = \|F(.,h(.))\|h(x) < \infty$, by the dominated convergence theorem it follows that $|Tw(x) - Tw'(x)| \to 0$ as $\|w - w'\|_\infty \to 0$. Since by Proposition 4.1, $T(S_\lambda)$ is relatively compact in $C(\overline{D} \cup \{\infty\})$, it follows that $\|Tw - Tw'\|_\infty \to 0$ as $\|w - w'\|_\infty \to 0$, and so T is continuous. Thus, we proved that T is a compact and continuous operator from S_λ into itself and so by the Schauder fi ed point theorem there exists $w \in S_\lambda$ such that $Tw = w$. □

Proof of Theorem 1.4. Assume that V is a nonnegative Borel measurable function and the result of problem $(\mathcal{E}')$ holds. Then there is $\lambda > 0$ such that, for all $x \in D$,

$$u(x) = \lambda h(x) + \int_D G(x,y)V(y)u^p(y)dy,$$

$$\frac{u}{h} \in C(\overline{D}), \qquad \frac{\lambda}{2} \leq \frac{u}{h} \leq \frac{3\lambda}{2} \qquad \text{and} \qquad \lim_{|x|\to\infty}\frac{u(x)}{h(x)} = \lambda.$$

It follows that

$$\lim_{|x|\to\infty}\int_D \frac{G(x,y)}{h(x)}V(y)h^p(y)dy = 0,$$

and so for $\varepsilon > 0$ there is $R > 0$ such that

$$\sup_{|x|\geq R}\int_D \frac{G(x,y)}{h(x)}V(y)h^p(y)dy < \varepsilon/2. \tag{4.40}$$

By the Dini-theorem, we also have,

$$\lim_{M\to\infty}\sup_{\overline{D}\cap(|x|\leq R)}\int_{D\cap(|y|\geq M)}\frac{G(x,y)}{h(x)}V(y)h^p(y)dy = 0, \tag{4.41}$$

and for all $x_0 \in \overline{D}$,

$$\lim_{r \to 0} \sup_{\overline{D} \cap (|x| \leq R)} \int_{D \cap (|x_0 - y| < r)} \frac{G(x,y)}{h(x)} V(y) h^p(y) dy = 0. \tag{4.42}$$

By (4.13) and (4.14), the potential Vh^{p-1} satisfie (1.4) and by (4.13) and (4.15), Vh^{p-1} satisfie (1.5). Thus, Vh^{p-1} is in the class $\mathcal{K}_n^{\infty}(D)$. □

References

[1] S. Athreya, On a singular semilinear elliptic boundary value problem and the boundary Harnack principle, *Potential Analysis*, **17** (2002), 293–301.

[2] Z. Q. Chen, R. J. Williams and Z. Zhao, On the existence of positive solutions of semilinear elliptic equations with Dirichlet boundary conditions, *Math. Annalen*, **298** (1994), 543–556.

[3] K. L. Chung and Z. Zhao, *From Brownian motion to Schrödinger's equation,* Springer-Verlag, New York 1995.

[4] W. Hansen, Uniform boundary Harnack principle and generalized triangle property, *J. Funct. Anal.*, **226** (2005), 452–484.

[5] L. L. Helms, *Introduction to potential theory*, Wiley-Interscience, 1969.

[6] D. Jerison and C. Kenig, Boundary behavior of harmonic functions in nontangentially accessible domains, *Advances in Math.*, **46** (1982), 80–147.

[7] C. Kenig and W. N. Ni, An exterior Dirichlet problem with applications to some non-linear equations arising in geometry, *Amer. J. Math.*, **106** (1984), 689–702.

[8] A. C. Lazer and P. J. Mackenna, On a singular nonlinear elliptic boundary value problem, *Proc. Amer. Math. Soc.*, **111 (3)** (1991), 721–730.

[9] F. H. Lin, On the elliptic equation $D_i[a_{ij}(x)D_jU] - k(x)U^p = 0$, *Proc. Amer. Math. Soc.*, **95** (1985), 219–226.

[10] H. Mâagli and L. Mâatoug, Positive solutions of nonlinear elliptic equations in unbouded domains in $\mathbb{R}^2$, *Potential Anal.*, **19** (2003), 261–279.

[11] W. M. Ni, On the elliptic equation $\Delta v + K(x)u^{(n+2)/(n-2)} = 0$, its generalizations and applications in geometry, *Indiana Univ. Math. J.* **31** (1982), 493–529.

[12] S. Port and C. Stone, *Brownian motion and classical potential theory*, Prob. Math. Statist., Academic Press, New York, 1978.

[13] L. Riahi, A $3G$-Theorem for Jordan domains in $\mathbb{R}^2$, *Colloq. Math.*, **101** (2004), 1–7.

[14] L. Riahi, The $3G$-inequality for general Schrödinger operators on Lipschitz domains, *Manuscripta Math.*, **116** (2005), 211–227.

[15] L. Riahi, Singular solutions of a semi-linear elliptic equation on nonsmooth domains, *International J. Evolution Eq.*, **2,** 4 (2006), 55–64.

[16] M. Selmi, Inequalities for Green function in a Dini-Jordan domain in $\mathbb{R}^2$, *Potential Anal.*, **13** (2000), 81–102.

[17] U. Ufuktepe and Z. Zhao, Positive solutions of nonlinear elliptic equations in the euclidian plane, *Proc. Am. Math. Soc.,* **126**, 12 (1998), 3681–3692.

[18] Z. Zhao, On the existence of positive solutions of nonlinear elliptic equations – A probabilistic potential theory approach, *Duke Math. J.*, **69**, 2 (1993), 247–258.

[19] Z. Zhao, Positive solutions of nonlinear second order ordinary differential equations, *Proc. Am. Math. Soc.*, **121** (1994), 465–469.

In: Evolution Equations: Theories, Solutions and Functions ISBN 978-1-61324-090-8
Editor: Gaston M. N'Guerekata

Chapter 14

EXISTENCE AND UNIQUENESS OF ENTROPY SOLUTIONS TO NONLINEAR ELLIPTIC PROBLEMS WITH VARIABLE GROWTH

Stanislas Ouaro[1,*] ***and Sado Traoré***[2,†]
[1] Laboratoire d'Analyse Mathématique des Equations (LAME),
UFR. Sciences Exactes et Appliquées, Université de Ouagadougou,
03 BP 7021 Ouaga 03, Ouagadougou, Burkina Faso
[2] Laboratoire d'Analyse Mathématique des Equations (LAME),
Institut des Sciences Exactes et Appliquées,
Université de Bobo Dioulasso,
01 BP 1091 Bobo-Dioulasso 01, Bobo Dioulasso, Burkina Faso

Abstract

We study the boundary value problem $u - div(a(x, \nabla u)) = f$ in Ω, $u = 0$ on $\partial\Omega$, where Ω is a smooth bounded domain in $\mathbb{R}^N$ and $div(a(x, \nabla u))$ is a $p(x)$-Laplace type operator. We obtain existence and uniqueness of an entropy solution for L^1 data f and for a variable exponent not necessarily continuous.

2000 Mathematics Subject Classification: 35J20, 35J25, 35D30, 35B38, 35J60.

Keywords and phrases: Generalized Lebesgue-Sobolev spaces; weak energy solution; entropy solution; $p(x)$-Laplace operator; electrorheological fluids

1. Introduction

The study of partial differential equations and variational problems involving $p(x)-$growth conditions has received specifi attention in the recent decades. This is a consequence of the fact that such equations can be used to model phenomena which arise in mathematical physics. Electrorheological fluid and elastic mechanics are two examples of physical field

*E-mail address: souaro@univ-ouaga.bf
†E-mail address: sado@univ-ouaga.bf

which benefi from such kind of studies. In that context, we refer to Diening [8], Ruzicka [25], and the references therein.

Most materials can be modelled with sufficien accuracy using classical Lebesgue and Sobolev spaces L^p and $W^{1,p}$, where p is a fi ed constant. For electrorheological fluids this is not adequate, but rather the exponent p should be able to vary. This situation leads us to the study of variable exponent Lebesgue and Sobolev spaces, $L^{p(.)}$ and $W^{1,p(.)}$ where $p(.)$ is a real-valued function.

The variable exponent Lebesgue Spaces $L^{p(.)}$, where $p(.)$ is a real-valued function, appeared in the literature for the firs time in a 1931 article of W. Orlicz [23]. In the 1950's, this study was carried on by Nakano [22] who made the firs systematic study of spaces with variable exponent. Later, Polish and Czechoslovak mathematicians investigated the modular function spaces (see for example Musielak [21], Kovacik and Rakosnik [16]). Variable exponent Lebesgue spaces on the real line have been independently developed by Russian researchers. In that context, we refer to the work of Tsenov [28], Sharapudinov [27] and Zhikov [29, 30].

This paper is motivated by phenomena which are described by nonlinear boundary value problems of the type

$$\begin{cases} u - div(a(x, \nabla u)) = f \text{ in } \Omega, \\ u = 0 \text{ on } \partial\Omega, \end{cases} \tag{1.1}$$

where $\Omega \subset \mathbb{R}^N$ $(N \geq 3)$ is a bounded domain with smooth boundary and $f \in L^1(\Omega)$.

A problem like (1.1) was studied firs by Sachon and Urbano [26]. More precisely, they studied the problem

$$\begin{cases} -div(a(x, \nabla u)) = f \text{ in } \Omega, \\ u = 0 \text{ on } \partial\Omega, \end{cases} \tag{1.2}$$

under the following assumption on the exponent:
$p(.) : \Omega \to (1, N)$ is a measurable function such that

$$\exists C > 0 : |p(x) - p(y)| \leq \frac{C}{-\ln|x-y|}, \text{ for } |x-y| < \frac{1}{2}. \tag{1.3}$$

Another tentative in this direction was done by Bendahmane and Wittbold in [2], where they proved the existence and uniqueness of renormalized solution of the problem

$$\begin{cases} -div(|\nabla u|^{p(x)-2}\nabla u) = f \text{ in } \Omega, \\ u = 0 \text{ on } \partial\Omega, \end{cases} \tag{1.4}$$

under the following assumption on the right hand side and the exponent:

$$f \in L^1(\Omega), \;\; p(.) : \overline{\Omega} \to (1, +\infty) \text{ and } 2 - \frac{1}{N} < p_- \leq p_+ < N, \tag{1.5}$$

where $p_- := ess \inf\limits_{x\in\Omega} p(x)$ and $p_+ := ess \sup\limits_{x\in\Omega} p(x)$.

Since the domain is bounded, then the notion of renormalized solution define in [2] is an equivalent notion of that of entropy solution used in this paper.

Note that assumption (1.3) allows Sanchon and Urbano in particular to exploit some useful properties of Sobolev spaces in variable exponents like Sobolev embedding and Poincaré type inequality to derive a priori estimates for the proof of the existence of entropy solutions. Nonetheless, it is quite natural to consider the problem with $p(.)$ that can even be discontinuous; this is the case e.g. of a stratifie medium. For this reason, in our paper the assumption on the exponent is such that it can even be discontinous. By this choice, the main difficult is that the usual Sobolev $W_0^{1,p}$ framework has to be replaced with the space $W_0^{1,p(.)}$ which does not have nice properties like Sobolev embedding and Poincaré inequality. Another difference with [26] is on the assumptions on the vector fiel a (see assumptions 2.2 and 2.5 below). Our assumptions permit us in particular to avoid the diffi culties of the pure functional-analytic approach for the proof of existence of weak solutions of (1.1) in which the properties of the spaces $W_0^{1,p(.)}$ and of its dual have to be carefully described since some of these properties were not yet clearly understood by the experts in the fiel of generalized Sobolev spaces. The paper contains four sections. In section 2, we introduce the assumptions and we recall the definition of Lebesgue and Sobolev spaces with variable exponent and some of their properties. In section 3, we prove existence and uniqueness of weak energy solution of (1.1) when the right-hand side f has enough integrability i.e., $f \in L^\infty(\Omega)$. Using the results of section 3, we study in section 4, the question of existence and uniqueness of entropy solution of (1.1) for $f \in L^1(\Omega)$. The L^∞ framework in section 3 is used to simplify the proof of existence of entropy solutions in section 4 since we need bounded sequences to approach L^1-functions. But more generally, for $f \in L^{(p_-)'}(\Omega)$, we have existence and uniqueness of weak solutions to (1.1) and the proof is similar to that of L^∞-data.

2. Assumptions and Preliminaries

In this paper, we study problem (1.1) for a variable exponent $p(.)$ which can even be discontinuous, more precisely, we assume that

$$\begin{cases} p(.) : \Omega \longrightarrow \mathbb{R} \text{ is a measurable function such that} \\ 1 < p_- \leq p_+ < +\infty, \end{cases} \tag{2.6}$$

For the vector field $a(.,.)$, we assume that $a(x,\xi) : \Omega \times \mathbb{R}^N \longrightarrow \mathbb{R}^N$ is Carathéodory and is the continuous derivative with respect to ξ of the mapping $A : \Omega \times \mathbb{R}^N \to \mathbb{R}$, $A = A(x,\xi)$, i.e. $a(x,\xi) = \nabla_\xi A(x,\xi)$ such that:

- The following equality holds

$$A(x,0) = 0 \tag{2.7}$$

  for almost every $x \in \Omega$.

- There exists a positive constant C_1 such that

$$|a(x,\xi)| \leq C_1(j(x) + |\xi|^{p(x)-1}) \tag{2.8}$$

for almost every $x \in \Omega$ and for every $\xi \in \mathbb{R}^N$, where j is a nonnegative function in $L^{p'(.)}(\Omega)$, with $1/p(x) + 1/p'(x) = 1$.

- The following inequalities hold

$$(a(x,\xi) - a(x,\eta)) \,.\, (\xi - \eta) > 0, \tag{2.9}$$

for almost every $x \in \Omega$ and for every $\xi, \eta \in \mathbb{R}^N$, with $\xi \neq \eta$, and

$$\frac{1}{C}\,|\xi|^{p(x)} \leq a(x,\xi).\xi \leq Cp(x)A(x,\xi) \tag{2.10}$$

for almost every $x \in \Omega$, $C > 0$ and for every $\xi \in \mathbb{R}^N$.

Remark 2.1. Assumption (2.4) and relation $a(x,\xi) = \nabla_\xi A(x,\xi)$ imply in particular that $A(x,\xi)$ is convex with respect to the second variable. Indeed, by the Hadamard's formula, we have

$$A(x, \xi + \delta) - A(x,\xi) = \delta. \int_0^1 a(x, \xi + t\delta)dt. \tag{2.11}$$

Using (2.6), we deduce

$$\begin{aligned}
&A(x, r\xi + (1-r)\psi) - rA(x,\xi) - (1-r)A(x,\psi)\\
&= r(A(x, r\xi + (1-r)\psi) - A(x,\xi)) + (1-r)(A(x, r\xi + (1-r)\psi) - A(x,\psi))\\
&= r(1-r)(\psi - \xi)\int_0^1 a(x, \xi + t(1-r)(\psi - \xi))dt\\
&\quad +(1-r)r(\xi - \psi)\int_0^1 a(x, \psi + tr(\xi - \psi))dt\\
&= r(1-r)(\psi - \xi)\int_0^1 (a(x, \xi + t(1-r)(\psi - \xi)) - a(x, \psi + tr(\xi - \psi)))dt;
\end{aligned}$$

since $(\xi + t(1-r)(\psi - \xi)) - (\psi + tr(\xi - \psi)) = (\xi - \psi)(1 - t(r-1) - tr) = (1 + t(1-2r))(\xi - \psi)$, and $1 + t(1-2r) \geq 1 + 1(1-2) = 0$, then $(\xi - \psi)$ is colinear to $(\xi + t(1-r)(\psi - \xi)) - (\psi + tr(\xi - \psi))$ and has the same direction. Using (2.4), we get

$$A(x, r\xi + (1-r)\psi) - rA(x,\xi) - (1-r)A(x,\psi) \leq 0.$$

Assumptions (2.2)-(2.5) allow us to exploit minimization methods to get existence of weak energy solution of problem (1.1) (cf. [19]) in the context of assumption (2.1). As examples of models with respect to above assumptions for problem (1.1), we can give the following:

(i) Set $A(x,\xi) = (1/p(x))\,|\xi|^{p(x)}$, $a(x,\xi) = |\xi|^{p(x)-2}\,\xi$ where $p(x) \geq 2$. Then we get the $p(x)$-Laplace operator

$$div(|\nabla u|^{p(x)-2}\,\nabla u).$$

(ii) Set $A(x,\xi) = (1/p(x))\left[\left(1 + |\xi|^2\right)^{p(x)/2} - 1\right]$, $a(x,\xi) = \left(1 + |\xi|^2\right)^{(p(x)-2)/2}\xi$, where $p(x) \geq 2$. Then we obtain the generalized mean curvature operator

$$div\left(\left(1 + |\nabla u|^2\right)^{(p(x)-2)/2}\nabla u\right).$$

As the exponent $p(.)$ appearing in (2.3) and (2.5) depend on the variable x, we must work with Lebesgue and Sobolev spaces with variable exponent.

We defin the Lebesgue space with variable exponent $L^{p(.)}(\Omega)$ as the set of all measurable function $u : \Omega \to \mathbb{R}$ for which the convex modular

$$\rho_{p(.)}(u) := \int_{\Omega} |u|^{p(x)}\, dx$$

is finite If the exponent is bounded, i.e., if $p_+ < \infty$, then the expression

$$|u|_{p(.)} := \inf\left\{\lambda > 0 : \rho_{p(.)}(u/\lambda) \leq 1\right\}$$

define a norm in $L^{p(.)}(\Omega)$, called the Luxembourg norm. The space $(L^{p(.)}(\Omega), |.|_{p(.)})$ is a separable Banach space. Moreover, if $p_- > 1$, then $L^{p(.)}(\Omega)$ is uniformly convex, hence refl xive, and its dual space is isomorphic to $L^{p'(.)}(\Omega)$, where $\frac{1}{p(x)} + \frac{1}{p'(x)} = 1$. Finally, we have the Hölder type inequality:

$$\left|\int_{\Omega} uv dx\right| \leq \left(\frac{1}{p_-} + \frac{1}{(p')_-}\right) |u|_{p(.)} |v|_{p'(.)}, \tag{2.12}$$

for all $u \in L^{p(.)}(\Omega)$ and $v \in L^{p'(.)}(\Omega)$.

Now, let

$$W^{1,p(.)}(\Omega) := \left\{u \in L^{p(.)}(\Omega) : |\nabla u| \in L^{p(.)}(\Omega)\right\},$$

which is a Banach space equipped with the norm

$$\|u\|_{1,p(.)} := |u|_{p(.)} + |\nabla u|_{p(.)}.$$

Next, we defin $W_0^{1,p(.)}(\Omega)$ as the closure of $C_0^\infty(\Omega)$ in $W^{1,p(.)}(\Omega)$ under the norm

$$\|u\| := |\nabla u|_{p(.)}.$$

The space $\left(W_0^{1,p(.)}(\Omega), \|u\|\right)$ is a separable and refl xive Banach space.

An important role in manipulating the generalized Lebesgue-Sobolev spaces is played by the modular $\rho_{p(.)}$ of the space $L^{p(.)}(\Omega)$. We have the following result (cf. [13]):

Lemma 2.2. *If* $u_n, u \in L^{p(.)}(\Omega)$ *and* $p_+ < +\infty$ *then the following relations hold:*

(i) $|u|_{p(.)} > 1 \Rightarrow |u|_{p(.)}^{p_-} \leq \rho_{p(.)}(u) \leq |u|_{p(.)}^{p_+}$;

(ii) $|u|_{p(.)} < 1 \Rightarrow |u|_{p(.)}^{p_+} \leq \rho_{p(.)}(u) \leq |u|_{p(.)}^{p_-}$;

(iii) $|u|_{L^{p(.)}(\Omega)} < 1$ (*respectively* $= 1; > 1$) $\Leftrightarrow \rho_{p(.)}(u) < 1$ (*respectively* $= 1; > 1$);

(iv) $|u_n|_{L^{p(.)}(\Omega)} \to 0$ (*respectively* $\to +\infty$) $\Leftrightarrow \rho_{p(.)}(u_n) \to 0$ (*respectively* $\to +\infty$);

(v) $\rho_{p(.)}\left(u/|u|_{L^{p(.)}(\Omega)}\right) = 1.$

Let us introduce the following notation: given two bounded measurable functions $p(.), q(.) : \Omega \to \mathbb{R}$, we write

$$q(.) \ll p(.) \text{ if } ess \inf_{x\in\Omega}(p(x) - q(x)) > 0.$$

3. Existence and Uniqueness of Weak Energy Solution for $f \in L^\infty(\Omega)$

In this section, we study the weak energy solution of (1.1).

Definition 3.1. A weak solution of (1.1) is a function $u \in W_0^{1,1}(\Omega)$ such that $a(., \nabla u) \in (L^1_{loc}(\Omega))^N$ and

$$\int_\Omega a(x, \nabla u).\nabla\varphi dx + \int_\Omega u\varphi dx = \int_\Omega f(x)\varphi dx, \text{ for all } \varphi \in C_0^\infty(\Omega). \tag{3.13}$$

A weak energy solution is a weak solution such that $u \in W_0^{1,p(.)}(\Omega)$.

The main result of this section is the following:

Theorem 3.2. *Assume* (2.1)-(2.5) *and* $f \in L^\infty(\Omega)$. *Then there exists a unique weak energy solution of* (1.1).

Proof. $*$ **Existence**. Let E denote the generalized Sobolev space $W_0^{1,p(.)}(\Omega)$.
Defin the energy functional $I : E \to \mathbb{R} \cup \{+\infty\}$ by

$$I(u) = \int_\Omega A(x, \nabla u)dx + \frac{1}{2}\int_\Omega u^2 dx - \int_\Omega fudx.$$

We firs establish some basic properties of I.

Proposition 3.3. *The functional I is well-defined on E and $I \in C^1(E \cap L^2(\Omega), \mathbb{R})$ with the derivative given by*

$$\langle I'(u), \varphi\rangle = \int_\Omega a(x, \nabla u).\nabla\varphi dx + \int_\Omega u\varphi dx - \int_\Omega f\varphi dx,$$

for all $u, \varphi \in E \cap L^2(\Omega)$.

To prove Proposition 3.3, we defin the functonal $\Lambda : E \to \mathbb{R}$ by

$$\Lambda(u) = \int_\Omega A(x, \nabla u)dx, \text{ for all } u \in E.$$

Lemma 3.4. (i) *The functional Λ is well-defined on E.*

(ii) *The functional Λ is of class $C^1(E,\mathbb{R})$ and*

$$\langle \Lambda'(u),\varphi\rangle = \int_\Omega a(x,\nabla u).\nabla\varphi dx,$$

for all $u,\varphi \in E$.

Proof of Lemma 3.4. (i) For any $x\in\Omega$ and $\xi\in\mathbb{R}^N$, we have

$$A(x,\xi) = \int_0^1 \frac{d}{dt}A(x,t\xi)dt = \int_0^1 a(x,t\xi).\xi dt.$$

Then by (2.3),

$$A(x,\xi) \le C_1\int_0^1 \left(j(x) + |\xi|^{p(x)-1}\, t^{p(x)-1}\right)|\xi|\, dt \le C_1 j(x)\,|\xi| + \frac{C_1}{p(x)}\,|\xi|^{p(x)}.$$

The above inequality and (2.5) imply

$$0 \le \int_\Omega A(x,\nabla u)dx \le C_1\int_\Omega j(x)\,|\nabla u|\, dx + \frac{C_1}{p_-}\int_\Omega |\nabla u|^{p(x)}\, dx,\ \text{ for all } u\in E.$$

Using (2.7) and Lemma 2.2, we deduce that Λ is well-define on E.

(ii) Existence of the Gâteaux derivative. Let $u,\varphi\in E$. Fix $x\in\Omega$ and $0<|r|<1$. Then due to the mean value theorem (see also [20]), there exists $\nu\in[0,1]$ such that

$$\begin{aligned}|a(x,\nabla u(x)+\nu r\nabla\varphi(x))|\,|\nabla\varphi(x)| &= |A(x,\nabla u(x)+r\nabla\varphi(x)) - A(x,\nabla u(x))|\,/\,|r|\\ &\le \left[C_1 j(x) + C_1\left(|\nabla u(x)|+|\nabla\varphi(x)|\right)^{p(x)-1}\right]|\nabla\varphi(x)|.\end{aligned}$$

Next, by (2.7), we have

$$\int_\Omega C_1 j(x)|\nabla\varphi(x)|dx \le \beta|C_1 j|_{p'(x)}.|\nabla\varphi|_{p(x)}$$

and

$$\int_\Omega \left(|\nabla u(x)|+|\nabla\varphi(x)|\right)^{p(x)-1}|\nabla\varphi(x)|dx \le \alpha\left|\left(|\nabla u(x)|+|\nabla\varphi(x)|\right)^{p(x)-1}\right|_{p'(x)}.|\nabla\varphi|_{p(x)}.$$

The above inequalities imply

$$C_1\left[j(x) + \left(|\nabla u(x)|+|\nabla\varphi(x)|\right)^{p(x)-1}\right]|\nabla\varphi(x)| \in L^1(\Omega).$$

It follows from the Lebesgue theorem that

$$\langle \Lambda'(u),\varphi\rangle = \int_\Omega a(x,\nabla u).\nabla\varphi dx.$$

Assume now that $u_n \to u$ in E. Let us defin $\psi(x,u) = a(x,\nabla u)$. Using assumption (2.3), theorems 4.1 and 4.2 in [16], we deduce that $\psi(x,u_n) \to \psi(x,u)$ in $(L^{p'(x)}(\Omega))^N$. By (2.7), we obtain

$$\left|\langle \Lambda'(u_n) - \Lambda'(u), \varphi\rangle\right| \le C\left|\psi(x,u_n) - \psi(x,u)\right|_{p'(x)} |\nabla\varphi|_{p(x)}$$

and so

$$\left\|\Lambda'(u_n) - \Lambda'(u)\right\| \le C\left|\psi(x,u_n) - \psi(x,u)\right|_{p'(x)} \to 0, \text{ as } n \to \infty.$$

The proof of Lemma 3.4 is complete. □

By Lemma 3.4, it is clear that Proposition 3.3 holds and then, the proof of Proposition 3.3 is also complete.

Lemma 3.5. *The functional Λ is weakly lower semi-continuous.*

Proof of Lemma 3.5. Using corollary III.8 in [6], it is enough to show that Λ is lower semi-continuous. For this purpose, fi $u \in E$ and $\epsilon > 0$. Since Λ is convex (by remark 2.1), we deduce that for any $v \in E$, the following inequality holds

$$\int_\Omega A(x,\nabla v)dx \ge \int_\Omega A(x,\nabla u)dx + \int_\Omega a(x,\nabla u).(\nabla v - \nabla u)dx.$$

Using (2.3) and (2.7), we have

$$\int_\Omega A(x,\nabla v)dx \ge \int_\Omega A(x,\nabla u)dx - \int_\Omega |a(x,\nabla u)|\,|\nabla v - \nabla u|\,dx$$

$$\ge \int_\Omega A(x,\nabla u)dx - C_1\int_\Omega j(x)\,|\nabla(v-u)|\,dx - C_1\int_\Omega |\nabla u|^{p(x)-1}\,|\nabla(v-u)|\,dx$$

$$\ge \int_\Omega A(x,\nabla u)dx - C_2\,|j|_{p'(x)}\,|\nabla(v-u)|_{p(x)} - C_3\left||\nabla u|^{p(x)-1}\right|_{p'(x)}|\nabla(v-u)|_{p(x)}$$

$$\ge \int_\Omega A(x,\nabla u)dx - C_4\,\|v-u\| \ge \int_\Omega A(x,\nabla u)dx - \epsilon,$$

for all $v \in E$ with $\|v-u\| < \delta = \epsilon/C_4$, where C_2, C_3 and C_4 are positive constants.

We conclude that Λ is weakly lower semi-continuous. The proof of Lemma 3.5 is complete. □

Proposition 3.6. *The functional I is bounded from below, coercive and weakly lower semi-continuous.*

Proof of Proposition 3.6. Using (2.5), we have

$$I(u) = \int_\Omega A(x,\nabla u)dx + \frac{1}{2}\int_\Omega u^2dx - \int_\Omega fudx \ge \int_\Omega \frac{1}{p(x)}|\nabla u|^{p(x)}\,dx - \int_\Omega fudx$$

$$\ge \int_\Omega \frac{1}{p(x)}|\nabla u|^{p(x)}\,dx - \|f\|_{(p_-)'}\,\|u\|_{p_-} \ge \frac{1}{p_+}\int_\Omega |\nabla u|^{p(x)}\,dx - \|f\|_{(p_-)'}\,\|u\|_{p_-},$$

where $\|u\|_{p_-} = \left(\int_\Omega |u|^{p_-}\, dx \right)^{\frac{1}{p_-}}$.

For the coerciveness of I, we want to show that $\lim_{\|u\| \to +\infty} I(u) = +\infty$. For this, we can assume that $\|u\| = |\nabla u|_{p(x)} > 1$ and then, by Lemma 2.2 we obtain from above inequalities that

$$I(u) \geq \frac{1}{p_+} |\nabla u|_{p(x)}^{p_-} - C \|u\|_{p_-} \geq \frac{1}{p_+} \|u\|^{p_-} - C' \|u\|,$$

since E is continuously embedded in $L^{p_-}(\Omega)$.

As $p_- > 1$, then I is coercive. It is obvious that I is bounded from below. By Lemma 3.5, Λ is weakly lower semi-continuous. We show that I is weakly lower semi-continuous. Let $(u_n) \subset E$ be a sequence which converges weakly to u in E. Since Λ is weakly lower semi-continuous, we have

$$\Lambda(u) \leq \liminf_{n \to +\infty} \Lambda(u_n). \qquad (3.14)$$

On the other hand, (u_n) has a subsequence converging strongly in L^{p_-} and a.e. Then, by Fatou's lemma we get $\int_\Omega u^2 dx \leq \liminf_{n \to +\infty} \int_\Omega u_n^2 dx$. Combining this latter inequality with (3.2), we get:

$$I(u) \leq \liminf_{n \to +\infty} I(u_n).$$

Therefore, I is weakly lower semi-continuous. The proof of Proposition 3.6 is complete. □

Since I is proper, lower semi-continuous and coercive, I has a minimizer which is a weak energy solution of (1.1). The proof of existence is then complete.

$*$ **Uniqueness**. Let u_1, u_2 be two weak energy solutions of (1.1). Then

$$\int_\Omega (u_1 - u_2)^2 dx = - \int_\Omega (a(x, \nabla u_1) - a(x, \nabla u_2)) . (\nabla u_1 - \nabla u_2) dx \leq 0. \qquad (3.15)$$

From (3.3), we deduce that

$$\int_\Omega (u_1 - u_2)^2 dx = 0,$$

from where it follows

$$u_1 = u_2.$$

□

4. Entropy Solutions

In this section, we study the problem (1.1) for a right hand side $f \in L^1(\Omega)$. In the L^1 setting, the suitable notion of solution for the study of (1.1) is the notion of entropy solution.

We firs defin the troncation function T_t by $T_t(s) := \max\{-t, \min\{t, s\}\}$.

Definition 4.1. A measurable function u is an entropy solution to problem (1.1) if, for every $t > 0$, $T_t(u) \in W_0^{1,p(.)}(\Omega)$ and

$$\int_\Omega uT_t(u-\varphi)dx + \int_\Omega a(x,\nabla u).\nabla T_t(u-\varphi)dx \leq \int_\Omega f(x)T_t(u-\varphi)dx \qquad (4.16)$$

for all $\varphi \in W_0^{1,p(.)}(\Omega) \cap L^\infty(\Omega)$.

Remark 4.2. A function u such that $T_t(u) \in W_0^{1,p(.)}(\Omega)$ for all $t > 0$ does not necessarily belong in $W_0^{1,1}(\Omega)$. However, it is possible to defin its weak gradient, still denoted by ∇u.

Our main result in this section is the following:

Theorem 4.3. *Assume* (2.1)-(2.5) *and* $f \in L^1(\Omega)$. *There exists a unique entropy solution* u *to problem* (1.1).

Proof. $*$ **A priori estimates**. We start with the existence of the weak gradient for every measurable function u such that $T_t(u) \in W_0^{1,p(.)}(\Omega)$ for all $t > 0$.

Proposition 4.4. *If* u *is a measurable function such that* $T_t(u) \in W_0^{1,p(.)}(\Omega)$ *for all* $t > 0$, *then there exists a unique measurable function* $v : \Omega \to \mathbb{R}^N$ *such that*

$$v\chi_{\{|u|<t\}} = \nabla T_t(u) \text{ for a.e. } x \in \Omega, \text{ and for all } t > 0,$$

where χ_B *denotes the characteristic function of a measurable set* B. *Moreover, if* u *belongs to* $W_0^{1,1}(\Omega)$, *then* v *coincides with the standard distributional gradient of* u.

Proof of Proposition 4.4. As $T_t(u) \in W_0^{1,p(.)}(\Omega) \subset W_0^{1,p_-}(\Omega)$ for all $t > 0$, then by Theorem 1.5 in [1], the result follows. □

Proposition 4.5. *Assume* (2.1)-(2.5) *and* $f \in L^1(\Omega)$. *Let* u *be an entropy solution of* (1.1), *then*

$$\int_\Omega |\nabla T_t(u)|^{p(x)}\, dx \leq Ct\,\|f\|_1\,, \text{ for all } t > 0. \qquad (4.17)$$

and

$$\|u\|_1 \leq meas(\Omega) + \|f\|_1\,. \qquad (4.18)$$

Proof of Proposition 4.5. Taking $\varphi = 0$ in the entropy inequality (4.1) and using (2.5), we obtain

$$\begin{aligned}\frac{1}{C}\int_{\{|u|\leq t\}} |\nabla T_t(u)|^{p(x)}\, dx &\leq \int_\Omega uT_t(u)dx + \int_{\{|u|\leq t\}} a(x,\nabla u).\nabla u dx \\ &\leq \int_\Omega f(x)T_t(u)dx \leq t\,\|f\|_1\,, \text{ for all } t > 0.\end{aligned}$$

Thanks to inequality above, we have

$$\begin{aligned}\int_{\Omega} uT_t(u)dx \le t\,\|f\|_1 &\Rightarrow \int_{\{|u|\le t\}} uT_t(u)dx + \int_{\{|u|>t\}} uT_t(u)dx \le t\,\|f\|_1 \\ &\Rightarrow \int_{\{|u|>t\}} uT_t(u)dx \le t\,\|f\|_1 \\ &\Rightarrow \int_{\{|u|>t\}} |u|\,dx \le \|f\|_1\,, \text{ for all } t>0.\end{aligned}$$

As $\int_{\Omega} |u|\,dx = \int_{\{|u|\le 1\}} |u|\,dx + \int_{\{|u|>1\}} |u|\,dx$, then $\|u\|_1 \le meas(\Omega) + \|f\|_1\,.$ □

Proposition 4.6. *Assume* (2.1)-(2.5) *and* $f \in L^1(\Omega)$. *Let* u *be an entropy solution of* (1.1), *then*

$$meas\,\{|\nabla u| > h\} \le const(\|f\|_1)/h^{p_-/2} \textit{ for } h \ge 1. \tag{4.19}$$

Proof of Proposition 4.6. Note that

$$\int_{\{|u|>t\}} tdx \le \int_{\{|u|>t\}} |u|\,dx \le \int_{\Omega} |u|\,dx.$$

We deduce from inequality above, thanks to (4.3):

$$\int_{\{|u|>t\}} tdx \le meas(\Omega) + \|f\|_1\,.$$

Therefore,

$$\begin{aligned}\int_{\left\{|\nabla u|^{\frac{p(.)}{2}}>t\right\}} tdx &\le \int_{\left\{|\nabla u|^{\frac{p(.)}{2}}>t\right\}\cap\{|u|\le t\}} tdx + \int_{\{|u|>t\}} tdx \\ &\le \int_{\{|u|\le t\}} t\left(\frac{|\nabla u|^{\frac{p(.)}{2}}}{t}\right)^2 dx + \|f\|_1 + meas(\Omega) \le (C+1)\,\|f\|_1 + meas(\Omega).\end{aligned}$$

We deduce from the preceding estimates that

$$meas\left(\left\{|\nabla u|^{\frac{p(.)}{2}} > t\right\}\right) \le \frac{1}{t}((C+1)\,\|f\|_1 + meas(\Omega)).$$

Next, as $\left\{|\nabla u|^{\frac{p(.)}{2}} > t\right\} \supset \left\{|\nabla u|^{\frac{p_-}{2}} > t\right\}$ for $t \ge 1$, it follows

$$meas\left(\left\{|\nabla u|^{\frac{p_-}{2}} > t\right\}\right) \le \frac{1}{t}((C+1)\,\|f\|_1 + meas(\Omega)).$$

Finally, taking $h = t^{\frac{2}{p_-}}$ in the inequality above, we obtain

$$meas\,(\{|\nabla u| > h\}) \le ((C+1)\,\|f\|_1 + meas(\Omega))/h^{p_-/2}.$$ □

Proposition 4.7. *Assume* (2.1)-(2.5) *and* $f \in L^1(\Omega)$. *Let* u *be an entropy solution of* (1.1), *then*

$$meas\,\{|u| > h\} \leq const(\|f\|_1)/h^{p_- -1}.$$

Proof of Proposition 4.7. Taking $\varphi = 0$ in the entropy inequality (4.1) and using (2.5), we obtain

$$\begin{aligned}\int_{\{|u|\leq h\}} |\nabla T_h(u)|^{p(x)}\,dx &\leq \int_\Omega uT_h(u)dx + \int_{\{|u|\leq h\}} a(x,\nabla u).\nabla u dx \\ &\leq \int_\Omega f(x)T_h(u)dx \leq h\,\|f\|_1\end{aligned}$$

for all $h > 0$. Next,

$$\int_{\{|u|\leq h\}} |\nabla T_h(u)|^{p(x)}\,dx \leq Mh \Rightarrow \int_{\{|u|\leq h\}} |\nabla T_h(u)|^{p_-}\,dx \leq C(1+h).$$

By the Poincaré inequality in constant exponent, we obtain

$$\|T_h(u)\|_{L^{p_-}(\Omega)} \leq D(1+h)^{\frac{1}{p_-}}.$$

The above inequality implies that

$$\int_\Omega |T_h(u)|^{p_-}\,dx \leq D^{p_-}(1+h),$$

hence

$$meas\,\{|u| > h\} \leq D^{p_-}\frac{1+h}{h^{p_-}}.$$

□

$*$ **Uniqueness of entropy solution.** Let $h > 0$ and u, v be two entropy solutions of (1.1). We write the entropy inequality (4.1) corresponding to the solution u, with $T_h v$ as test function, and to the solution v, with $T_h u$ as test function. Upon addition, we get

$$\left\{\begin{aligned}&\int_{\{|u-T_hv|\leq t\}} a(x,\nabla u).\nabla(u-T_hv)dx + \int_{\{|v-T_hu|\leq t\}} a(x,\nabla v).\nabla(v-T_hu)dx \\ &\quad + \int_\Omega uT_t(u-T_hv)dx + \int_\Omega vT_t(v-T_hu)dx \\ &\leq \int_\Omega f(x)(T_t(u-T_hv)+T_t(v-T_hu))dx.\end{aligned}\right. \tag{4.20}$$

Defin

$$E_1 := \{|u-v|\leq t, |v|\leq h\},\ E_2 := E_1 \cap \{|u|\leq h\},\ \text{and } E_3 := E_1 \cap \{|u| > h\}.$$

We start with the firs integral in (4.5). By (2.5), we have

$$\begin{cases} \int_{\{|u-T_hv|\le t\}} a(x,\nabla u).\nabla(u-T_hv)dx = \\ \int_{\{|u-T_hv|\le t\}\cap(\{|v|\le h\}\cup\{|v|>h\})} a(x,\nabla u).\nabla(u-T_hv)dx = \\ \int_{\{|u-T_hv|\le t,|v|\le h\}} a(x,\nabla u).\nabla(u-T_hv)dx \\ +\int_{\{|u-T_hv|\le t,|v|>h\}} a(x,\nabla u).\nabla(u-T_hv)dx \\ =\int_{\{|u-v|\le t,|v|\le h\}} a(x,\nabla u).\nabla(u-v)dx + \int_{\{|u-h|\le t,|v|>h\}} a(x,\nabla u).\nabla u dx \\ \ge \int_{\{|u-v|\le t,|v|\le h\}} a(x,\nabla u).\nabla(u-v)dx = \int_{E_1} a(x,\nabla u).\nabla(u-v)dx \\ =\int_{E_1\cap(\{|u|\le h\}\cup\{|u|>h\})} a(x,\nabla u).\nabla(u-v)dx = \int_{E_2} a(x,\nabla u).\nabla(u-v)dx \\ +\int_{E_3} a(x,\nabla u).\nabla(u-v)dx = \int_{E_2} a(x,\nabla u).\nabla(u-v)dx + \\ \int_{E_3} a(x,\nabla u).\nabla u dx - \int_{E_3} a(x,\nabla u).\nabla v dx \ge \int_{E_2} a(x,\nabla u).\nabla(u-v)dx \\ -\int_{E_3} a(x,\nabla u).\nabla v dx. \end{cases} \tag{4.21}$$

Using (2.3) and (2.7), we estimate the last integral in (4.6) as follows

$$\begin{cases} \left|\int_{E_3} a(x,\nabla u).\nabla v dx\right| \le C_1 \int_{E_3} \left(j(x) + |\nabla u|^{p(x)-1}\right) |\nabla v|\, dx \\ \le C\left(|j|_{p'(.)} + \left||\nabla u|^{p(x)-1}\right|_{p'(.),\{h<|u|\le h+t\}}\right) |\nabla v|_{p(.),\{h-t<|v|\le h\}}, \end{cases} \tag{4.22}$$

where $\left||\nabla u|^{p(x)-1}\right|_{p'(.),\{h<|u|\le h+t\}} = \left\||\nabla u|^{p(x)-1}\right\|_{L^{p'(.)}(\{h<|u|\le h+t\})}$.

The quantity $C\left(|j|_{p'(.)} + \left||\nabla u|^{p(x)-1}\right|_{p'(.),\{h<|u|\le h+t\}}\right)$ is finite since $u \in W_0^{1,p(.)}(\Omega)$ and $j \in L^{p'(.)}$; then by Proposition 4.7, the last expression converges to zero as h tends to infinit . Therefore, from (4.6) and (4.7), we obtain

$$\int_{\{|u-T_hv|\le t\}} a(x,\nabla u).\nabla(u-T_hv)dx \ge I_h + \int_{E_2} a(x,\nabla u).\nabla(u-v)dx, \tag{4.23}$$

where I_h converges to zero as h tends to infinit . We may adopt the same procedure to treat the second integral in (4.5) to obtain

$$\int_{\{|v-T_h u|\leq t\}} a(x,\nabla v).\nabla(v-T_h u)dx \geq J_h - \int_{E_2} a(x,\nabla v).\nabla(u-v)dx, \quad (4.24)$$

where J_h converges to zero as h tends to infinit .

Next, we consider the term

$$K_h = \int_\Omega uT_t(u-T_h v)dx + \int_\Omega vT_t(v-T_h u)dx.$$

Note that

$$\begin{aligned} K_h = & \int_{\{|u|\leq h,|v|\leq h\}} (u-v)\,T_t(u-v)dx + \int_{\{|u|\leq h,|v|>h\}} uT_t(u-T_h v)dx \\ & + \int_{\{|u|>h,|v|\leq h\}} uT_t(u-T_h v)dx + \int_{\{|u|>h,|v|>h\}} uT_t(u-T_h v)dx \\ & + \int_{\{|u|>h,|v|\leq h\}} vT_t(v-T_h u)dx + \int_{\{|u|\leq h,|v|>h\}} vT_t(v-T_h u)dx \\ & + \int_{\{|u|>h,|v|>h\}} vT_t(v-T_h u)dx \\ \geq & \int_{\{|u|\leq h,|v|\leq h\}} (u-v)\,T_t(u-v)dx - 2t\int_{\{|v|>h\}} |u|\,dx - 2t\int_{\{|u|>h\}} |v|\,dx \\ & + \int_{\{|u|>h,|v|\leq h\}} uT_t(u-T_h v)dx + \int_{\{|v|>h,|u|\leq h\}} vT_t(v-T_h u)dx. \end{aligned}$$

Since u and v are entropy solutions and $f \in L^1(\Omega)$, then $u \in L^1(\Omega)$ and $v \in L^1(\Omega)$. We deduce from Proposition 4.7 that both $meas\,\{|u| > h\}$ and $meas\,\{|v| > h\}$ tend to zero as h goes to infinit and then

$$-2t\int_{\{|v|>h\}} |u|\,dx - 2t\int_{\{|u|>h\}} |v|\,dx \to 0 \text{ has } h \to 0.$$

Note also that

$$\begin{aligned} \int_{\{|u|\leq h,|v|\leq h\}} (u-v)\,T_t(u-v)dx + \int_{\{|u|>h,|v|\leq h\}} uT_t(u-T_h v)dx \\ + \int_{\{|v|>h,|u|\leq h\}} vT_t(v-T_h u)dx \geq 0, \end{aligned}$$

then

$$\lim_{h\to 0} K_h \geq 0. \quad (4.25)$$

Next, consider the right hand side of inequality (4.5). Noting that

$$T_t(u-T_h v) + T_t(v-T_h u) = 0 \text{ in } \{|u| \leq h, |v| \leq h\};$$

we obtain

$$
\begin{aligned}
&\left|\int_{\Omega} f(x)(T_t(u - T_h v) + T_t(v - T_h u))dx\right| \\
&= \left|\int_{\{|u|>h\}} f(x)(T_t(u - T_h v) + T_t(v - T_h u))dx \right.\\
&\left. + \int_{\{|u|\le h\}} f(x)(T_t(u - T_h v) + T_t(v - T_h u))dx\right| \\
&= \left|\int_{\{|u|>h\}} f(x)(T_t(u - T_h v) + T_t(v - T_h u))dx \right.\\
&\left. + \int_{\{|u|\le h, |v|>h\}} f(x)(T_t(u - T_h v) + T_t(v - T_h u))dx\right| \\
&\le 2t\left(\int_{\{|u|>h\}} |f|\, dx + \int_{\{|v|>h\}} |f|\, dx\right).
\end{aligned}
$$

Due to Proposition 4.7, both $meas\,\{|u| > h\}$ and $meas\,\{|v| > h\}$ tend to zero as h goes to infinit , then by the inequality above, the right hand side of inequality (4.5) tends to zero as h goes to infinit . From this assertion, (4.5), (4.8), (4.9) and (4.10), we obtain, letting $h \to +\infty$,

$$
\int_{\{|u-v|\le t\}} (a(x, \nabla u) - a(x, \nabla v)).\nabla(u - v)dx \le 0, \text{ for all } t > 0.
$$

By assertion (2.4), we conclude that $\nabla u = \nabla v$, a.e. in Ω.

Finally, from Poincaré inequality, we have

$$
\int_{\Omega} |T_t(u - v)|^{p-}\, dx \le C\int_{\Omega} |\nabla(T_t(u - v))|^{p-}\, dx = \int_{\{|u-v|\le t\}} |\nabla(u - v)|^{p-}\, dx = 0, \text{ for all } t > 0;
$$

and hence $u = v$, a.e. in Ω.

$*$ **Existence of entropy solutions**. Let $(f_n)_n$ be a sequence of bounded functions, strongly converging to $f \in L^1(\Omega)$ and such that

$$
\|f_n\|_1 \le \|f\|_1\,, \text{ for all } n \in \mathbb{N}. \tag{4.26}
$$

We consider the problem

$$
\begin{cases} u_n - div(a(x, \nabla u_n)) = f_n \text{ in } \Omega, \\ u_n = 0 \text{ on } \partial\Omega. \end{cases} \tag{4.27}
$$

It follows from Theorem 3.2 that problem (4.12) has a unique weak energy solution $u_n \in W_0^{1,p(.)}(\Omega)$. Our interest is to prove that as n goes to infinit , these approximate solutions u_n tend to a measurable function u which is an entropy solution of the limit problem (1.1). We start the proof of existence by proving that the sequence $(u_n)_{n\in\mathbb{N}}$ of solutions of problem (4.12) converges in measure to a measurable function u.

Proposition 4.8. *Assume* (2.1)-(2.5), $f \in L^1(\Omega)$ *and* (4.11). *Let* $u_n \in W_0^{1,p(.)}(\Omega)$ *be the solution of* (4.12). *The sequence* $(u_n)_{n\in\mathbb{N}}$ *is Cauchy in measure. In particular, there exists a measurable function* u *and a subsequence still denoted* u_n *such that* $u_n \to u$ *in measure.*

Proof of Proposition 4.8. Let $s > 0$ and defin

$$E_1 := \{|u_n| > t\},\ E_2 := \{|u_m| > t\},\ \text{and } E_3 := \{|T_t(u_n) - T_t(u_m)| > s\},$$

where $t > 0$ is to be fi ed. We note that

$$\{|u_n - u_m| > s\} \subset E_1 \cup E_2 \cup E_3,$$

and hence

$$meas\,\{|u_n - u_m| > s\} \leq meas\,(E_1) + meas\,(E_2) + meas\,(E_3)\,. \tag{4.28}$$

Let $\epsilon > 0$. Using (4.11) and the uniform bound given by Proposition 4.7, we choose $t = t(\epsilon)$ such that

$$meas(E_1) \leq \epsilon/3 \text{ and } meas(E_2) \leq \epsilon/3. \tag{4.29}$$

On the other hand, by (4.2), we get

$$\int_\Omega |\nabla T_t(u_n)|^{p_-}\,dx \leq M(1+t),\ \text{for all } n \geq 0, \tag{4.30}$$

therefore, by Sobolev embedding, we can assume that $(T_t(u_n))_{n\in\mathbb{N}}$ is a Cauchy sequence in $L^{p_-}(\Omega)$. Consequently, there exists a measurable function u and a subsequence still denoted $T_t(u_n)$ such that

$$T_t(u_n) \to T_t(u),\ \text{in } L^{p_-}(\Omega) \text{ and a.e.}$$

Thus,

$$meas(E_3) \leq \int_\Omega \left(\frac{|T_t(u_n) - T_t(u_m)|}{s}\right)^{p_-} dx \leq \frac{\epsilon}{3},$$

for all $n, m \geq n_0(s,\epsilon)$.

Finally, from (4.13), (4.14) and the last estimate, we obtain that

$$meas\,\{|u_n - u_m| > s\} \leq \epsilon,\ \text{for all } n, m \geq n_0(s,\epsilon), \tag{4.31}$$

i.e., $(u_n)_{n\in\mathbb{N}}$ is a Cauchy sequence in measure. □

Next, in order to prove that the sequence $(\nabla u_n)_{n\in\mathbb{N}}$ converges in measure to the weak gradient of u, we need two technical lemmas.

Lemma 4.9 (cf. [26, lemma 5.4]). *Let* $(v_n)_{n\in\mathbb{N}}$ *be a sequence of measurable functions. If* v_n *converges in measure to* v *and is uniformly bounded in* $L^{p(.)}(\Omega)$ *for some* $1 \ll p(.) \in L^\infty(\Omega)$, *then* $v_n \to v$ *strongly in* $L^1(\Omega)$.

The second technical lemma is a standard fact in measure theory (cf. [14]).

Lemma 4.10. *Let $(X, \mathcal{M}, \mu)$ be a measure space such that $\mu(X) < +\infty$. Consider a measurable function $\gamma : X \to [0, +\infty]$ such that*

$$\mu(\{x \in X : \gamma(x) = 0\}) = 0.$$

Then, for every $\epsilon > 0$, there exists $\delta > 0$ such that

$$\mu(A) < \epsilon, \textit{ for all } A \in \mathcal{M} \textit{ with } \int_A \gamma d\mu < \delta.$$

We can now prove the convergence in measure of the weak gradient.

Proposition 4.11. *Assume* (2.1)-(2.5), $f \in L^1(\Omega)$ *and* (4.11). *Let $u_n \in W_0^{1,p(.)}(\Omega)$ be the weak energy solution of* (4.12). *The following assertions hold:*
(i) ∇u_n *converges in measure to the weak gradient of u.*
(ii) *For all $t > 0$, $a(x, \nabla T_t(u_n))$ converges to $a(x, \nabla T_t(u))$ in $\left(L^1(\Omega)\right)^N$ strongly and in $\left(L^{p'(.)}(\Omega)\right)^N$ weakly.*

Proof of Proposition 4.11. (i) We claim that $(\nabla u_n)_{n\in\mathbb{N}}$ is Cauchy in measure. Indeed, let $s > 0$, and consider

$$E_1 := \{|\nabla u_n| > h\} \cup \{|\nabla u_m| > h\}, \; E_2 := \{|u_n - u_m| > t\},$$

and

$$E_3 := \{|\nabla u_n| \le h, |\nabla u_m| \le h, |u_n - u_m| \le t, |\nabla u_n - \nabla u_m| > s\},$$

where h and t will be chosen later. We note that

$$\{|\nabla u_n - \nabla u_m| > s\} \subset E_1 \cup E_2 \cup E_3. \tag{4.32}$$

Let $\epsilon > 0$. By Proposition 4.6, we may choose $h = h(\epsilon)$ large enough such that $meas(E_1) \le \epsilon/3$ for all $n, m \ge 0$. On the other hand, by Proposition 4.8 (see (4.16)), we have that $meas(E_2) \le \epsilon/3$ for all $n, m \ge n_0(t, \epsilon)$. Moreover, by assumption (2.4), there exists a real valued function $\gamma : \Omega \to [0, +\infty]$ such that $meas\{x \in \Omega : \gamma(x) = 0\} = 0$ and

$$(a(x,\xi) - a(x,\xi').(\xi - \xi')) \ge \gamma(x), \tag{4.33}$$

for all $\xi, \xi' \in \mathbb{R}^N$ such that $|\xi|, |\xi'| \le h, |\xi - \xi'| \ge s$, for a.e. $x \in \Omega$ (see [5]). Let $\delta = \delta(\epsilon)$ be given from Lemma 4.11, replacing ϵ and A by $\epsilon/3$ and E_3, respectively. Using (4.18), the equation and (4.11), we obtain

$$\int_{E_3} \gamma dx \le \int_{E_3} (a(x, \nabla u_n) - a(x, \nabla u_m)) . \nabla(u_n - u_m) dx \le 2\|f\|_1 t < \delta,$$

by choosing $t = \delta/(4\|f\|_1)$. From Lemma 4.10 again, it follows that $meas(E_3) < \epsilon/3$. Thus, using (4.17) and the estimates obtained for E_1, E_2 and E_3, it follows that $meas(\{|\nabla u_n - \nabla u_m| \ge s\}) \le \epsilon$, for all $n, m \ge n_0(s, \epsilon)$, and then the claim is proved. □

As a consequence, $(\nabla u_n)_{n\in\mathbb{N}}$ converges in measure to some measurable function v. In order to end the proof of Proposition 4.11, we need the following lemma:

Lemma 4.12. *(a) For a.e. $t \in \mathbb{R}$, $\nabla T_t(u_n)$ converges in measure to $v\chi_{\{|u|<t\}}$;*
(b) for a.e. $t \in \mathbb{R}$, $\nabla T_t(u) = v\chi_{\{|u|<t\}}$;
(c) $\nabla T_t(u) = v\chi_{\{|u|<t\}}$ holds for all $t \in \mathbb{R}$.

Proof of Lemma 4.12. • Proof of (a).

We know that $\nabla u_n \to v$ in measure. Thus $\chi_{\{|u|<t\}}\nabla u_n \to \chi_{\{|u|<t\}}v$ in measure.

Now, let us show that $\left(\chi_{\{|u_n|<t\}} - \chi_{\{|u|<t\}}\right)\nabla u_n \to 0$ in measure. For that, it is sufficien to show that $\left(\chi_{\{|u_n|<t\}} - \chi_{\{|u|<t\}}\right) \to 0$ in measure. Now, for all $\delta > 0$,

$$\left\{\left|\chi_{\{|u_n|<t\}} - \chi_{\{|u|<t\}}\right| |\nabla u_n| > \delta\right\} \subset \left\{\left|\chi_{\{|u_n|<t\}} - \chi_{\{|u|<t\}}\right| \neq 0\right\}$$
$$\subset \{|u| = t\} \cup \{u_n < t < u\} \cup \{u < t < u_n\} \cup \{u_n < -t < u\} \cup \{u < -t < u_n\}.$$

Thus,

$$\begin{cases} meas\left\{\left|\chi_{\{|u_n|<t\}} - \chi_{\{|u|<t\}}\right| |\nabla u_n| > \delta\right\} \leq meas\{|u| = t\} + meas\{u_n < t < u\} \\ +meas\{u < t < u_n\} + meas\{u_n < -t < u\} + meas\{u < -t < u_n\}. \end{cases} \tag{4.34}$$

Note that $meas\{|u| = t\} \leq meas\{t - h < u < t + h\} \to 0$ as $h \to 0$ for a.e. t, since u is a fi ed function. Next,

$$meas\{u_n < t < u\} \leq meas\{t < u < t + h\} + meas\{|u - u_n| \leq h\}, \text{ for all } h > 0.$$

Due to Proposition 4.8 and since $meas\{t < u < t + h\} \to 0$ as $h \to 0$, we have for all fi ed $h > 0$, $meas\{|u - u_n| \leq h\} \to 0$ as $n \to +\infty$. Thus for all $\epsilon > 0$, one can fin N such that for all $n > N$, $meas\{u_n < t < u\} < \epsilon/2 + \epsilon/2 = \epsilon$ by choosing h and then N. Each of the other terms in the right-hand side of (4.19) can be treated in the same way as for $meas\{u_n < t < u\}$. Thus, $meas\left\{\left|\chi_{\{|u_n|<t\}} - \chi_{\{|u|<t\}}\right| |\nabla u_n| > \delta\right\} \to 0$ as $n \to +\infty$. Finally, since $\nabla T_t(u_n) = \nabla u_n \chi_{\{|u_n|<t\}}$, the claim (a) follows.

• Proof of (b).

Let ψ_t be the weak $W_0^{1,p(.)}$-limit of $T_t(u_n)$, then it is also the strong L^1-limit of $T_t(u_n)$. But, as T_t is a Lipschitz function, the convergence in measure of u_n to u implies the convergence in measure of $T_t(u)$. Thus by the uniqueness of the limit in measure, ψ_t is identifie with $T_t(u)$, and we conclude that $\nabla T_t(u_n) \to \nabla T_t(u)$ weakly in $L^{p(\cdot)}(\Omega)$.

The previous convergence also ensures that $\nabla T_t(u_n)$ converges to $\nabla T_t(u)$ weakly in $L^1(\Omega)$. On the other hand, by (a), $\nabla T_t(u_n)$ converges to $v\chi_{\{|u|<t\}}$ in measure. By lemma 4.9, since $\nabla T_t(u_n)$ is uniformly bounded in $L^{p_-}(\Omega)$, the convergence is actually strong in $L^1(\Omega)$; thus it is also weak in $L^1(\Omega)$. By the uniqueness of a weak L^1-lmit, $v\chi_{\{|u|<t\}}$ coincides with $\nabla T_t(u)$.

• Proof of (c)

Let $0 < t < s$, and s be such that $v\chi_{\{|u|<t\}}$ coincides with $\nabla T_t(u)$. Then

$$\nabla T_t(u) = \nabla T_t(T_s(u)) = \nabla T_s(u)\chi_{\{|T_s(u)|<t\}} = v\chi_{\{|u|<t\}}\chi_{\{|u|<t\}} = v\chi_{\{|u|<t\}}. \qquad \square$$

Now, we can end the proof of Proposition 4.11. Combining Lemma 4.12(c) and Proposition 4.4, (i) follows.

Next, according to lemmas 4.9 and 4.12, for all $t > 0$, $a(x, \nabla T_t(u_n))$ converges to $a(x, \nabla T_t(u))$ in $\left(L^1(\Omega)\right)^N$ strongly and $a(x, \nabla T_t(u_n))$ converges to $\chi_t \in (L^{p'(.)}(\Omega))^N$ in $(L^{p'(.)}(\Omega))^N$ weakly. Since each of the convergences implies the weak L^1-convergence, χ_t can be identifie with $a(x, \nabla T_t(u))$; thus $a(x, \nabla T_t(u)) \in (L^{p'(.)}(\Omega))^N$. The proof of (ii) is then complete.

Now, let us end the proof of the existence of entropy solution. To this end, fi $t > 0$, $\varphi \in W_0^{1,p(.)}(\Omega) \cap L^\infty(\Omega)$, and choose $T_t(u_n - \varphi)$ as test function in (3.1), with u replaced by u_n, to obtain

$$\int_\Omega u_n T_t(u_n - \varphi)dx + \int_\Omega a(x, \nabla u_n).\nabla\left(T_t(u_n - \varphi)\right)dx = \int_\Omega f_n(x)T_t(u_n - \varphi)dx. \quad (4.35)$$

Note that this choice can be made using a standard density argument. We now pass to the limit in the previous identity. For the right hand side in (4.20), the convergence is obvious since f_n converges strongly in L^1 to f and $T_t(u_n - \varphi)$ converges weakly-$*$ in L^∞, and a.e., to $T_t(u - \varphi)$.

Next, we write the second term of the left hand side in (4.20) as

$$\int_{\{|u_n - \varphi| \le t\}} a(x, \nabla u_n).\nabla u_n dx - \int_{\{|u_n - \varphi| \le t\}} a(x, \nabla u_n).\nabla \varphi dx \quad (4.36)$$

and note that $\{|u_n - \varphi| \le t\}$ is a subset of $\{|u_n| \le t + \|\varphi\|_\infty\}$. Hence, taking $s = t + \|\varphi\|_\infty$, we rewrite the second integral in (4.21) as

$$\int_{\{|u_n - \varphi| \le t\}} a(x, \nabla T_s(u_n)).\nabla \varphi dx.$$

Since $a(x, \nabla T_s(u_n))$ is uniformly bounded in $(L^{p'(.)}(\Omega))^N$ (by (4.15) and assumption (2.3)), by Proposition (4.11)-(i), we have that it converges weakly to $a(x, \nabla T_s(u))$ in $(L^{p'(.)}(\Omega))^N$. Therefore, the last integral converges to

$$\int_{\{|u - \varphi| \le t\}} a(x, \nabla T_s(u)).\nabla \varphi dx.$$

The firs integral in (4.21) is nonnegative by (2.5), and it converges a.e. by Proposition 4.11. It follows from Fatou's Lemma that

$$\int_{\{|u - \varphi| \le t\}} a(x, \nabla u).\nabla u dx \le \liminf_{n \to +\infty} \int_{\{|u_n - \varphi| \le t\}} a(x, \nabla u_n).\nabla u_n dx.$$

Finally, we may study the convergence of the firs term of left hand side in (4.20). We rewrite it as

$$\int_\Omega u_n T_t(u_n - \varphi)dx = \int_\Omega (u_n - \varphi)T_t(u_n - \varphi)dx + \int_\Omega \varphi T_t(u_n - \varphi)dx. \quad (4.37)$$

Note that $\int_\Omega \varphi T_t(u_n - \varphi)dx \to \int_\Omega \varphi T_t(u - \varphi)dx$ as $n \to +\infty$ since $T_t(u_n - \varphi)$ converges weakly-$*$ in L^∞, and a.e., to $T_t(u - \varphi)$ and $\varphi \in L^\infty(\Omega) \subset L^1(\Omega)$.

Next, the term $(u_n-\varphi)T_t(u_n-\varphi)$ is nonnegative and by (4.15), we deduce that $T_t(u_n-\varphi)$ converges strongly in $L^{p_-}(\Omega)$ and $a.e.$, to $T_t(u-\varphi)$ and since $meas(\Omega)<+\infty$, then

$$T_t(u_n-\varphi) \text{ converges in measure to } T_t(u-\varphi). \tag{4.38}$$

Again, since $meas(\Omega)<+\infty$, then thanks to (4.23) and proposition 4.8, we have

$$(u_n-\varphi)T_t(u_n-\varphi) \text{ converges in measure to } (u-\varphi)T_t(u-\varphi). \tag{4.39}$$

Due to (4.23), there exists a subsequence still denoted by $((u_n-\varphi)T_t(u_n-\varphi))$ such that

$$(u_n-\varphi)T_t(u_n-\varphi) \text{ converges } a.e. \text{ to } (u-\varphi)T_t(u-\varphi).$$

We obtain by using Fatou's Lemma that

$$\int_\Omega (u-\varphi)T_t(u-\varphi)dx \leq \liminf_{n\to+\infty}\int_\Omega (u_n-\varphi)T_t(u_n-\varphi)dx. \tag{4.40}$$

Combining results, we obtain

$$\int_\Omega uT_t(u-\varphi)dx + \int_\Omega a(x,\nabla u).\nabla T_t(u-\varphi)dx \leq \int_\Omega f(x)T_t(u-\varphi)dx,$$

as n goes to infinit in (4.20) i.e., u is an entropy solution of (1.1). □

Acknowledgment

The authors want to express their deepest thanks to the editor and anonymous referee's for comments and suggestions on the paper.

References

[1] A. Alvino, L. Boccardo and V. Ferone, L. Orsina, G. Trombetti, Existence results for non-linear elliptic equations with degenerate coercivity, *Ann. Mat. Pura Appl.* **182** (2003), 53–79.

[2] M. Bendahmane and P. Wittbold, Renormalized solutions for nonlinear elliptic equations with variable exponents and L^1 data, *Nonlinear Anal., Theory Methods Appl.* **70** (2009), No. 2 (A), 567-583.

[3] Ph. Bénilan, L. Boccardo, T. Gallouët, R. Gariepy, M. Pierre and J. L. Vazquez, An L^1-theory of existence and uniqueness of solutions of nonlinear elliptic equations, *Ann. Sc. Norm. Super. Pisa, Cl. Sci.* **22** (1995), 241–273.

[4] L. Boccardo and T. Gallouët, Nonlinear elliptic and parabolic equations involving measure data, *J. Funct. Anal.* **87** (1989), 149–169.

[5] L. Boccardo and T. Gallouët, Nonlinear elliptic equations with right hand side measures, *Comm. Partial Differential Equations* **17** (1992), 641–655.

[6] H. Brezis, *Analyse Fonctionnelle: Théorie et Applications,* Paris, Masson, 1983.

[7] L. Diening, Riesz potential and Sobolev embeddings on generalized Lebesgue and Sobolev spaces $L^{p(\cdot)}$ and $W^{1,p(\cdot)}$, *Math. Nachr.* **268** (2004), 31–43.

[8] L. Diening, Theoretical and numerical results for electrorheological fluids *Ph.D. thesis,* University of Freiburg, Germany, 2002.

[9] D. E. Edmunds and J. Rakosnik, Density of smooth functions in $W^{k,p(x)}(\Omega)$, *Proc. R. Soc. A* **437** (1992), 229–236.

[10] D. E. Edmunds and J. Rakosnik, Sobolev embeddings with variable exponent, *Sudia Math.* **143** (2000), 267–293.

[11] D. E. Edmunds and J. Rakosnik, Sobolev embeddings with variable exponent, II, *Math. Nachr.* **246-247** (2002), 53–67.

[12] X. Fan and Q. Zhang, Existence of solutions for $p(x)$-Laplacian Dirichlet problem, *Nonlinear Anal.* **52** (2003), 1843–1852.

[13] X. Fan and D. Zhao, On the spaces $L^{p(x)}(\Omega)$ and $W^{m,p(x)}(\Omega)$, *J. Math. Anal. Appl.* **263** (2001), 424–446.

[14] P. Halmos, *Measure Theory,* D. Van Nostrand Company, New York, 1950.

[15] A. El Hamidi, Existence results to elliptic systems with nontandard growth conditions, *Publication Département de Mathématiques, Université de La Rochelle,* Vol. **12**, pp. 1–15.

[16] O. Kovacik and J. Rakosnik, On spaces $L^{p(x)}$ and $W^{1,p(x)}$, *Czech. Math. J.* **41** (1991), 592–618.

[17] M. Krasnosel'skii, *Topological Methods in the Theory of Nonlinear Integral Equations,* Pergamon Press, New York, 1964.

[18] A. Kufner, J. Oldrich and S. Fucik, *Functions Spaces,* Noordhoff International Publishing, Leyden, 1977.

[19] J. Leray and J. L. Lions, Quelques résultats de Visik sur les problèmes elliptiques nonlinéaires par les méthodes de Minty et Browder, *Bull. Soc. Math. France* **93** (1965), 97–107.

[20] M. Mihailescu and V. Radulescu, A multiplicity result for a nonlinear degenerate problem arising in the theory of electrorheological fluids *Proc. R. Soc. A* **462** (2006), 2625–2641.

[21] J. Musielak, Orlicz Spaces and Modular Spaces, *Lecture Notes in Mathematics,* vol. 1034, springer, Berlin, 1983.

[22] H. Nakano, *Modulared Semi-Ordered Linear Spaces,* Maruzen Co., Ltd., Tokyo, 1950.

[23] W. Orlicz, Über konjugierte Exponentenfolgen, *Studia Math.* **3** (1931), 200–212.

[24] K. R. Rajagopal and M. Ruzicka, Mathematical Modeling of Electrorheological materials, *Contin. Mech. Thermodyn.* **13** (2001), 59–78.

[25] M. Ruzicka, *Electrorheological Fluids: Modelling and Mathematical Theory,* Lecture Notes in Mathematics, vol. 1748, Springer-Verlag, Berlin, 2002.

[26] M. Sanchon and J. M. Urbano, Entropy solutions for the $p(x)$-Laplace Equation, *Trans. Amer. Math. Soc.* **361** (2009), no. 12, 6387–6405.

[27] I. Sharapudinov, On the topology of the space $L^{p(t)}([0,1])$, *Math. Zametki* **26** (1978), 613–632.

[28] I. Tsenov, Generalization of the problem of best approximation of a function in the space L^s, *Uch. Zap. Dagestan Gos. Univ.* **7** (1961), 25–37.

[29] V. Zhikov, Averaging of functionals in the calculus of variations and elasticity. *Math. USSR Izv.* **29** (1987), 33–66.

[30] V. Zhikov, On passing to the limit in nonlinear variational problem. *Math. Sb.* **183** (1992), 47–84.

INDEX

A

B

C

D

E

F

G

H

I

J

L

M

N

W

Y

Z

MATHEMATICS RESEARCH DEVELOPMENTS

EVOLUTION EQUATIONS

THEORIES, SOLUTIONS AND FUNCTIONS

Mathematics Research Developments

Additional books in this series can be found on Nova's website
under the Series tab.

Additional E-books in this series can be found on Nova's website
under the E-book tab.